Noncommutative Cosmology

Noncommutative Cosmology

Matilde Marcolli

California Institute of Technology, USA
Perimeter Institute for Theoretical Physics, Canada
University of Toronto, Canada

World Scientific

NEW JERSEY · LONDON · SINGAPORE · BEIJING · SHANGHAI · HONG KONG · TAIPEI · CHENNAI · TOKYO

Published by

World Scientific Publishing Co. Pte. Ltd.

5 Toh Tuck Link, Singapore 596224

USA office: 27 Warren Street, Suite 401-402, Hackensack, NJ 07601

UK office: 57 Shelton Street, Covent Garden, London WC2H 9HE

Library of Congress Cataloging-in-Publication Data

Names: Marcolli, Matilde, author.

Title: Noncommutative cosmology / by Matilde Marcolli (California Institute of Technology, USA).

Description: New Jersey : World Scientific, 2018. | Includes bibliographical references.

Identifiers: LCCN 2017042860| ISBN 9789813202832 (hardcover : alk. paper) |
     ISBN 9789813202849 (pbk : alk. paper)

Subjects: LCSH: Noncommutative differential geometry. | Mathematical physics. | Cosmology.

Classification: LCC QC20.7.G44 M37 2018 | DDC 523.101/51636--dc23

LC record available at https://lccn.loc.gov/2017042860

British Library Cataloguing-in-Publication Data

A catalogue record for this book is available from the British Library.

For any available supplementary material, please visit
http://www.worldscientific.com/worldscibooks/10.1142/10335#t=suppl

Preface

This book explores the use of mathematical methods from Noncommutative Geometry to construct new geometric models for classical and quantum cosmology and for (Euclidean) quantum gravity. A theme we will be discussing at length is a class of modified gravity models based on Dirac operators and the spectral action functional. We will also discuss the use of Algebraic Geometry and Arithmetic in cosmological models.

The book is based on a series of lectures given by the author at the summer school "Geometric, algebraic and topological methods for Quantum Field Theory", organized by Alexander Cardona, Hernán Ocampo, Sylvie Paycha and Andrés Reyes at Villa de Leyva, Colombia, July 4–22, 2011; additional material was introduced based on the graduate courses "Topics in Geometry and Physics" (Ma193b) and "Mathematical Physics" (Ma148b) taught by the author at Caltech in 2012 and 2016, and on several talks given by the author in the Cosmology Seminar at the Perimeter Institute for Theoretical Physics.

The research work surveyed in this book is based on the results of several extensive collaborations the author has had over the past few years. Many of the topics discussed in the book were developed in collaborations with Yuri Manin and with Elena Pierpaoli, on a range of different aspects of the relation between geometry and cosmology. Other parts of the book reflect past and ongoing work with Wentao Fan and Farzad Fathizadeh and with Walter van Suijlekom. Other results reported in the volume arise from work carried out by the author with several of her students: Adam Ball, Domenic Denicola, Christopher Duston, Christopher Estrada, Yeorgia Kafkoulis, Daniel Kolodrubetz, Nicolas Tedeschi, Ahmed Zainy al-Yasry, and Nick Zolman who worked on various parts of this project, and especially Branimir Ćaćić and Kevin Teh, who contributed essentially to several crucial aspects of this work on noncommutative cosmology.

The volume is structured as follows: the first chapter recalls general background on Noncommutative Geometry and on the spectral action functional. Part of this chapter also discusses particle physics models based on the spectral action and summarizes previous work of the author with Ali Chamseddine and Alain Connes. The chapter also briefly reviews results of Beenakker, van den Broek, and van Suijlekom on particle physics models with supersymmetry and the spectral action, and work of the author and Zolman on the spectral action and supersymmetry algebras, as well as the work of Chamseddine, Connes and van Suijlekom on spectral action models of grand unified theories. It also briefly surveys work of Boyle and Farnsworth on the fused algebra approach to the noncommutative geometry of the Standard Model of particle physics. The second chapter is based on joint work with Elena Pierpaoli on Eary Universe models based on the spectral action functional of gravity coupled to matter and its large energies asymptotic expansion. It also discusses recent results on the Higgs mass problem in spectral action models, based on results of Chamseddine–Connes and of the author and Chris Estrada. The third chapter is based on results of the author with Elena Pierpaoli, Branimir Ćaćić, and Kevin Teh on the problem of Cosmic Topology in the context of models of gravity based on the spectral action and associated slow-roll inflation models. The fourth chapter is based on joint work with Yuri Manin on the use of algebro-geometric techniques in cosmology and associated models of conformally cyclic cosmologies and of eternal inflation scenarios based on algebro-geometric blow-ups. The fifth chapter is based on work with Yuri Manin and on work with Chris Estrada on mixmaster universe models and their arithmetic properties, and on their noncommutative deformations and possible Early Universe scenario. The case of the $SU(2)$ Bianchi IX gravitational instantons is discussed based on joint work with Yuri Manin. In the sixth chapter the discussion of the $SU(2)$ Bianchi IX gravitational instantons is further expanded, based on joint work with Wentao Fan and Farzad Fathizadeh, with explicit computations of the coefficients of the spectral action and a result on the arithmetic structure of the spectral action, based on modular forms. The seventh chapter summarizes work in collaboration with Farzad Fathizadeh on the occurrence of mixed Tate motives and periods in the asymptotic expansion of the spectral action for Robertson–Walker spacetimes. The eighth chapter is based on joint work of the author with Adam Ball and with Nicolas Tedeschi, on multifractal models of cosmology treated with the methods of noncommutative geometry. In particular it covers the spectral action model of gravity and slow-roll inflation scenarios for Packed Swiss Cheese Cosmologies, and a version of the "Eternal Symmetree" p-adic model of eternal inflation. The ninth chapter deals with an approach to Euclidean quantum cosmology, in the style of Hartle–Hawking, based on the spectral action functional. The chapter includes a survey of results of Chris Duston on exotic smoothness in Euclidean quantum cosmology, results of the author and Domenic Denicola and Ahmed Zainy al-Yasry, and of the author and Walter van Suijlekom, on topological spin foam models and gauge spin networks models. It also surveys recent work of John Barrett and Lisa Glaser on random finite noncommutative geometries.

Matilde Marcolli

California Institute of Technology

Perimeter Institute for Theoretical Physics

University of Toronto

Acknowledgments

This work was supported in part by NSF grants DMS-0901221, DMS-1007207, DMS-1201512, DMS-1707882, PHY-1205440. The author also thanks the Mathematical Sciences Research Institute in Berkeley for hospitality and support during some of the writing of this book in 2013. Part of this work was done at the Perimeter Institute for Theoretical Physics, over the course of several visits between 2014 and 2016. The Perimeter Institute is supported by the Government of Canada through Industry Canada and by the Province of Ontario through the Ministry of Economic Development and Innovation. The author thanks all her collaborators and students listed in the Preface above, for making this book possible and for the very rewarding joint work, especially among them Yuri Manin, Elena Pierpaoli, Farzad Fathizadeh, Branimir Ćaćić and Kevin Teh. The author also thanks Latham Boyle and Shane Farnsworth for many useful discussions, and especially the latter for reading an earlier draft of the manuscript and providing extensive comments. She also thanks Leon Kot for his involvement and invaluable help with the writing of this volume.

Fig. 0.1   This book in the making, Perimeter Institute, August 2016.

Contents

Preface   v

Acknowledgments   ix

1. Gravity and Matter in Noncommutative Geometry   1

   1.1   Spectral triples   4

        1.1.1   Dimension in noncommutative geometry   6

        1.1.2   Manifolds   7

        1.1.3   Almost-commutative geometries   10

        1.1.4   Cartesian products   11

        1.1.5   Finite spectral triples   12

        1.1.6 θ-deformations   15

        1.1.7   Fractals   17

   1.2   The Spectral Action functional   19

        1.2.1   Asymptotic expansion   20

        1.2.2   The spectral action as modified gravity   22

        1.2.3   Riemannian versus Lorentzian geometries   23

        1.2.4   Einstein–Hilbert action with cosmological term . .   24

        1.2.5   Conformal gravity   24

        1.2.6   Gauss–Bonnet gravity   25

        1.2.7   Poisson summation   26

        1.2.8   Spectral action and expansion on fractals   28

   1.3   Particle Physics models   30

        1.3.1   Gravity coupled to matter   30

        1.3.2   The Standard Model: νMSM   31

        1.3.3   The moduli space of Dirac operators   34

        1.3.4   Bosons and inner fluctuations   36

1.3.5	Asymptotic expansion of the spectral action . . .	39
1.3.6	Coefficients of the gravitational terms	42
1.3.7	Fused algebra approach	43
1.3.8	Supersymmetric theories	45
1.3.9	Adinkras, SUSY algebras, and the spectral action	51
1.3.10	Grand Unified theories	57

2. **Renormalization Group Flows and Early Universe Models** — **61**

2.1	RGE flows	62
	2.1.1 RGE flow from the Minimal Standard Model . . .	62
	2.1.2 RGE flow from the νMSM	65
	2.1.3 Geometric constraints at unification	67
	2.1.4 Maximal mixing and initial condition at unification	68
	2.1.5 Sensitive dependence and fine tuning	69
2.2	Gravitational terms	69
2.3	Early universe models	71
	2.3.1 Effective gravitational constant	73
	2.3.2 Effective cosmological constant	74
	2.3.3 Antigravity in the early universe	75
	2.3.4 Gravity balls	76
	2.3.5 Primordial black holes with gravitational memory	77
	2.3.6 Emergent Hoyle–Narlikar cosmologies	78
	2.3.7 Slow-roll inflation	79
2.4	Higgs mass estimates	80
	2.4.1 Scalar fields and the Higgs mass problem	81
	2.4.2 Asymptotic safety and anomalous dimensions . .	84

3. **Cosmic Topology** — **89**

3.1	The problem of cosmic topology	90
3.2	The spectral action and cosmic topology	92
	3.2.1 Slow-roll potential and slow-roll parameters . . .	93
	3.2.2 Spherical space forms	96
	3.2.3 Flat tori and Bieberbach manifolds	102
	3.2.4 A heat kernel view	106
	3.2.5 Gravity coupled to matter and slow-roll potential	109
	3.2.6 Engineering inflation via Dirac spectra	113

4. Algebro-geometric models in Cosmology   115

    4.1 Spacetimes and complex geometry 115
         4.1.1 Complexified spacetimes and Grassmannians . . . 116
         4.1.2 Twistor spaces 116
    4.2 Blowup models 117
         4.2.1 Gluing spacetimes 117
         4.2.2 Conformally cyclic cosmological models 119
         4.2.3 Eternal inflation via trees of projective spaces . . 120
    4.3 Time and elliptic curves 121
    4.4 Noncommutativity and gluing of spacetimes 122

5. Mixmaster Cosmologies   125

    5.1 Kasner metrics and mixmaster universe models 125
         5.1.1 The shift of the continued fraction expansion . . . 126
         5.1.2 Continued fractions and the mixmaster universe . 127
         5.1.3 Continued fractions and modular curves 129
         5.1.4 Kasner times and geodesic lengths 132
    5.2 Modular curves, C^*-algebras, and mixmaster models . . . 134
    5.3 Noncommutative mixmaster cosmologies 139
         5.3.1 Mixmaster universes with torus sections 139
         5.3.2 Noncommutative θ-deformations 141
         5.3.3 Spectral action and inflation scenario 142
    5.4 Bianchi IX $SU(2)$-gravitational instantons 146
         5.4.1 Painlevé VI equation 147
         5.4.2 Gravitational instantons and Painlevé 148
    5.5 Noncommutativity in the early universe 150

6. The Spectral Action on Bianchi IX Cosmologies   155

    6.1 Pseudodifferential calculus and parametrix method 156
    6.2 Wodzicki residues method 160
    6.3 Rationality result 162
    6.4 Gravitational instantons and the spectral action 163
    6.5 The spectral action and modular forms 165

7. Motives and Periods in Cosmology   171

    7.1 Robertson–Walker metrics 173
    7.2 The a_2 term period 175
    7.3 The periods of the higher order terms a_{2n} 177

7.4 The mixed motives of Robertson–Walker gravity 179
    7.4.1 Triangulated category of motives 180
    7.4.2 Grothendieck classes of RW spacetimes 181
    7.4.3 Mixed Tate motives of RW spacetimes 182

8. Fractal and Multifractal Structures in Cosmology   187

    8.1 Packed Swiss Cheese Cosmology and the spectral action . 187
        8.1.1 Apollonian sphere packings 188
        8.1.2 Length spectrum and zeta functions 191
        8.1.3 Models of fractal spacetimes 195
        8.1.4 The spectral action functional on fractal
            cosmologies 197
        8.1.5 Slow-roll inflation potentials with fractality 201
    8.2 A *p*-adic model of eternal inflation 201
        8.2.1 Bruhat–Tits tree and Bethe tree 202
        8.2.2 Multifractals, symbolic dynamics, and stochastics 203

9. Noncommutative Quantum Cosmology   209

    9.1 Hartle-Hawking quantum cosmology 209
        9.1.1 Path integral and wave function of the universe . 210
        9.1.2 Hamiltonian constraint, Wheeler-DeWitt equation 212
        9.1.3 Minisuperspace models 214
        9.1.4 Exotic smoothness 216
    9.2 Categories and algebras of geometries 218
        9.2.1 Cobordisms: equivalences and 2-categories 222
        9.2.2 Vertical composition and Hartle-Hawking gravity 227
        9.2.3 Horizontal composition and Connes–Chern
            character 228
        9.2.4 Almost-commutative cobordisms 230
    9.3 Topological spin networks and foams 233
        9.3.1 Spin networks and monodromies 233
        9.3.2 Spin foams and monodromies 237
        9.3.3 2-categories and convolution algebras 239
        9.3.4 Quantized area operator and dynamics 240
    9.4 Discretized almost-commutative geometries 243
        9.4.1 Categorical data and finite spectral triples 243
        9.4.2 Gauge networks 245
        9.4.3 Spectral action on a lattice 247

9.4.4   Continuum limit and the Wilson action 249

9.5   Random finite noncommutative geometries 250

Bibliography   255

Index   271

Chapter 1

Gravity and Matter in Noncommutative Geometry

The main idea behind noncommutative geometry (see [Connes (1994)] for an extensive treatment of the subject) lies in the correspondence between spaces and algebras of functions. When one thinks of spaces as topological spaces, with some reasonable assumption about the topology (locally compact and Hausdorff), and of functions as continuous functions, the equivalence is expressed in terms of the Gelfand–Naimark correspondence. Assigning a locally compact Hausdorff topological space X is the same thing as assigning its C^*-algebra of continuous functions $C_0(X)$, in the sense that the space can be reconstructed from the algebra (with points corresponding to characters, that is, C^*-algebra homomorphisms to $\mathbb{C}$) and isomorphic algebras correspond to homeomorphic spaces. All C^*-algebras of the form $C_0(X)$, for some locally compact Hausdorff topological space X, are in particular *commutative* algebras, since the product is the ordinary pointwise product of functions, which is clearly commutative. However, the world of C^*-algebras is rich with many interesting *noncommutative* examples. These cannot be directly interpreted as algebras of functions on an ordinary space. However, geometric properties of a space that can be formulated purely in terms of its algebra of functions may continue to make sense when one drops the commutativity assumption. One can then think of a noncommutative C^*-algebra as the algebra of continuous functions on a *noncommutative space*. Unlike ordinary spaces, which can be defined in two a priori independent ways, as spaces or as algebras of functions, and then shown to be equivalent by the Gelfand–Naimark correspondence, noncommutative spaces can only be defined through their (noncommutative) algebra of functions. For simplicity, when not stated otherwise, we will assume our topological spaces X are compact and Hausdorff, with algebra of functions $C(X)$. For a general introduction to C^*-algebras, we refer the reader to [Davidson (1996)].

Thus, the main idea, and programmatic proposal, of noncommutative geometry is first of all to reformulate as much as possible of ordinary (differential) geometry on a space (manifold) X in terms of the C^*-algebra $C(X)$ and a dense subalgebra $C^\infty(X)$. Differential forms, bundles, connections, cohomology: these are all important geometric notions that continue to make sense (when properly formulated) without commutativity, hence they can be applied to both commutative and noncommutative spaces. In order to describe physics on possibly noncommutative spaces, one needs a good framework for the analog of *Riemannian geometry*, as well as suitable *action functionals* for noncommutative spaces.

It turns out that noncommutative geometry is a very good framework for theories of (modified) gravity coupled to matter. The main idea behind gravity and particle physics models based on noncommutative geometry is that "all forces become gravity" on an noncommutative space. In other words, it is only from the point of view of a slice of the geometry consisting of an ordinary spacetime manifold that we see a difference between gravity and the other forces, while from the point of view of the overall (noncommutative) geometry they are all seen together as gravity. As we will see, the main construction is not unlike the idea of "extra dimensions" many people are familiar with from string theory, except for the fact that the extra dimensions in these models are not only small, but also noncommutative, while the extended dimensions of spacetime maintain their commutative nature. Moreover, the theory is not based on extended objects like strings or branes, but on a very general action functional for gravity, the *spectral action*, which can be defined on an arbitrary noncommutative space (with an analog of a Riemannian structure). In the case of an ordinary manifold, this action functional recovers the usual Einstein–Hilbert action with cosmological constant, hence it incorporates General Relativity. However, already in the ordinary commutative case, one also obtains some correction terms, which correspond to modified gravity models, which include conformal gravity and Gauss–Bonnet gravity. When applied to a class of noncommutative spaces called *almost commutative geometries*, the spectral action delivers, in addition to these gravitational term, an action functional for matter non-minimally coupled to gravity, which involves a conformal coupling of the Higgs and scalar fields to the curvature R. Depending on the geometry, this produces different types of particle physics models, including an extension of the Standard Model by right handed neutrinos with Majorana masses, grand unified Pati–Salaam models, and some supersymmetric extensions of the Standard Model. The literature on these models is very extensive, and we will be giving precise references in the relevant chapters that follow. For the moment, we refer the reader to the general overviews given in [Beenakker, van den Broek, van Suijlekom (2015)], [Connes and Marcolli (2008)], [van Suijlekom (2015)].

There are many different sources of noncommutative spaces. For example, a large class of them arises when one considers "bad quotients", that is, quotients of smooth manifolds by group actions (or other equivalence relations) where the quotient is no longer a smooth manifold. Noncommutative geometry provides a method for replacing the bad quotient by a noncommutative space, which still behaves (in the sense of noncommutative geometry) like a smooth object. However, in the specific context of cosmological models we consider here, this will not be the main class of noncommutative spaces we work with. Our setting will involve noncommutative spaces that are very close to being ordinary manifolds, or more precisely a class of spaces known as almost-commutative geometries, which are locally a product of a smooth manifold and a finite noncommutative geometry. We will also consider noncommutative geometries that arise as deformations of ordinary manifolds, along actions of tori that become noncommutative. Methods of noncommutative geometry can also be used to treat ordinary spaces that are not smooth manifolds (fractals, for example) as if they were smooth. We will use such methods to address gravity models of multifractal structures in cosmology.

In this first chapter we review the main setting of (metric) noncommutative geometry, based on the notion of *spectral triple*, and the natural action functional for gravity on noncommutative spaces, the *spectral action*. We will then survey recent results, which show how this formalism can be used to derive different types of modified gravity models and various particle physics models with matter non-minimally coupled to gravity.

1.1   Spectral triples

The basic setting for metric noncommutative geometry (which extends Riemannian and spin geometry to the noncommutative world) is provided by the theory of *spectral triples*, introduced in [Connes (1995)].

A spectral triple $(\mathcal{A}, \mathcal{H}, D)$ is characterized by a simple list of formal properties:

- $\mathcal{A}$ is an involutive algebra;
- it has a representation $\pi : \mathcal{A} \to \mathcal{L}(\mathcal{H})$ by bounded linear operators on a separable Hilbert space $\mathcal{H}$;
- D is a self adjoint linear operator on $\mathcal{H}$, with dense domain;
- the operator D has compact resolvent, $(1 + D^2)^{-1/2} \in \mathcal{K}$;
- the commutators $[\pi(a), D]$ are bounded, for all $a \in \mathcal{A}$;
- the spectral triple $(\mathcal{A}, \mathcal{H}, D)$ is *even* if there is a $\mathbb{Z}/2$-grading γ on $\mathcal{H}$ which satisfies

$$[\gamma, \pi(a)] = 0, \quad \forall a \in \mathcal{A}, \qquad D\gamma = -\gamma D.$$

There are interesting examples of spectral triples on noncommutative spaces that satisfy only these properties. However, spectral triples that more closely resemble the behavior of classical commutative compact Riemannian spin manifolds have additional properties.

The first type of additional condition one considers about spectral triples is a *real structure*. This consists of an anti-linear isometry J on the Hilbert space $\mathcal{H}$ that satisfies:

- there are signs $\varepsilon, \varepsilon', \varepsilon'' \in \{\pm 1\}$ such that

$$J^2 = \varepsilon, \quad JD = \varepsilon' DJ, \quad \text{and} \quad J\gamma = \varepsilon'' \gamma J;$$

- order zero condition: $[\pi(a), \pi(b)^0] = 0$, for all $a, b \in \mathcal{A}$, where $\pi(b)^0 = J\pi(b)^* J^{-1}$ for all $b \in \mathcal{A}$;
- order one condition: $[[D, \pi(a)], \pi(b)^0] = 0$, for all $a, b \in \mathcal{A}$.

The order zero condition implies that $\mathcal{H}$ is an $\mathcal{A}$-bimodule with commuting left and right action given by $\pi(a)$ and $\pi(b)^0$. The order one condition corresponds to the property of geometric Dirac operators of being first order elliptic differential operators.

In the formulation above we have left the choice of signs $\varepsilon, \varepsilon', \varepsilon''$ unconstrained. It is customary, if one wants to enforce a closer resemblance to the properties of ordinary manifolds, to further require that only certain combinations of signs are allowed. These are the ones listed in the following table, and one assigns to each of them a dimension modulo 8, which corresponds to the KO-dimension in the case of ordinary manifolds:

n	0	1	2	3	4	5	6	7
ε	1	1	-1	-1	-1	-1	1	1
ε'	1	-1	1	1	1	-1	1	1
ε''	1		-1		1		-1	

Recall that the K-theory of a manifold X is a generalized cohomology theory that is 2-periodic, with the even $K^0(X)$ obtained by considering the Grothendieck group of stable isomorphism classes of complex vector bundles with the Whitney sum and the tensor product operations, and the odd $K^1(X)$ obtained via suspension. Equivalently, $K^0(X) = [X, BU \times \mathbb{Z}]$, the set of homotopy classes of maps, with BU the classifying space of the unitary groups $U = \lim_n U(n)$, and $K^1(X) = [X, U]$. The 2-periodicity arises from the homotopy equivalence $BU \times \mathbb{Z} \simeq \Omega^2(BU \times \mathbb{Z})$ with its double loop space. In a similar way KO-theory is an 8-periodic generalized cohomology theory based on real vector bundles, with the real Bott periodicity corresponding to the homotopy equivalence $BO \times \mathbb{Z} \simeq \Omega^8(BO \times \mathbb{Z})$, where BO is the classifying space of the special orthogonal groups.

However, one may want more generally to allow for real structures with other (non-manifold-like) combinations of signs. This would include, for instance the "exotic" cases considered in [Dabrowski, Dossena (2011)] in relation to products of finite spectral triples, see also the detailed discussion in [Ćaćić (2013)]. In order to distinguish these cases, we will say that real structures with signs as in the table above are "manifold-like" and other combinations of signs are "non-manifold-like", or "exotic", in the sense that these combinations of signs do not occur in the KO-dimension table of ordinary manifolds. The more general table of KO-dimensions is then as in [Dabrowski, Dossena (2011)]:

n	0_+	0_-	1	2_+	2_-	3	4_+	4_-	5	6_+	6_-	7
ε	1	1	1	-1	1	-1	-1	-1	-1	1	-1	1
ε'	1	-1	-1	1	-1	1	1	-1	-1	1	-1	1
ε''	1	1		-1	-1		1	1		-1	-1	

1.1.1 *Dimension in noncommutative geometry*

In ordinary geometry there are many different (but equivalent) ways of defining the dimension of a manifold. In the case of a smooth compact manifold, the dimension is usually understood as the number of independent linear directions in the tangent space at any given point. The same dimension, however, can also be read off the rate of growth of the eigenvalues of the Laplacian (or of the Dirac operator) by Weyl's law

$$N(\lambda) = \frac{Vol(X)}{(4\pi)^{n/2}\Gamma(\frac{n}{2}+1)} \lambda^n + o(\lambda^n),$$

for $\dim X = n$, where $N(\lambda)$ is the counting function

$$N(\lambda) = \#\{j : \sqrt{\lambda_j} \leq \lambda\},$$

with $0 \leq \lambda_0 \leq \lambda_1 \leq \cdots \leq \lambda_j \cdots$ the Laplace spectrum. Additionally, the dimension can be read off deRham cohomology, as the highest non-vanishing degree. While cohomology is integer graded and topological K-theory is $\mathbb{Z}/2\mathbb{Z}$-graded, by Bott periodicity, real K-theory, KO-theory, is graded by integers modulo 8, by real Bott periodicity. For a compact spin manifold, the fundamental class of X corresponds to a class in K-homology, which, in the description of K-homology in terms of elliptic operators [Atiyah (1970)], can be described in terms of the Dirac operator, and a KO-orientation class in real KO-theory. Thus, one also has, for ordinary manifolds a notion of KO-dimension, which is equal to $\dim(X)$ mod 8. These different ways of describing the dimension of a smooth manifold generalize to noncommutative geometries, but for a noncommutative space they no longer necessarily agree.

Moreover, there is an additional more refined notion of dimension, which is no longer given by a single number but by a set of complex numbers: the dimension spectrum, which we define more precisely in the following paragraph. Even when applied to an ordinary manifold, this contains additional points besides the usual topological dimension of the manifold. It is customary to say that a noncommutative space manifests itself in several dimensions, given by the points of its dimension spectrum. At each of these points, as we discuss in more detail below, there is a corresponding notion of volume and integration. The dimension spectrum plays an important role in the asymptotic expansion of the action functional for noncommutative geometries.

Summarizing, the main different notions of dimension for a spectral triple $(\mathcal{A}, \mathcal{H}, D)$ are:

- *metric dimension*: it is determined by the rate of growth of the eigenvalues of the Dirac operator D. The metric dimension is equal to n if the operator $|D|^{-s}$ is of trace class on the half plane $\{s \in \mathbb{C} : \Re(s) > n\}$;
- the *KO-dimension* (an integer modulo 8) is determined by the signs of the commutation relations of J, γ, and D;
- The *dimension spectrum* is the set of poles in $\mathbb{C}$ of the zeta functions $\zeta_{b,D}(s) := \mathrm{Tr}(b|D|^{-s})$ associated to the spectral triple, where b is an element in the algebra generated by the $\delta^m(\pi(a))$ and $\delta^m([D, \pi(a)])$ for all $a \in \mathcal{A}$, and $m \in \mathbb{N}$, where $\delta(T) = [|D|, T]$, and the δ^m are the iterates of δ.

For general noncommutative spaces there is no reason why these different notions of dimension would continue to be as closely related as in the commutative case. In particular, is it shown in [Ćaćić (2013)] that a compact oriented Riemannian manifold X admits almost commutative spectral triples with arbitrary KO-dimension. The metric dimension of the almost commutative geometry is the same as that of the base manifold, hence this shows that one can have complete independence of metric and KO-dimension. Moreover, the dimension spectrum, as a subset of $\mathbb{C}$, can contain non-integer as well as non-real points. This is the case, for example, for fractal geometries, viewed as spectral triples.

1.1.2 *Manifolds*

The first and prototypical case of spectral triples is, of course, provided by compact smooth manifolds, endowed with a geometric structure. The latter include the cases of a Riemannian manifold, a Riemannian spin manifold, or and more generally a geometric structure coming from a Hermitian vector bundle with an essentially self-adjoint first-order elliptic differential operator. In the case of a Riemannian spin manifold, the data $(C^\infty(X), L^2(X, S), \not{D}_X)$, where S is the spinor bundle and $\not{D}_X$ the Dirac operator, determine a spectral triple, the Dirac spectral triple. Similarly, if X is a compact oriented Riemannian manifold of dimension n, the data $(C^\infty(X), L^2(X, \Lambda T^*_{\mathbb{C}} X), d + d^*, \gamma)$ with γ the $\mathbb{Z}/2\mathbb{Z}$-grading by parity of degree, also determine a spectral triple, the Hodge–de Rham spectral triple. More interestingly, the manifold, along with the geometric structure, can be reconstructed from the spectral triple, provided a certain set of axioms that we recall below are satisfied. These axioms describe stronger conditions on the data of the spectral triples than the basic definition we considered above, and are not always satisfied in more general cases.

The following statement of Connes' reconstruction theorem involves an orientability condition defined in terms of Hochschild cocycles. Since we will not need to explicitly use this condition and Hochschild and cyclic homology, we will not include here an explicit introduction to these topics, but we refer the reader to the book [Khalkhali (2009)] for a quick and very readable introduction.

The reconstruction theorem [Connes (2013)] for commutative spectral triples shows that a commutative unital Frechet pre-C^*-algebra $\mathcal{A}$ is isomorphic to $C^\infty(X)$ for some compact oriented n-dimensional smooth manifold X, if and only if there is a representation $\pi : \mathcal{A} \to \mathcal{L}(\mathcal{H})$ of $\mathcal{A}$ as bounded operators on a Hilbert space and a self-adjoint linear operator D on $\mathcal{H}$ such that the data $(\mathcal{A}, \mathcal{H}, D)$ is a spectral triple satisfying the properties:

- *metric dimension*: $(\mathcal{A}, \mathcal{H}, D)$ has metric dimension n;
- *order one*: $[[D, \pi(a)], \pi(b)] = 0$ holds for all $a, b \in \mathcal{A}$;
- *(weak) orientability*: there is a Hochschild cocycle $c \in Z_n(\mathcal{A}, \mathcal{A})$ such that $\pi_D(c)$, defined by

$$\pi_D(a^0 \otimes a^1 \otimes \cdots \otimes a^n) = a^0[D, a^1] \cdots [D, a^n],$$

  is self-adjoint, unitary, satisfying $\pi(a)\pi_D(c) = \pi_D(c)\pi(a)$ and $[D, \pi(a)]\pi_D(c) = (-1)^{n+1}\pi_D(c)[D, \pi(a)]$, for all $a \in \mathcal{A}$.
- *absolute continuity*: $\mathcal{H}^\infty = \cap_m \mathrm{Dom}(D^m)$ is a finite projective $\mathcal{A}$-module with an $\mathcal{A}$-valued inner product $(\cdot|\cdot)$, related to the inner product $\langle \cdot, \cdot \rangle$ of the Hilbert space by

$$\langle \xi, a\eta \rangle = \int a\,(\xi|\eta)\,|D|^{-n},$$

  for all $a \in \mathcal{A}$ and all $\xi, \eta \in \mathcal{H}^\infty$, with $\int$ the Dixmier trace;
- *regularity*: all operators $\pi(a)$ for $a \in \mathcal{A}$ and all the commutators $[D, \pi(a)]$ are in the domain of all the powers δ^m, $m \in \mathbb{N}$, of the unbounded derivations $\delta(T) = [|D|, T]$;
- *strong regularity*: endomorphisms of the $\mathcal{A}$-module $\mathcal{H}^\infty$ are in the domain of the δ^m.

The (weak) orientability axiom stated here can be strengthened to a (strong) orientability condition requiring that the operator $\pi_D(c)$ additionally satisfies $\pi_D(c)\,D + D\,\pi_D(c) = 0$ when n is even and $\pi_D(c) = 1$ when n is odd. One also says that the spectral triple is of Dirac type if $[D, \pi(a)]^2 \in \mathcal{A}$

for all $a \in \mathcal{A}$, see the discussion in [Ćaćić (2012)], [Ćaćić (2013)]. It was proved in [Connes (2013)] that a strongly orientable spectral triples satisfying the other properties listed above necessarily has $\mathcal{A} = C^\infty(X)$ for some compact oriented manifold X. Theorem 11.2 of [Gracia-Bondía, Várilly, Figueroa (2001)] then shows that the rest of the data of the spectral triple are of the form $(L^2(X, E), D)$ for some Hermitian bundle E, with D an essentially self-adjoint elliptic first order differential operator on E.

If in addition the weak closure $\mathcal{A}''$ (the double commutant of $\mathcal{A}$ in $\mathcal{B}(\mathcal{H})$) acts on $\mathcal{H}$ with multiplicity $2^{\lfloor n/2 \rfloor}$, then X is a compact spin^c manifold of dimension n. This spectral multiplicity fixes the Hilbert space representation of the von Neumann algebra $\mathcal{A}''$ and expresses a weaker form of Poincaré duality, see [Connes (2013)]. For such an algebra $\mathcal{A} \simeq C^\infty(X)$, the data $(\mathcal{A}, \mathcal{H}, D)$ are then of the form $(C^\infty(X), L^2(X, E), D\!\!\!/)$ where E is a Hermitian vector bundle on X and $D\!\!\!/$ is an essentially self-adjoint first-order elliptic differential operator on E, see [Gracia-Bondía, Várilly, Figueroa (2001)]. With the additional hypothesis that $\mathcal{A}''$ acts on $\mathcal{H}$ with multiplicity $2^{\lfloor n/2 \rfloor}$, the Hermitian bundle is a spinor bundle with $D\!\!\!/$ a Dirac type operator.

Without assuming strong orientability, and under a Dirac type condition $[D, a]^2 \in \mathcal{A}$ for $a \in \mathcal{A}$, it was shown in [Ćaćić (2012)] that the data $(\mathcal{A}, \mathcal{H}, D)$ are of the form $(C^\infty(X), L^2(X, E), D)$ with X a compact oriented *Riemannian* n-manifold, E a Hermitian vector bundle, and D is a Dirac-type operator.

These results say that, in the commutative case, spectral triples with sufficiently strong properties do indeed correspond to smooth manifolds with geometric structures, which include both Riemannian geometry and spin geometry.

The existence of a real structure J, in the case where E is a spinor bundle, corresponds to an isomorphism of E with its dual bundle, as Clifford modules. The signs $\varepsilon, \varepsilon', \varepsilon''$ are determined in this case by the Clifford algebra representation. Recall that the Clifford algebra $Cl(V, g)$ on an n-dimensional vector space V with a nondegenerate quadratic form g of signature (p, q), $p + q = n$, is obtained from the tensor algebra $T^\bullet(V)$ by imposing the relations $vw + wv = 2g(v, w)$. As a vector space, it is isomorphic to the exterior algebra $\Lambda^\bullet(V)$, but not as algebra. We then write $Cl_n^+ = Cl(\mathbb{R}^n, g_{n,0})$ and $Cl_n^- = Cl(\mathbb{R}^n, g_{0,n})$. These Clifford algebras satisfy the periodicity condition $Cl_{n+8}^\pm = Cl_n^\pm \otimes M_{16}(\mathbb{R})$. We have $Cl_n^\pm \subset \mathbb{C}l_n = Cl_n^\pm \otimes_\mathbb{R} \mathbb{C}$ and the following identifications:

n	Cl_n^+	Cl_n^-	Cl_n	S_n
1	$\mathbb{R} \oplus \mathbb{R}$	$\mathbb{C}$	$\mathbb{C} \oplus \mathbb{C}$	$\mathbb{C}$
2	$M_2(\mathbb{R})$	$\mathbb{H}$	$M_2(\mathbb{C})$	$\mathbb{C}^2$
3	$M_2(\mathbb{C})$	$\mathbb{H} \oplus \mathbb{H}$	$M_2(\mathbb{C}) \oplus M_2(\mathbb{C})$	$\mathbb{C}^2$
4	$M_2(\mathbb{H})$	$M_2(\mathbb{H})$	$M_4(\mathbb{C})$	$\mathbb{C}^4$
5	$M_2(\mathbb{H}) \oplus M_2(\mathbb{H})$	$M_4(\mathbb{C})$	$M_4(\mathbb{C}) \oplus M_4(\mathbb{C})$	$\mathbb{C}^4$
6	$M_4(\mathbb{H})$	$M_8(\mathbb{R})$	$M_8(\mathbb{C})$	$\mathbb{C}^8$
7	$M_8(\mathbb{C})$	$M_8(\mathbb{R}) \oplus M_8(\mathbb{R})$	$M_8(\mathbb{C}) \oplus M_8(\mathbb{C})$	$\mathbb{C}^8$
8	$M_{16}(\mathbb{R})$	$M_{16}(\mathbb{R})$	$M_{16}(\mathbb{C})$	$\mathbb{C}^{16}$

Both the real Clifford algebra and its complexification act on the spinor representation S_n. The existence of an antilinear $J : S_n \to S_n$ with $J^2 = 1$ and $[J, \pi(a)] = 0$ for all elements a of the real algebra, determines a real subbundle $\{v \in E : Jv = v\}$, while the existence of an antilinear $J : E \to E$ with $J^2 = -1$ and $[J, \pi(a)] = 0$ implies the existence of a quaternionic structure on E, compatible with the action of the real algebra. In both cases, for a faithful, irreducible representation, the real algebra is characterized as elements a of the complex algebra with $[J, \pi(a)] = 0$ or as elements of $End(E)$ such that $JaJ^* = a$. This provides the possible real structures J with the signs as in the classical table of signs of KO-dimension discussed at the beginning of §1.1, see [Landsman (2010)] for more details. The relation of this table of signs to KO-homology is then explained in §9.5 of [Gracia-Bondía, Várilly, Figueroa (2001)]. The Dirac operator of a compact spin manifold defines a K-homology class $[\slashed{D}] \in K_0(X)$. If the signs obtained as above correspond to a value of the KO-dimension, then the K-homology class in turn determines a class in KO-homology (which is graded modulo 8, by real Bott periodicity) in the corresponding degree. We refer the reader to [Gracia-Bondía, Várilly, Figueroa (2001)] and to [Lawson, Michelsohn (1989)] for more details and for a general introduction to spin geometry and KO-theory.

1.1.3   *Almost-commutative geometries*

The next class of interesting spectral triples, which play a crucial role in the construction of particle physics models, consists of the *almost commutative geometries*.

We adopt the general framework of [Ćaćić (2012)], [Ćaćić (2013)]. An almost commutative geometry from this viewpoint consists of a spectral triple $(\tilde{\mathcal{A}}, \mathcal{H}, D)$ together with a central unital ∗-subalgebra $\mathcal{A}_c$ of $\tilde{\mathcal{A}}$ such that

- $(\mathcal{A}_c, \mathcal{H}, D)$ is a Dirac-type n-dimensional commutative spectral triple satisfying the axioms of [Connes (2013)] listed above;
- $\tilde{\mathcal{A}}$ is a finitely generated projective unital $\mathcal{A}_c$-module-*-subalgebra of $\mathrm{End}_{\mathcal{A}_c}(\mathcal{H}^\infty)$;
- the order one condition $[[D, b], a] = 0$ holds for all $a \in \tilde{\mathcal{A}}$ and $b \in \mathcal{A}_c$.

A bundle of algebras $\mathcal{B}$ is a locally trivial bundle of finite-dimensional C^*-algebras over a compact smooth manifold X. A Hermitian vector bundle E on X is a $\mathcal{B}$-module (a representation of $\mathcal{B}$) if there is a monomorphism of bundles of algebras $\mathcal{B} \to \mathrm{End}(E)$. A Clifford $\mathcal{B}$-module is a Clifford module $E \to X$ with a representation of $\mathcal{A}$ that commutes with the Clifford action. A *concrete* almost-commutative spectral triple is of the form $(C^\infty(X, \mathcal{B}), L^2(X, E), D)$, with $\mathcal{B}$ a bundle of algebras, E a Clifford $\mathcal{B}$-module, X a compact oriented Riemannian manifold, and D a Dirac-type operator.

The main reconstruction theorem for almost-commutative geometries [Ćaćić (2012)] states that almost commutative geometries are always of this concrete type.

An especially simple case of almost-commutative geometries are the Cartesian ones, which are just products of a commutative spectral triple arising from a compact Riemannian manifold and a *finite spectral triple*, defined as in §1.1.5 below.

Particle physics models were usually built out of Cartesian almost commutative geometries, although it was shown in the work [Boeijink, van Suijlekom (2011)] that more general almost-commutative geometries involving nontrivial bundles also have relevance to particle physics models. Other interesting examples of almost commutative geometries that involve nontrivial bundles include the rational noncommutative tori. Examples of almost commutative geometries also occur in the approach to quantum error correcting codes in [Marcolli, Perez (2012)].

1.1.4 *Cartesian products*

In view of the special case of Cartesian almost commutative geometries, which we will be focusing on later in this chapter, it is worth recalling how the data of spectral triples behave under Cartesian products.

Let $(\mathcal{A}_i, \mathcal{H}_i, D_i)$ for $i = 1, 2$ be two spectral triples, then the product is of the form

- both spectral triples are even, with $\mathbb{Z}/2\mathbb{Z}$-gradings γ_i:

$$(\mathcal{A}_1 \otimes \mathcal{A}_2, \mathcal{H}_1 \otimes \mathcal{H}_2, D_1 \otimes 1 + \gamma_1 \otimes D_2), \quad \text{with grading} \quad \gamma_2 \otimes \gamma_2;$$

- the first is even with γ_1 grading and the second is odd:

$$(\mathcal{A}_1 \otimes \mathcal{A}_2, \mathcal{H}_1 \otimes \mathcal{H}_2, D_1 \otimes 1 + \gamma_1 \otimes D_2);$$

- the first is odd and the second is even with γ_2 grading:

$$(\mathcal{A}_1 \otimes \mathcal{A}_2, \mathcal{H}_1 \otimes \mathcal{H}_2, D_1 \otimes \gamma_2 + 1 \otimes D_2);$$

- both are odd:

$$(\mathcal{A}_1 \otimes \mathcal{A}_2, \mathcal{H}_1 \otimes \mathcal{H}_2 \otimes \mathbb{C}^2, D_1 \otimes 1 \otimes \sigma_1 + 1 \otimes D_2 \otimes \sigma_2), \quad \text{graded by } 1 \otimes 1 \otimes \sigma_3,$$

with σ_j the Pauli matrices

$$\sigma_1 = \begin{pmatrix} 0 & 1 \\ 1 & 0 \end{pmatrix}, \quad \sigma_2 = \begin{pmatrix} 0 & -i \\ i & 0 \end{pmatrix}, \quad \sigma_3 = \begin{pmatrix} 1 & 0 \\ 0 & -1 \end{pmatrix}.$$

1.1.5 *Finite spectral triples*

A *finite spectral triple* (or finite noncommutative geometry) is a triple $F = (\mathcal{A}_F, \mathcal{H}_F, D_F)$ where $\mathcal{A}_F$ is a *finite dimensional* algebra. This means that F has metric dimension zero, although, as we already discussed, its KO-dimension can be an arbitrary integer modulo 8.

It is convenient to allow for the possibility of both complex and real algebras. This is not a significant discrepancy with respect to the usual setting of spectral triples that only uses complex algebras. Indeed, one can reformulate the construction purely in terms of complex algebras, along the lines discussed in §13 of Chapter 1 of [Connes and Marcolli (2008)].

The data of a finite spectral triple $F = (\mathcal{A}_F, \mathcal{H}_F, D_F)$ are given by

- a finite dimensional (real) C^*-algebra $\mathcal{A}$, which by Wedderburn theorem can always be written as

$$\mathcal{A} = \oplus_{i=1}^{N} M_{n_i}(\mathbb{K}_i)$$

where $\mathbb{K}_i = \mathbb{R}$ or $\mathbb{C}$ or the quaternions $\mathbb{H}$,

$$\mathbb{H} = \{ \begin{pmatrix} \alpha & \beta \\ -\bar{\beta} & \bar{\alpha} \end{pmatrix} : \alpha, \beta \in \mathbb{C} \},$$

- a representation π of $\mathcal{A}$ on a finite dimensional (complex) Hilbert space $\mathcal{H}_F$;

- an antilinear involution J on $\mathcal{H}_F$ with signs $\epsilon, \epsilon', \epsilon''$ as discussed at the beginning of §1.1;
- a bimodule structure on $\mathcal{H}_F$ with the left and right action of $a, b \in \mathcal{A}$ given by $\pi(a)$ and $\pi(b)^0 = J\pi(b)^* J^{-1}$, satisfying the condition condition $[\pi(a), \pi(b)^0] = 0$ for the real structure J;
- a self adjoint liner operator $D^* = D$ satisfying the order one condition $[[D, \pi(a)], \pi(b)^0] = 0$.

The compact resolvent and bounded commutators conditions for Dirac operators of spectral triples are automatically satisfied since the algebra and Hilbert space are finite dimensional and D is just a matrix. Finite spectral triples are purely linear algebra objects.

There is a convenient graphical method for classifying finite spectral triples, consisting of Krajewski diagrams, [Krajewski (1995)]. Suppose given the algebra of a finite spectral triple as a sum $\mathcal{A} = \sum_{i=1}^N M_{n_i}(\mathbb{C})$ of matrix algebras. A faithful representation $\mathcal{H} = \oplus_{i=1}^N \mathbb{C}^{n_1} \otimes V_i$, has multiplicity specified by the vector space V_i of dimension r_i. The components of the Dirac operator are given by linear maps

$$D_{ij} : \mathbb{C}^{n_1} \otimes V_i \to \mathbb{C}^{n_j} \otimes V_j,$$

with $D_{ij} = D_{ji}^*$. To such data one can associate a *decorated graph* with sets of vertices and edges (V, E) (looping edges and multiple edges are allowed), with $\#V = N$. One decorates each vertex v by non-negative integers n_i (rank) and r_i (multiplicity). The graph has an edge between the vertices $v(n_i, r_i)$ and $v(n_j, r_j)$ if $D_{ij} \neq 0$.

For example, the finite spectral triple $(M_n(\mathbb{C}), \mathbb{C}^n, D = D_e + D_e^*)$ corresponds to a graph given by a single loop with one vertex decorated by n and one edge decorated by D_e, as shown in the figure.

It is possible to also incorporate the information about the real structure in this graphical description. Suppose again that the algebra has the form

$\mathcal{A} = \oplus_{i=1}^{N} M_{n_i}(\mathbb{C})$. The Hilbert space $\mathcal{H}$ is now a bimodule: a representation of $\mathcal{A} \otimes \mathcal{A}^o$,

$$\mathcal{H} = \oplus_{i,j=1}^{N} \mathbb{C}^{n_i} \otimes \mathbb{C}^{n_j o} \otimes V_{ij},$$

with irreducible representations $\mathbb{C}^{n_i} \otimes \mathbb{C}^{n_j o}$ and multiplicities V_{ij}. The real structure is an anti-unitary $J : \mathcal{H} \to \mathcal{H}$ with $J^2 = \pm 1$. When $J^2 = 1$, there is an orthonormal basis $\{e_k^{ij}\}$ of $\mathbb{C}^{n_i} \otimes \mathbb{C}^{n_j o} \otimes V_{ij}$ with $Je_k^{ij} = e_k^{ij}$. When $J^2 = -1$, there is an orthonormal basis $\{e_k^{ij}, f_k^{ji}\}$ with $e_k^{ij} \in \mathbb{C}^{n_i} \otimes \mathbb{C}^{n_j o} \otimes V_{ij}$ and $f_k^{ji} \in \mathbb{C}^{n_j} \otimes \mathbb{C}^{n_i o} \otimes V_{ij}$, with $Je_k^{ij} = f_k^{ji}$ and $Jf_k^{ji} = -e_k^{ij}$. As before, the Dirac operator is specified by components $D_{ij,kl}^* = D_{kl,ij}$, with

$$D_{ij,kl} : \mathbb{C}^{n_i} \otimes \mathbb{C}^{n_j o} \otimes V_{ij} \to \mathbb{C}^{n_k} \otimes \mathbb{C}^{n_l o} \otimes V_{kl}.$$

One needs to impose the compatibility with J and the order-one condition, $JD = \pm DJ$ and $[[D, a]b^0] = 0$. The edges in the diagram are only horizontal or vertical or looping edges at a vertex: indeed, the order one condition, when a, b are diagonal implies $[[D, a]b^0]_{ij,kl} = 0$, hence $D_{ij,kl} = 0$ when $i \neq j$ and $k \neq l$, so that only vertical and horizontal arrows or looping edges are possible. The compatibility with J relates $D_{ij,kl}$ with $D_{ji,lk}$, as a diagonal symmetry of the diagram.

In the case where one also considers real algebras, by Wedderburn theorem

$$\mathcal{A} = \oplus_{i=1}^{N} M_{n_i}(\mathbb{F}_i), \quad \mathbb{F}_i = \mathbb{C}, \mathbb{R}, \text{ or } \mathbb{H},$$

with $\mathbb{H}$ the quaternions, where $\mathbb{H} \otimes_{\mathbb{R}} \mathbb{C} = M_2(\mathbb{C})$ and $M_k(\mathbb{H}) \otimes_{\mathbb{R}} \mathbb{C} = M_{2k}(\mathbb{C})$. Representations are $\mathbb{R}$-linear $\pi : \mathcal{A} \to \mathcal{L}(\mathcal{H})$ to a complex $\mathcal{H}$ and are in one-to-one correspondence with complex representations of $\mathcal{A} \otimes_{\mathbb{R}} \mathbb{C}$. Krajewski diagrams in this case are modified by adding vertex labels of the form $(n_i, n_j, v_{ij}, \mathbb{F}_i, \mathbb{F}_j)$, while edges are still labelled by the Dirac components $D_{ij,kl}$. Irreducible finite real spectral triples are classified in [Chamseddine, Connes (2008)].

When one fixes the data $\mathcal{A}_F$, $\mathcal{H}_F$, and the real structure J of the finite spectral triple, and looks for the set of all possible matrices D, up to unitary equivalence, that satisfies the properties $D^* = D$ and $[[D, \pi(a)], \pi(b)^0] = 0$, one obtains a *moduli space* of Dirac operators. These moduli spaces were studied in general terms in [Ćaćić (2011)]. We will discuss more in details one case for a specific particle physics model, taken from [Chamseddine, Connes, Marcolli (2007)].

1.1.6 θ-deformations

The best known and most extensively studied examples of noncommutative spaces are certainly the *noncommutative tori*.

As a C^*-algebra, a noncommutative torus with modulus θ is a *rotation algebra*, with angle θ, namely the universal C^*-algebra $\mathcal{A}_\theta$ generated by two unitaries U and V subject to the commutation relation

$$VU = e^{2\pi i\theta}UV.$$

The C^*-algebra $\mathcal{A}_\theta$ is regarded as the algebra of continuous functions on the noncommutative space $\mathbb{T}_\theta$ (the noncommutative torus of modulus θ). The dense subalgebra given by

$$\mathcal{A}_\theta^\infty = \{a \in \mathcal{A}_\theta \,|\, a = \sum_{n,m} c_{n,m} U^n V^m, \ c_{n,m} \in \mathbb{C} \ \text{ rapid decay sequence}\}$$

can be seen as the algebra of smooth functions on $\mathbb{T}_\theta$.

The C^*-algebra $\mathcal{A}_\theta$ can be conveniently written as a crossed product algebra $\mathcal{A}_\theta = C(S^1) \rtimes_\theta \mathbb{Z}$ of the commutative algebra of functions over the circle, with respect to the action of $\mathbb{Z}$ generated by the rotation by angle θ. Another useful equivalent way of describing the algebra $\mathcal{A}_\theta$ is as a twisted group algebra, in the following sense. The group ring $\mathbb{C}[\mathbb{Z}^2]$ can be completed to the group C^*-algebra $C^*(\mathbb{Z}^2)$. By Fourier transform, this is isomorphic to $C(T^2) \simeq C^*(\mathbb{Z}^2)$, with T^2 the Pontrjagin dual of $\mathbb{Z}^2$. The algebra $C^*(\mathbb{Z}^2)$ has unitary commuting generators $U = \delta_{(0,1)}$ and $V = \delta_{(1,0)}$, seen as elements in the group ring, acting on $\ell^2(\mathbb{Z}^2)$ by $Uf(n,m) = f(n,m+1)$ and $Vf(n,m) = f(n+1,m)$. The twisted group ring $\mathbb{C}[\mathbb{Z}^2, \sigma]$, where $\sigma : \mathbb{Z}^2 \times \mathbb{Z}^2 \to U(1)$ is given by $\sigma((n,m),(k,r)) = \exp(\pi i\theta(nr - mk))$, has generators $Uf(n,m) = e^{-\pi i\theta n} f(n,m+1)$ and $Vf(n,m) = e^{-\pi i\theta m} f(n+1,m)$. This determines a new C^*-algebra, called the twisted group C^*-algebra $C^*(\mathbb{Z}^2, \sigma)$, which is still associative, but no longer commutative. In fact, the commutation relation becomes $VU = e^{2\pi i\theta}UV$, so that $C^*(\mathbb{Z}^2, \sigma) = \mathcal{A}_\theta$.

Clearly, when $\theta = 0$, the algebra is simply $\mathcal{A}_{\theta=0} = C(T^2)$, the algebra of functions on an ordinary 2-torus, with U and V the functions $e^{2\pi ix}$ and $e^{2\pi iy}$ on the two circles, and with the smooth algebra corresponding to smooth functions $C^\infty(T^2)$ written in their Fourier modes. When $\theta \neq 0$, the properties and behavior of noncommutative tori are significantly different, in the case where the angle θ is rational and in the case where it is irrational. In the rational case, these are again examples of noncommutative spaces that belong to the class of the almost commutative geometries (though not Cartesian ones) that we discussed in the previous section. However, in the irrational case, these are more essentially noncommutative objects: the irrational rotation algebras are simple, hence they cannot be almost commutative. In fact, this property is a sign that they are very far away from the commutative world.

These noncommutative spaces have been studied extensively since their introduction in the early '80s, both from the geometric [Connes (1980)] and the C^*-algebraic point of view [Rieffel (1981)], see also [Pimsner, Voiculescu (1980)], [Popa, Rieffel (1980)], for some of the early literature.

Noncommutative tori have found many diverse applications, ranging from the Quantum Hall Effect [Bellissard (1994)] to String Theory compactifications [Connes, Douglas, Schwarz (1998)]. They are as ubiquitous in noncommutative geometry as elliptic curves are in algebraic geometry. In fact they bear many similarities and intriguing relations to elliptic curves, see for example [Manin (2004)], [Marcolli (2005)].

Higher rank noncommutative tori are constructed similarly to the rank 2 case: given an $n \times n$ antisymmetric matrix $\Theta = (\theta_{ij})$, the rank n noncommutative torus $\mathcal{A}_\Theta$ is defined by the universal C^*-algebra generated by n unitaries U_i, subject to the commutation relations

$$U_j U_k = e^{2\pi i \theta_{jk}} U_k U_j.$$

We refer the reader to [Elliott (1984)], [Rieffel (1988)] for some early results on properties of higher rank noncommutative tori.

Besides being very interesting noncommutative spaces in themselves, noncommutative tori also provide a way of deforming certain classes of Riemannian manifolds to noncommutative spaces, in such a way that the deformation is *isospectral*, in the sense that the Hilbert space and Dirac operator of the spectral triple remain the same in the deformation. This method, introduced in [Connes, Landi (2001)], is known as θ-deformations.

Let X be a compact Riemannian spin manifold, and let

$$(C^\infty(X), L^2(X, S), \not{D}_X)$$

be its Dirac spectral triple. Suppose that the manifold X admits an action of a torus T^2 by isometries, $\alpha : T^2 \to \mathrm{Isom}(X, g_X)$, where g_X is the Riemannian metric. Then it is shown in [Connes, Landi (2001)] that the action α determines a noncommutative deformation $\mathcal{A}_\theta^\infty$ of the algebra of smooth functions $C^\infty(X)$, so that $(\mathcal{A}_\theta^\infty, L^2(X, S), \not{D}_X)$ is still a spectral triple. In turn, this determines the corresponding C^*-algebra $\mathcal{A}_\theta$ deforming $C(X)$.

The algebra $\mathcal{A}_\theta^\infty$ is obtained by decomposing elements of the original algebra into Fourier modes (weighted components) with respect to the torus action, and replacing the commutative pointwise product with a noncommutative product as in the noncommutative torus of modulus θ. Geometrically this can be thought of as deforming all the torus orbits of maximal dimension into noncommutative tori. More precisely, given an operator $\pi(f) \in \mathcal{L}(\mathcal{H}_X)$, for $\mathcal{H}_X = L^2(X, S)$ and $f \in C^\infty(X)$ represented on $\mathcal{H}_X$ as a multiplication operator acting on spinors, one decomposes it into weighted components according to the action of T^2,

$$\alpha_t(\pi(f_{n,m})) = e^{2\pi i(nt_1 + mt_2)} \pi(f_{n,m}).$$

The deformed product is given component-wise by setting

$$f_{n,m} \star_\theta h_{k.r} = e^{\pi i\theta(nr - mk)} f_{n,m} h_{k,r},$$

where, as before, we identify $\mathcal{A}_\theta = C^*(\mathbb{Z}^2, \sigma)$.

More general forms of θ-deformations, of possibly noncommutative spectral triples with isometric torus actions, have been studied in [Yamashita (2010)]. An abstract characterization of θ-deformations of commutative spectral triples, along with a corresponding reconstruction theorem, were obtained in [Ćaćić (2014)].

1.1.7 *Fractals*

Fractals are typically defined as spaces of non-integer dimension. The more precise notion of dimension in this context is the Hausdorff dimension (or some more easily computable approximations such as the box counting dimension). The Hausdorff dimension is associated to a one parameter family of Hausdorff measures, and it corresponds to the value of the parameter at which the Hausdorff measure of the set jumps from being infinity to being zero. We refer the reader to [Falconer (1986)] for a rigorous introduction to the subject of fractal geometry. Since for smooth manifolds the Hausdorff dimension agrees with the usual notion of dimension, fractals of non-integer dimension are necessarily non-smooth. Examples of fractals include, for instance, graphs of some nowhere differentiable functions. Thus, although they are still ordinary commutative spaces, they cannot be treated as smooth objects with the usual notions of commutative geometry. However, fractals can carry interesting structures of spectral triples, which allow them to be treated, to some extent, as smooth spaces from the point of view of noncommutative geometry.

A very simple example of a spectral triple associated to a fractal is the case of the middle third Cantor set X. This example is discussed in [Connes (1994)], [Connes and Marcolli (2008b)]. Consider the C^*-algebra $C(X)$ and the dense involutive subalgebra $\mathcal{A} \subset C(X)$ of locally constant functions on Λ. Let $\mathcal{J} = \{J_k\}$ be countable collection of bounded open intervals in $\mathbb{R} \setminus X$, and let $\{\ell_k\}$ be the set of their lengths, $\ell_k = \ell(J_k)$, listed in non-increasing order. Let $E = \{x_k^{\pm}\}$ be the set of endpoints, $J_k = (x_k^-, x_k^+)$. Construct the Hilbert space $\mathcal{H} = \ell^2(E)$. The algebra $C(X)$ is represented on $\mathcal{H}$ by

$$\pi(f)\xi(x) = f(x)\xi(x), \quad \forall f \in C(X), \ \xi \in \mathcal{H}, \ x \in E.$$

We define a Dirac operator D by defining its restriction D_k to each subspace $\mathcal{H}_k \subset \mathcal{H}$ spanned by $\xi(x_k^{\pm})$ as

$$D_k \begin{pmatrix} \xi(x_k^-) \\ \xi(x_k^+) \end{pmatrix} = \ell_k^{-1} \begin{pmatrix} \xi(x_k^+) \\ \xi(x_k^-) \end{pmatrix}.$$

This has polar decomposition $D = |D|\, F$, where the absolute value $|D|$ has spectrum $\{\ell_k^{-1}\}$ and the sign is given by $F_k = \begin{pmatrix} 0 & 1 \\ 1 & 0 \end{pmatrix}$. The data $(\mathcal{A}, \mathcal{H}, D)$ define a spectral triple, which is finitely summable, with

$$\mathrm{Tr}(|D|^{-s}) = 2\zeta_{L,X}(s), \tag{1.1}$$

where

$$\zeta_{L,X}(s) := \sum_k \ell_k^{-s} = \sum_{k \geq 1} 2^k 3^{-sk} \tag{1.2}$$

is the geometric zeta function of the fractal, in the sense of [Lapidus, van Frankenhuijsen (2013)]. The metric dimension of this spectral triple is equal to the Hausdorff dimension of the Cantor set

$$\dim_H(X) = \frac{\log 2}{\log 3},$$

while the dimension spectrum consists of all points

$$\mathrm{DimSp}(\mathcal{A}, \mathcal{H}, D) = \{ \frac{\log 2}{\log 3} + \frac{2\pi i n}{\log 3}, \ n \in \mathbb{Z} \}.$$

In the case of Cantor sets, the spectral action has a different form of asymptotic expansion, with respect to the case of a manifold or an almost commutative geometry, and for instance one can see the appearance of $\log(t)$ terms in the heat kernel expansion. We refer the reader to the recent results of [Pearson, Bellissard (2009)] and [Kellendonk, Savinien (2012)],

[Kellendonk, Savinien (2012b)]. We will discuss these oscillatory logarithmic terms in the spectral action in further details in §1.2.8 below.

More elaborate examples of spectral triples associated to fractals like Sierpinski gaskets were constructed more recently in [Christensen, Ivan, Lapidus (2008)], [Christensen, Ivan, Schrohe (2012)], [Cipriani et al. (2014)]. We will return to discuss the results of [Christensen, Ivan, Lapidus (2008)] and [Christensen, Ivan, Schrohe (2012)], as they will provide a model for the construction of gravity action functionals based on the spectral action that account for the possibility of fractal and multifractal structures in cosmology.

1.2 The Spectral Action functional

In the noncommutative geometry setting, one wants a good notion of an action functional, that can be associated to an arbitrary finitely summable spectral triple (commutative or noncommutative), which depends only on the data $(\mathcal{A}, \mathcal{H}, D)$ of the spectral triple (with the additional datum of the real structure J, if one is given). A suitable construction of such an action functional, *the spectral action*, was proposed in [Chamseddine, Connes (1997)]. The spectral action is defined as

$$\mathrm{Tr}(f(D/\Lambda)), \tag{1.3}$$

where D is the Dirac operator of the spectral triple, Λ is a mass (energy) scale that makes D/Λ dimensionless, and f is a non-negative even smooth function, which regularizes the otherwise divergent trace. One should think of the function f as a rapidly decaying function, which is a smooth approximation to a cutoff function. The cutoff would simply count the eigenmodes of the Dirac operator only up to a certain cutoff scale, and the smooth approximation allows for nicer analytic properties of the functionals. Indeed, for analytic purposes, it is convenient to require that the function f is the Laplace transform of a measure m on $\mathbb{R}_+$,

$$f(x) = \int_{\mathbb{R}_+} e^{-tx^2} \, dm(t).$$

One thinks of (1.3) as a functional defined over a space of possible Dirac operators, a space which, in particular, encodes the information on the possible choices of different Riemannian metrics. Thus, as we will discuss more in detail below, in the case of an ordinary manifold, the functional (1.3) is related to an action functional for gravity. One can also include in the expression (1.3) additional bosonic fields, which come from *fluctuations* of the Dirac operator. This leads to a version of (1.3) written as

$$\mathcal{S}_\Lambda[D,A] = \mathrm{Tr}(f(D_A/\mathbf{\Lambda})), \tag{1.4}$$

where

$$D_A = D + A + \varepsilon' \, J \, A \, J^{-1}$$

with $A = A^*$ a self-adjoint operator of the form

$$A = \sum a_j[D,b_j], \quad a_j, b_j \in \mathcal{A}.$$

In specific particle physics models, the fluctuations A will correspond to gauge fields, Higgs fields, and supersymmetric partners of the standard model fermions.

At first, an action functional written in the form (1.3) or (1.4) looks very different from the usual action functionals of physical theories, which are based on an integral of a local expressions in the fields, the Lagrangian density. The functional (1.3) depends only on the *spectrum* of the Dirac operator D, which depends on global information about the underlying geometry. However, as we discuss in §1.2.1 below, the spectral action has an asymptotic expansion that recovers the usual form of the familiar physical action functionals, and which will allow us to compare it with various modified gravity models.

1.2.1 *Asymptotic expansion*

A finitely summable spectral triple $(\mathcal{A}, \mathcal{H}, D)$ has *simple dimension spectrum*, if the zeta functions $\zeta_{b,D}(s) = \mathrm{Tr}(b\,|D|^{-s})$ have only simple poles. Also, as we discussed above in the axioms of the reconstruction theorem, a spectral triple is *regular* if all the operators $\pi(a)$ and $[D, \pi(a)]$ belong to the domain of all the δ^m, $m \in \mathbb{N}$, with $\delta(T) = [|D|, T]$. For regular spectral triples with simple dimension spectrum, it was proved in [Chamseddine, Connes (1997)] that the spectral action has an asymptotic expansion, for $\Lambda \to \infty$, with terms labeled by points in the dimension spectrum and coefficients that are residues of the zeta functions at those points.

To give a sketch of the argument, let e^{-tD^2} be the heat kernel of the Dirac operator D of a regular spectral triple with simple dimension spectrum. Then there is a heat kernel expansion at $t \to 0$ of the form

$$\mathrm{Tr}(e^{-tD^2}) \sim \sum_\alpha a_\alpha t^\alpha. \tag{1.5}$$

The heat kernel expansion is related to the zeta function $\zeta_D(s) = \text{Tr}(|D|^{-s})$ via Mellin transform

$$|D|^{-s} = \frac{1}{\Gamma\left(\frac{s}{2}\right)} \int_0^\infty e^{-tD^2} t^{s/2-1} \, dt$$

with

$$\int_0^1 t^{\alpha+s/2-1} \, dt = \frac{1}{\alpha + s/2}.$$

Thus, a non-zero terms a_α in (1.5), with $\alpha < 0$, corresponds to a *pole* of $\zeta_D(s)$ at $s = -2\alpha$ with residue

$$\text{Res}_{s=-2\alpha}\, \zeta_D(s) = \frac{2\, a_\alpha}{\Gamma(-\alpha)}.$$

The fact that no $\log t$ terms are present in the expansion implies regularity at 0 for ζ_D with $\zeta_D(0) = a_0$. In fact, we have

$$\frac{1}{\Gamma\left(\frac{s}{2}\right)} \sim \frac{s}{2} \qquad \text{as} \quad s \to 0.$$

The contribution to the zeta function $\zeta_D(0)$ from pole part at $s = 0$ of

$$\int_0^\infty \text{Tr}(e^{-tD^2})\, t^{s/2-1} \, dt$$

is given by

$$a_0 \int_0^1 t^{s/2-1} \, dt = a_0 \frac{2}{s}.$$

Thus, as proved in [Chamseddine, Connes (1997)], for a regular spectral triple with simple dimension spectrum, there is an asymptotic expansion for $\Lambda \to \infty$ of the spectral action (1.3) of the form

$$\text{Tr}(f(D/\Lambda)) \sim \sum_{\beta \in \text{DimSp}^+} f_\beta \Lambda^\beta \fint |D|^{-\beta} + f(0)\,\zeta_D(0) + o(1), \qquad (1.6)$$

where $f_\beta = \int_0^\infty f(v)\, v^{\beta-1} \, dv$ are the momenta of f, and the summation runs over the part $\text{DimSp}^+ \subset \text{DimSp}(\mathcal{A}, \mathcal{H}, D)$ of the dimension spectrum contained in the positive axis $\mathbb{R}_+^*$. The coefficient is given by

$$\fint |D|^{-\beta} = \text{Res}_{s=\beta}\, \zeta_D(s),$$

and it is interpreted as the noncommutative integration in dimension β:

$$\fint a\, |D|^{-\beta} := \text{Res}_{s=\beta}\, \zeta_{a,D}(s) = \text{Res}_{s=\beta}\, \text{Tr}(a\, |D|^{-s}), \quad \forall a \in \mathcal{A}.$$

This asymptotic expansion in powers of Λ, for large energies $\Lambda \to \infty$ applies equally to both (1.3) and (1.4). The spectral action (1.4) has another, very useful, asymptotic expansion in the fields A, for which we refer the reader to §7.2.2 of [van Suijlekom (2015)] and to [van Suijlekom (2011)], see also [Chamseddine, Connes (2006)].

1.2.2   *The spectral action as modified gravity*

We first consider the case where of a spectral triple $(\mathcal{A}, \mathcal{H}, D)$ that is the standard Dirac spectral triple $(C^\infty(X), L^2(X, S), D\!\!\!/_X)$ of a compact Riemannian 4-dimensional spin or spin^c manifold. Note that, while not all smooth compact 4-dimensional manifolds admit a spin structure, all of them admit a spin^c structure, see [Lawson, Michelsohn (1989)]. In this case, we see that the asymptotic expansion (1.6) of the spectral action is of the form

$$\mathrm{Tr}(f(D/\Lambda)) \sim f_2\Lambda^2 \int |D|^{-2} + f_4\Lambda^4 \int |D|^{-4} + f(0)\,\zeta_D(0) + o(1),$$

with terms in dimension 0, 2, and 4. The coefficients of these terms can be computed explicitly from the heat kernel expansion of $D\!\!\!/_X^2$ and the Seeley-deWitt coefficients, see Chapter 1 of [Connes and Marcolli (2008)]. We refer the reader to [Kalau, Walze (1995)], [Kastler (1995)], and [Ackermann (1996)] for a detailed discussion of this case. See also [Gilkey (1995)] for the heat kernel expansion for $D\!\!\!/_X^2$ and more general elliptic operators on manifolds. The resulting terms are then of the form

$$f_2\Lambda^2 \int |D|^{-2} = \frac{96\, f_2\Lambda^2}{24\pi^2} \int R\,\sqrt{g}\, d^4x \qquad (1.7)$$

$$f_4\Lambda^4 \int |D|^{-4} = \frac{48 f_4\Lambda^4}{\pi^2} \int \sqrt{g}\, d^4x \qquad (1.8)$$

$$f(0)\,\zeta_D(0) = \frac{f_0}{10\pi^2} \int \left(\frac{11}{6} R^* R^* - 3 C_{\mu\nu\rho\sigma} C^{\mu\nu\rho\sigma}\right) \sqrt{g}\, d^4x, \qquad (1.9)$$

with the Gauss–Bonnet term

$$R^* R^* = \frac{1}{4} \epsilon^{\mu\nu\rho\sigma} \epsilon_{\alpha\beta\gamma\delta} R^{\alpha\beta}_{\ \ \mu\nu} R^{\gamma\delta}_{\ \ \rho\sigma}, \qquad (1.10)$$

and the Weyl curvature tensor

$$
\begin{aligned}
C_{\lambda\mu\nu\kappa} = R_{\lambda\mu\nu\kappa} &- \frac{1}{2}(g_{\lambda\nu}R_{\mu\kappa} - g_{\lambda\kappa}R_{\mu\nu} - g_{\mu\nu}R_{\lambda\kappa} + g_{\mu\kappa}R_{\lambda\nu}) \\
&+ \frac{1}{6} R^\alpha_{\ \alpha}(g_{\lambda\nu}g_{\mu\kappa} - g_{\lambda\kappa}g_{\mu\nu}).
\end{aligned}
\qquad (1.11)
$$

By direct inspection of this asymptotic expansion, we see that, in addition to the first two terms, which recover the Einstein–Hilbert action with cosmological constant, we also have modified gravity terms involving the Gauss–Bonnet form and the Weyl curvature. We discuss below the meaning of these terms from the point of view of modified gravity models.

We discuss below the role of the momenta f_0, f_2, f_4 and of the energy scale Λ in regard to the gravitational parameters for the classical Einstein–Hilbert action functional (gravitational and cosmological constant) and for the modified gravity terms. In fact all the next chapter will be dedicated to various different models describing the running of these coefficients with the renormalization group flow and their relevance to Early Universe models in cosmology. However, we will first need to enrich our gravity model with a coupling of gravity to matter, by changing the underlying geometry from commutative to almost commutative. This will also change the form of the coefficients of the gravitational terms.

1.2.3 *Riemannian versus Lorentzian geometries*

The theory of spectral triples we discussed in this chapter is tailor made to the geometric and analytic properties of Dirac operators on *compact Riemannian* manifolds. In particular, for the spectral action, defined as in (1.3) to make sense, one needs the operator D to have discrete spectrum with finite multiplicities, which is ensured by the compact resolvent condition in the spectral triple axioms. However, this property is not satisfied by Dirac operators on Lorentzian manifolds. Thus, one cannot simply extend the theory of spectral triples to the Lorentzian case. Several approaches to relate Lorentzian geometry to the setting of spectral triples have been proposed. Typically, they involve a formulation that uses Krein spaces instead of Hilbert spaces, which an indefinite signature version of inner product spaces, see [Bognár (1974)], and a Wick rotation operator relating the data to those of a Riemannian spectral triple. The reader may want to consult [van Suijlekom (2004)], [Marcolli (2008)], [van den Dungen, Paschke, Rennie (2013)], [Franco (2014)] for examples of such constructions.

It is well known that not all Riemannian manifolds arise as Wick rotations of Lorentzian manifolds, nor in general Riemannian manifolds would admit a Wick rotation to a Lorentzian manifold: there are topological obstructions that prevent it. A general Lorentzian metric need not have a section in a complexification of the spacetime manifold on which the metric is real and positive definite, and conversely. However, when that happens to be the case and we can Wick rotate between Riemannian and Lorentzian signature, one can define the spectral action functional as (1.3) on the Riemannian manifold and observe that the terms of the asymptotic expansion continue to make sense also for the Lorentzian metric. This gives us a more conservative approach, which does not require the development of a theory of spectral triples for Lorentzian manifolds, but which still provides us with an action functional for modified gravity. We will see later in the book, when we discuss quantum cosmology, that this approach is consistent with the Riemannian setting for quantum cosmology of [Hawking (1984)].

1.2.4   *Einstein–Hilbert action with cosmological term*

The terms in dimension 2 and dimension 4 of the asymptotic expansion of the spectral action on a 4-dimensional manifold correspond to an Einstein–Hilbert action functional with cosmological constant,

$$S_{EH} = \int \frac{1}{2\kappa}(R + 2\Lambda)\sqrt{g}\,d^4x$$

with $\kappa = 8\pi G$ (in units with $c = 1$) where G is the gravitational constant and Λ is the cosmological constant (notation not to be confused with the energy scale $\boldsymbol{\Lambda}$). We will also write the second term with the notation

$$\gamma \int \sqrt{g}\,d^4x, \quad \text{with } \gamma = \frac{\Lambda}{8\pi G}.$$

More precisely, the type of Einstein–Hilbert action we will be considering, arising from spectral action models of gravity coupled to matter, will be of the form

$$\frac{1}{16\pi\, G_{\text{eff}}(\boldsymbol{\Lambda})} \int R\sqrt{g}\,d^4x + \frac{\Lambda_{\text{eff}}(\boldsymbol{\Lambda})}{8\pi\, G_{\text{eff}}(\boldsymbol{\Lambda})} \int \sqrt{g}\,d^4x,$$

with an effective gravitational constant $G_{\text{eff}}(\boldsymbol{\Lambda})$ and an effective cosmological constant $\Lambda_{\text{eff}}(\boldsymbol{\Lambda})$ that can run with the energy scale $\boldsymbol{\Lambda}$ according to suitable renormalization group equations, which depends on the specific underlying almost commutative geometry.

1.2.5   *Conformal gravity*

Conformal gravity denotes a class of modified gravity theories that are invariant under conformal transformations of the metric. The simplest case has action functional that depends on the Weyl curvature in the form

$$S_{\text{conf}} = \int C_{\mu\nu\rho\sigma}C^{\mu\nu\rho\sigma}\sqrt{g}\,d^4x. \qquad (1.12)$$

The equations of motion for this action functional are the Bach equations

$$2\nabla_\mu\nabla_\sigma C^{\mu}{}_{\nu\rho}{}^{\sigma} + C^{\mu}{}_{\nu\rho}{}^{\sigma}R_{\mu\sigma} = 0,$$

with $R_{\mu\nu}$ the Ricci curvature. Solutions are conformally flat metrics.

Conformal gravity is a very promising candidate for a modified gravity theory, see the extensive discussion in [Mannheim (2012)], [Mannheim (2012b)], including cosmological implications.

As in the case of the Einstein–Hilbert and cosmological terms, the conformal gravity term in the asymptotic expansion of the spectral action occurs with a coefficient that may be running with the energy scale $\mathbf{\Lambda}$ and whose expression depends on the underlying almost commutative geometry. In particular, it becomes an interesting question to identify if, for specific geometries and specific energy scales (hence a specific cosmological timeline), this modified gravity term is subdominant with respect to the Einstein–Hilbert action, or whether it becomes significant.

In particular, as we will discuss in detail below, over almost commutative geometries, the conformal gravity action will acquire additional terms involving a conformal non-minimal coupling of a field (which may be the Higgs field or a scalar field) to gravity. This type of conformal gravity terms, together with the Weyl curvature term (1.12) arise in the context of Hoyle–Narlikar cosmologies. We will discuss in the next chapter the significance of this occurrence for Early Universe models.

1.2.6 *Gauss–Bonnet gravity*

In dimension $\dim(X) = 4$ the term (1.10) is topological, in the sense that it just integrates to the Euler characteristic $\chi(X)$. While, for $\dim(X) = 4$ this is not a dynamical term, in higher dimensions it is the term that gives rise to the type of modified gravity model that is usually referred to as Lovelock gravity or Gauss–Bonnet gravity.

The general form of Gauss–Bonnet gravity has an action functional of the form

$$S_{GB} = \int (\gamma_0 + \gamma_1 R + \gamma_2 (R^2 + R_{\mu\nu\rho\sigma} R^{\mu\nu\rho\sigma} - 4 R_{\mu\nu} R^{\mu\nu})),$$

while more general forms of Lovelock gravity also allow for higher order correction terms given by a series $\sum_k \gamma_k \mathcal{R}_k$. In dimension $\dim(X) = 4$, the second order term is the Gauss–Bonnet form, hence this correction term integrates to a constant multiple of the topological Euler characteristic $\chi(X)$ and is non-dynamical. However, it becomes non-trivial in extra dimensional models.

As in the case of the conformal gravity terms, the Gauss–Bonnet term appears in the asymptotic expansion of the spectral action with a coefficient that depends on the underlying almost commutative geometry and that may run with the energy scale, hence the same observations apply regarding the subdominant nature or the significance of this term.

1.2.7 *Poisson summation*

We conclude this general discussion on the spectral action functional by mentioning a method described in [Chamseddine, Connes (2010)] for computing the spectral action, that applies to homogeneous and isotropic geometries with sufficient symmetry that the spectrum of the Dirac operator can be explicitly computed, along with the multiplicities of the eigenvalues (which are in general non-trivial for highly symmetric geometries). In such cases one can try to compute the spectral action directly from the definition (1.3), which requires summing the series

$$\text{Tr}(f(D/\mathbf{\Lambda})) = \sum_{\lambda \in \text{Spec}(D)} m_\lambda \, f(\lambda/\mathbf{\Lambda}), \tag{1.13}$$

where $m_\lambda = \dim E_\lambda$ are the spectral multiplicities, the dimensions of the eigenspaces E_λ.

Let $h \in \mathcal{S}(\mathbb{R})$ be a rapidly decaying function. Then the Poisson summation formula states that summing the values of h along a lattice (an arithmetic progression) is the same as summing the values of the Fourier transform along the dual lattice. More precisely, we have

$$\sum_{k \in \mathbb{Z}} h(x + 2\pi k) = \frac{1}{2\pi} \sum_{n \in \mathbb{Z}} \hat{h}(n) e^{inx}. \tag{1.14}$$

In fact, the function $f(x) = \sum_{k \in \mathbb{Z}} h(x + 2\pi k)$ is 2π-periodic with Fourier coefficients

$$\hat{f}_n = \frac{1}{2\pi} \int_0^{2\pi} f(x) e^{-inx} dx = \frac{1}{2\pi} \sum_{k \in \mathbb{Z}} \int_0^{2\pi} h(x + 2\pi k) e^{-inx} dx$$

$$= \frac{1}{2\pi} \sum_{k \in \mathbb{Z}} \int_{2\pi k}^{2\pi(k+1)} g(x) e^{-inx} dx = \frac{1}{2\pi} \int_{\mathbb{R}} h(x) e^{-inx} dx = \frac{1}{2\pi} \hat{h}(n).$$

Thus, we have

$$\sum_{n \in \mathbb{Z}} h(x + \lambda n) = \frac{1}{\lambda} \sum_{n \in \mathbb{Z}} \exp\left(\frac{2\pi inx}{\lambda}\right) \widehat{h}(\frac{n}{\lambda})$$

with $\lambda \in \mathbb{R}_+^*$ and $x \in \mathbb{R}$, and with

$$\widehat{h}(x) = \int_{\mathbb{R}} h(u) \, e^{-2\pi iux} \, du.$$

Consider then the case of a geometry for which the Dirac spectrum has the following structure:

- the eigenvalues $\lambda \in \mathrm{Spec}(D)$ form an arithmetic progression;
- there is a real polynomial $P(x)$ that interpolates the values of the spectral multiplicities, namely such that

$$m_\lambda = P(\lambda).$$

In this very special case, if the test function f is in the space of rapid decay functions $\mathcal{S}(\mathbb{R})$, then the product $P(\lambda)f(\lambda/\Lambda) = h(\lambda)$ is also a rapidly decaying function and summing the series (1.13) is the same thing, by the Poisson summation formula, as summing the series of the Fourier transforms. One can then show that all the terms in this dual series, except the one evaluated at zero are smaller than any Λ^{-k} for arbitrary $k \in \mathbb{N}$, see [Chamseddine, Connes (2010)].

We will discuss applications of this technique later in the book, when we describe the spectral action approach to the problem of cosmic topology. We reproduce here the example of the 3-sphere S^3, from [Chamseddine, Connes (2010)], which will be the prototype example for the other, more complicated, topologies we discuss later. In the case where $X = S^3$ with the standard round metric of radius a, the Dirac spectrum is given by

$$\mathrm{Spec}(\slashed{D}_{S^3}) = \{\pm a^{-1}(\frac{1}{2} + n) : n \in \mathbb{Z}\},$$

with multiplicities $n(n+1)$, and applying the Poisson summation technique described above one obtains for the spectral action

$$\mathrm{Tr}(f(D/\Lambda)) = (\Lambda a)^3 \widehat{f}^{(2)}(0) - \frac{1}{4}(\Lambda a)\widehat{f}(0) + O((\Lambda a)^{-k})$$

with $\widehat{f}^{(2)}$ the Fourier transform of $v^2 f(v)$.

For a 4-dimensional Riemannian product geometry $S^3 \times S^1$, with the standard metrics on both factors, and the circe S^1 of size β, the spectral action is correspondingly of the form

$$\mathrm{Tr}(h(D^2/\Lambda^2)) = \pi\Lambda^4 a^3 \beta \int_0^\infty u\, h(u)\, du - \frac{1}{2}\pi\Lambda a \beta \int_0^\infty h(u)\, du + O(\Lambda^{-k}),$$

where the Poisson summation formula is applied to

$$g(u,v) = 2P(u)\, h(u^2(\Lambda a)^{-2} + v^2(\Lambda \beta)^{-2})$$

$$\widehat{g}(n,m) = \int_{\mathbb{R}^2} g(u,v)e^{-2\pi i(xu+yv)}\, du\, dv.$$

1.2.8   *Spectral action and expansion on fractals*

We discuss here the case of a fractal with exact self-similarity, namely such that the contraction ratios are all integer powers of a fixed scale $0 < r < 1$, such as the classical ternary Cantor set we discussed in §1.1.7. In these cases, the Dirac zeta function (1.1)

$$\mathrm{Tr}(|D^{-s}|) = 2\zeta_{L,X}(s) = 2 \sum_k \ell_k^{-s},$$

with $\{\ell_k\}$ the associated sequence of lengths (fractal string L), has the property that $\zeta_{L,X}(s)$ has poles off the real line, which lie on the vertical line with $\Re(s) = \sigma = \dim_H(X)$, the Hausdorff dimension, periodically spaced with period $2\pi/\log(1/r)$, that is, poles

$$s_m = \sigma + \frac{2\pi i m}{\log(1/r)}, \quad m \in \mathbb{Z}.$$

These non-real poles contribute to the heat kernel asymptotics, through a series of log-oscillatory terms, in the following way, [Dunne (2012)].

Proposition 1.2.1. *Let X be a fractal with a spectral triple for which the Dirac operator D has eigenvalues of $|D|$ growing exponentially like b^n for some $b > 1$, with spectral multiplicities that also grow exponentially, like a^n, for some $a > 1$. Then the spectral action $\mathrm{Tr}(f(D/\Lambda))$ has an expansion for large Λ of the form*

$$\mathrm{Tr}(f(D/\Lambda)) \sim \Lambda^\sigma \sum_{m \in \mathbb{Z}} \Lambda^{\frac{2\pi i m}{\log b}} f_{s_m} \tag{1.15}$$

where $\sigma = \frac{\log a}{\log b}$ and

$$s_m = \sigma + \frac{2\pi i m}{\log b},$$

where the coefficients f_{s_m} are given by

$$f_{s_m} = \frac{1}{\log b} \int_0^\infty f(u)\, u^{s_m} \frac{du}{u}. \tag{1.16}$$

Proof.   The proof follows from [Dunne (2012)], see also Proposition 3.1 of [Ball, Marcolli (2016)]. The zeta function is given by

$$\zeta_{L,X}(s) = \sum_n a^n b^{-sn} = (1 - a b^{-s})^{-1}$$

which has simple poles at

$$s_m = \frac{\log a}{\log b} + \frac{2\pi i m}{\log b}.$$

The trace of the heat kernel is given by

$$\text{Tr}(e^{-tD^2}) = \sum_n a^n e^{-tb^{2n}}.$$

As shown in [Dunne (2012)], the small time asymptotics for this kind of heat kernels behave line

$$\text{Tr}(e^{-tD^2}) \sim \frac{t^{-\frac{\log a}{2\log b}}}{2\log b} \sum_{m\in\mathbb{Z}} \Gamma\left(\frac{\log a}{2\log b} + \frac{\pi i m}{\log b}\right) \exp\left(-\frac{\pi i m}{\log b}\log t\right).$$

Under Mellin transform

$$|D|^{-s} = \frac{1}{\Gamma(s/2)} \int_0^\infty e^{-tD^2} t^{\frac{s}{2}-1}\, dt$$

this corresponds to

$$\text{Tr}(|D|^{-s}) = \sum_m \frac{\Gamma(s_m/2)}{\Gamma(s/2)\cdot(s-s_m)\cdot\log b} + \text{ holomorphic terms}.$$

For a test function written as Laplace transform as $k(u) = \int_0^\infty e^{-xu} h(x)dx$, so that $k(tD_X^2) = \int_0^\infty e^{-xtD_X^2} h(x)dx$, one then has

$$k(tD_X^2) \sim \sum_m \frac{\Gamma(s_m/2)}{2\log b} t^{-s_m/2} \int_0^\infty x^{-s_m/2} h(x)dx.$$

Since $\Re(s_m) = \sigma > 0$, we can write $x^{-s_m/2}$ as Mellin transform

$$x^{-s_m/2} = \frac{1}{\Gamma(s_m/2)} \int_0^\infty e^{-xv}\, v^{\frac{s_m}{2}-1}\, dv,$$

hence we obtain

$$\text{Tr}(k(tD_X^2)) \sim \sum_m \text{Res}_{s=s_m} \zeta_{D_X}(s)\, t^{-s_m/2} \int_0^\infty k(v) v^{\frac{s_m}{2}-1} dv.$$

Taking $f(u) = k(u^2)$, we find

$$\int_0^\infty k(v) v^{\frac{s_m}{2}-1} dv = 2 \int_0^\infty f(u) u^{s_m-1} du.$$

We then set $t = \Lambda^{-2}$ to obtain the expansion of the spectral action

$$\Lambda^\sigma \sum_{m\in\mathbb{Z}} \Lambda^{\frac{2\pi i m}{\log b}} \left(\int_0^\infty f(u) u^{s_m-1} du\right) \text{Res}_{s=s_m} \zeta_{L,X}(s).$$

Using the relation between Mellin and Fourier transform, we write

$$f_{s_m} = \frac{1}{\log b} \int_0^\infty f(u) u^\sigma e^{-2\pi i m \frac{\log u}{\log b}}\frac{du}{u} = \int_\mathbb{R} F(\lambda) e^{-2\pi i m\lambda} d\lambda = 2\pi\hat{F}(-2\pi m),$$

with $\lambda = \frac{\log u}{\log b}$ and $F(\lambda) = f(b^\lambda)b^{\lambda\sigma}$, and with $\hat{F}(\xi) = (2\pi)^{-1}\int_\mathbb{R} F(\lambda)e^{i\xi\lambda}d\lambda$ the Fourier transform. If the test function $f(u)$ is sufficiently rapidly decaying, the Fourier series $\sum_{m\in\mathbb{Z}} \Lambda^{\frac{2\pi i m}{\log b}} f_{s_m}$ converges to a smooth function of the circle variable $\theta = \frac{\log\Lambda}{\log b}$ modulo $2\pi\mathbb{Z}$. $\qquad\square$

1.3   Particle Physics models

In the previous section, we only considered the simplified model where the underlying geometry is an ordinary 4-dimensional smooth compact Riemannian spin manifold. We showed that the spectral action, on such a geometry, can be considered as a modified gravity model.

This model can be considerably refined, if the commutative geometry X is replaced by an almost commutative geometry. Then the same action functional, written in the form (1.4) and supplemented with an appropriate fermionic term, delivers a model of gravity couple to matter, where the matter content and the possible interactions are completely determined by the geometry.

We will work under the simplifying assumption that the almost commutative geometry is Cartesian, hence of the form $X \times F$, with the product defined as in §1.1.4, where X is a 4-dimensional smooth compact Riemannian spin manifold, as before, and F is a finite noncommutative geometry, as in §1.1.5. However, we refer the reader to [Boeijink, van Suijlekom (2011)] for an example of a model of gravity coupled to matter, over a non-Cartesian almost commutative geometry, where the non-trivial structure of the underlying geometry plays a significant role in the model.

1.3.1   *Gravity coupled to matter*

The general procedure, given a Cartesian almost commutative geometry $X \times F$, is to consider as action functional for gravity coupled to matter the spectral action (1.4) for the Dirac operator D of the product geometry, together with a fermionic term. In the case of a finite geometry F with KO-dimension 6, that we will be focusing on in discussing the extension of the Minimal Standard Model obtained in [Chamseddine, Connes, Marcolli (2007)], the fermionic term takes the form

$$\frac{1}{2}\langle J\tilde{\xi}, D_A\tilde{\xi}\rangle, \tag{1.17}$$

where J is the real structure of the product geometry, D_A is the Dirac operator on the product geometry, with inner fluctuations A, and $\mathcal{H}_{cl}^+$ is the space of classical spinors, seen as anticommuting Grassmann variables. These are anticommuting variables satisfying the basic integration rule

$$\int \xi\, d\xi = 1.$$

The Pfaffian of an antisymmetric quadratic form $\mathfrak{A}(\xi, \xi)$ in Grassmann variables ξ is given by the functional integral

$$Pf(\mathfrak{A}) = \int e^{-\frac{1}{2}\mathfrak{A}(\xi,\xi)} \, D[\xi].$$

The reason from viewing fermions as Grassmann variables is the fact that the bilinear form (1.17) is an antisymmetric form on

$$\mathcal{H}_{cl}^+ = \{\xi \in \mathcal{H}_{cl} \mid \gamma\xi = \xi\},$$

and as such the associated quadratic form $\mathfrak{A}(\xi, \xi)$ is zero on commuting variables but nonzero on Grassmann variables.

The action functional for the full model of gravity coupled to matter, for the product geometry $X \times F$, is then of the form

$$S_\Lambda[D, A, \tilde{\xi}] = \mathrm{Tr}(f(D_A/\Lambda)) + \frac{1}{2} \langle J\tilde{\xi}, D_A \tilde{\xi} \rangle.$$

In [Chamseddine, Connes, Marcolli (2007)] it was shown that this action recovers the complete action functional of an extension of the Minimal Standard Model with right handed neutrinos and Majorana mass terms, non-minimally coupled to gravity, for a specific choice of the finite geometry F. We will discuss this model more in detail in §1.3.2 below.

Since the main focus in this book is on models of cosmology, rather than on particle physics models, we are mostly interested in the bosonic part of the spectral action, given by the term $\mathrm{Tr}(f(D_A/\Lambda))$, hence later in the book we will no longer discuss the fermionic term (1.17).

1.3.2 *The Standard Model: νMSM*

The particle physics model constructed in [Chamseddine, Connes, Marcolli (2007)] starts with, as input ansatz, a finite dimensional real algebra given by the "left-right symmetric" algebra

$$\mathcal{A}_{LR} = \mathbb{C} \oplus \mathbb{H}_L \oplus \mathbb{H}_R \oplus M_3(\mathbb{C}) \tag{1.18}$$

with involution $(\lambda, q_L, q_R, m) \mapsto (\bar{\lambda}, \bar{q}_L, \bar{q}_R, m^*)$. The $\mathbb{C}$-subalgebra $\mathbb{C} \oplus M_3(\mathbb{C})$ corresponds to integer spin, while the $\mathbb{R}$-subalgebra $\mathbb{H}_L \oplus \mathbb{H}_R$ corresponds to half-integer spin. More general choices of initial ansatz that lead to the same model are given in [Chamseddine, Connes (2008)].

Given a bimodule $\mathcal{M}$ over $\mathcal{A}$, a unitary $u \in \mathcal{U}(\mathcal{A})$ acts in the adjoint representation as

$$\mathrm{Ad}(u)\xi = u\xi u^* \quad \forall \xi \in \mathcal{M}.$$

A bimodule $\mathcal{M}$ for $\mathcal{A}_{LR}$ is *odd* iff $s = (1, -1, -1, 1)$ acts by $Ad(s) = -1$. Odd bimodule for the real algebra (1.18) are also representations of the complex algebra $M_2(\mathbb{C}) \oplus M_6(\mathbb{C})$, where $M_2(\mathbb{C}) = \mathbb{H} \otimes_{\mathbb{R}} \mathbb{C}$ and $M_6(\mathbb{C}) = H \otimes_{\mathbb{R}} M_3(\mathbb{C})$.

The Hilbert space of the finite geometry is constructed by considering the sum $\mathcal{M}_F$ of all inequivalent irreducible odd $\mathcal{A}_{LR}$-bimodules, and then taking

$$\mathcal{H}_F = \mathcal{M}_F \oplus \mathcal{M}_F \oplus \mathcal{M}_F.$$

Here the choice of taking three identical copies of $\mathcal{M}_F$ corresponds to the fact that there are $N = 3$ generations of particles in the Standard Model. The number of generations is an input choice that is not derived from other data of the model.

The contragradient bimodule $\mathcal{M}^0$ of $\mathcal{M}$ is defined as

$$\mathcal{M}^0 = \{\bar{\xi} \; ; \; \xi \in \mathcal{M}\}, \quad a\bar{\xi}b = \overline{b^*\xi a^*}.$$

We also denote by $\mathcal{B}$ the $\mathbb{C}$-algebra

$$\mathcal{B} = (\mathcal{A}_{LR} \otimes_{\mathbb{R}} \mathcal{A}_{LR}^{op})_p = \oplus^{4 \, \text{times}} M_2(\mathbb{C}) \oplus M_6(\mathbb{C}),$$

where $p = \frac{1}{2}(1 - s \otimes s^0)$. In the following we write $\mathbf{1}$ and $\mathbf{3}$ for the representations of $\mathbb{C}$ and $M_3(\mathbb{C})$ on $\mathbb{C}$ and $\mathbb{C}^3$, respectively, and $\mathbf{2}$ for the representation of $\mathbb{H}$ on $\mathbb{C}^2$ by

$$q = \begin{pmatrix} \alpha & \beta \\ -\bar{\beta} & \bar{\alpha} \end{pmatrix}$$

for $q = \alpha + \beta j \in \mathbb{H}$ with $\alpha, \beta \in \mathbb{C}$. We also use the notation $| \uparrow \rangle$ and $| \downarrow \rangle$ for the eigenvectors

$$q(\lambda) = \begin{pmatrix} \lambda & 0 \\ 0 & \bar{\lambda} \end{pmatrix} \quad q(\lambda)| \uparrow \rangle = \lambda | \uparrow \rangle, \quad q(\lambda)| \downarrow \rangle = \bar{\lambda} | \downarrow \rangle.$$

The module $\mathcal{M}_F$ has the following properties:

- $\dim_{\mathbb{C}} \mathcal{M}_F = 32$.
- $\mathcal{M}_F = \mathcal{E} \oplus \mathcal{E}^0$, where

$$\mathcal{E} = \mathbf{2}_L \otimes \mathbf{1}^0 \oplus \mathbf{2}_R \otimes \mathbf{1}^0 \oplus \mathbf{2}_L \otimes \mathbf{3}^0 \oplus \mathbf{2}_R \otimes \mathbf{3}^0.$$

- $\mathcal{M}_F \cong \mathcal{M}_F^0$ by the antilinear map $J_F(\xi, \bar{\eta}) = (\eta, \bar{\xi})$, for $\xi, \eta \in \mathcal{E}$, which satisfies

$$J_F^2 = 1, \quad \xi b = J_F b^* J_F \xi \quad \forall \xi \in \mathcal{M}_F, \, b \in \mathcal{A}_{LR}.$$

- $\mathcal{M}_F$ is a sum irreducible representations of $\mathcal{B}$:

$$\mathbf{2}_L \otimes \mathbf{1}^0 \oplus \mathbf{2}_R \otimes \mathbf{1}^0 \oplus \mathbf{2}_L \otimes \mathbf{3}^0 \oplus \mathbf{2}_R \otimes \mathbf{3}^0$$

$$\oplus \mathbf{1} \otimes \mathbf{2}_L^0 \oplus \mathbf{1} \otimes \mathbf{2}_R^0 \oplus \mathbf{3} \otimes \mathbf{2}_L^0 \oplus \mathbf{3} \otimes \mathbf{2}_R^0.$$

- $\mathcal{M}_F$ has a grading $\gamma_F = c - J_F c J_F$, with $c = (0, 1, -1, 0) \in \mathcal{A}_{LR}$, which satisfies

$$J_F^2 = 1, \quad J_F \gamma_F = -\gamma_F J_F.$$

This combination of signs, $J_F^2 = +1$ and $J_F \gamma_F = -\gamma_F J_F$, determines uniquely the KO-dimension: the only case that matches it in the table of classical KO-dimensions is the case where the KO-dimension is equal to 6 modulo 8.

The decomposition into irreducible representations of $\mathcal{B}$ can be interpreted in terms of particle content of the model by identifying a basis of $\mathcal{M}_F$ with fermions in the following way:

- The vectors $\mathbf{2}_L \otimes \mathbf{1}^0$ give neutrinos $\nu_L \in |\uparrow\rangle_L \otimes \mathbf{1}^0$ and charged leptons $e_L \in |\downarrow\rangle_L \otimes \mathbf{1}^0$.
- The vectors $\mathbf{2}_R \otimes \mathbf{1}^0$ give right-handed neutrinos $\nu_R \in |\uparrow\rangle_R \otimes \mathbf{1}^0$ and charged leptons $e_R \in |\downarrow\rangle_R \otimes \mathbf{1}^0$.
- The vectors $\mathbf{2}_L \otimes \mathbf{3}^0$ give the u/c/t quarks $u_L \in |\uparrow\rangle_L \otimes \mathbf{3}^0$ and the d/s/b quarks $d_L \in |\downarrow\rangle_L \otimes \mathbf{3}^0$, with color indices.
- The vectors $\mathbf{2}_R \otimes \mathbf{3}^0$ give the u/c/t quarks $u_R \in |\uparrow\rangle_R \otimes \mathbf{3}^0$ and the d/s/b quarks $d_R \in |\downarrow\rangle_R \otimes \mathbf{3}^0$, with color indices.
- The vectors $\mathbf{1} \otimes \mathbf{2}_{L,R}^0$ give antineutrinos $\bar{\nu}_{L,R} \in \mathbf{1} \otimes |\uparrow\rangle_{L,R}^0$, and charged antileptons $\bar{e}_{L,R} \in \mathbf{1} \otimes |\downarrow\rangle_{L,R}^0$.
- The vectors $\mathbf{3} \otimes \mathbf{2}_{L,R}^0$ give antiquarks $\bar{u}_{L,R} \in \mathbf{3} \otimes |\uparrow\rangle_{L,R}^0$ and $\bar{d}_{L,R} \in \mathbf{3} \otimes |\downarrow\rangle_{L,R}^0$, with color indices.

The identification with fermions is confirmed by the computation of the hypercharges. This first requires restricting the subalgebra so that the order one condition for the geometry can be satisfied.

Given the data $(\mathcal{A}_{LR}, \mathcal{H}_F, J_F, \gamma)$, one looks for pairs $(\mathcal{A}, D)$ of a subalgebra $\mathcal{A} \subset \mathcal{A}_{LR}$ and an operator D on $\mathcal{H}_F$ such that $D = D^*$ and D satisfies the order one condition $[[D, a], b^0] = 0$, for all elements a, b of the subalgebra $\mathcal{A}$. One selects solutions $(\mathcal{A}, D)$ with the largest possible $\mathcal{A}$. It is shown in [Chamseddine, Connes, Marcolli (2007)] that the largest subalgebra $\mathcal{A}$ on which the order one condition holds is (up to automorphisms of $\mathcal{A}_{LR}$) of the form

$$\mathcal{A}_F = \{(\lambda, q_L, \lambda, m) \mid \lambda \in \mathbb{C}, \ q_L \in \mathbb{H}, \ m \in M_3(\mathbb{C})\} = \mathbb{C} \oplus \mathbb{H} \oplus M_3(\mathbb{C}).$$
$$(1.19)$$

We discuss in §1.3.3 the general form of the operators D (up to unitary equivalence) that solve these constraints, with $\mathcal{A}$ as in (1.19).

Thus, imposing the order one condition for the Dirac operator has the effect of breaking the left-right symmetry of the algebra. It would be interesting to construct a version of the model where the order one condition is achieved dynamically instead of being imposed as a constraint, so that the associated symmetry breaking would happen spontaneously.

The unitary group of the finite geometry F is given by

$$U(\mathcal{A}_F) = \{u \in \mathcal{A}_F \mid uu^* = u^*u = 1\},$$

while one defines the special unitary group as

$$SU(\mathcal{A}_F) = \{u \in U(\mathcal{A}_F) \mid \det(u) = 1\},$$

with the determinant in the representation on $\mathcal{H}_F$. Up to a finite abelian group, the algebra (1.19) gives a symmetry group

$$SU(\mathcal{A}_F) \simeq U(1) \times SU(2) \times SU(3).$$

The adjoint action of $U(1)$, in powers of $\lambda \in U(1)$, is given by

	$\lvert\uparrow\rangle \otimes \mathbf{1}^0$	$\lvert\downarrow\rangle \otimes \mathbf{1}^0$	$\lvert\uparrow\rangle \otimes \mathbf{3}^0$	$\lvert\downarrow\rangle \otimes \mathbf{3}^0$
$\mathbf{2}_L$	-1	-1	$1/3$	$1/3$
$\mathbf{2}_R$	0	-2	$4/3$	$-2/3$

hence it assigns the correct hypercharges to all the fermions, according to the previous identification of basis vectors with particles.

1.3.3 *The moduli space of Dirac operators*

For the finite geometry $(\mathcal{A}_F, \mathcal{H}_F, \gamma_F, J_F)$, with $\mathcal{A}_F$ as in (1.19), we now look at all the possible self-adjoint D_F on $\mathcal{H}_F$, commuting with J_F, anticommuting with γ_F and satisfying the order one condition $[[D, a], b^0] = 0$, for all $a, b \in \mathcal{A}_F$.

These are classified in [Chamseddine, Connes, Marcolli (2007)], after imposing the additional condition that D_F commutes with the subalgebra

$$\mathbb{C}_F \subset \mathcal{A}_F, \quad \mathbb{C}_F = \{(\lambda, \lambda, 0), \lambda \in \mathbb{C}\}. \qquad (1.20)$$

Dirac operators are then of the form

$$D_F = \begin{pmatrix} S & T^* \\ T & \bar{S} \end{pmatrix} \quad \text{with} \quad S = S_1 \oplus (S_3 \otimes 1_3),$$

where in the basis described in §1.3.2 above we have

$$S_1 = \begin{pmatrix} 0 & 0 & Y^*_{(\uparrow 1)} & 0 \\ 0 & 0 & 0 & Y^*_{(\downarrow 1)} \\ Y_{(\uparrow 1)} & 0 & 0 & 0 \\ 0 & Y_{(\downarrow 1)} & 0 & 0 \end{pmatrix}$$

same for S_3, with

$$S_3 = \begin{pmatrix} 0 & 0 & Y^*_{(\uparrow 3)} & 0 \\ 0 & 0 & 0 & Y^*_{(\downarrow 3)} \\ Y_{(\uparrow 3)} & 0 & 0 & 0 \\ 0 & Y_{(\downarrow 3)} & 0 & 0 \end{pmatrix}$$

with $Y_{(\downarrow 1)}, Y_{(\uparrow 1)}, Y_{(\downarrow 3)}, Y_{(\uparrow 3)}$ in $GL_3(\mathbb{C})$ and Y_R a symmetric matrix.

Thus, Dirac operators, up to unitary equivalence, are parameterized by points of a moduli space $\mathcal{C}_3 \times \mathcal{C}_1$ where
$\mathcal{C}_3$ consists of pairs $(Y_{(\downarrow 3)}, Y_{(\uparrow 3)}) = (Y_d, Y_u)$ modulo the relation

$$Y'_{(\downarrow 3)} = W_1 Y_{(\downarrow 3)} W_3^*, \quad Y'_{(\uparrow 3)} = W_2 Y_{(\uparrow 3)} W_3^*$$

with W_j unitary matrices:

$$\mathcal{C}_3 = (K \times K)\backslash(G \times G)/K$$

with $G = GL_3(\mathbb{C})$ and $K = U(3)$. This part of the moduli space has $\dim_{\mathbb{R}} \mathcal{C}_3 = 10 = 3 + 3 + 4$ where these parameters correspond to $3 + 3$ eigenvalues, 3 angles, and 1 phase. Similarly, the lepton part $\mathcal{C}_1$ consists of triplets $(Y_{(\downarrow 1)}, Y_{(\uparrow 1)}, Y_R) = (Y_e, Y_\nu, M)$, with Y_R symmetric, modulo the relation

$$Y'_{(\downarrow 1)} = V_1 Y_{(\downarrow 1)} V_3^*, \quad Y'_{(\uparrow 1)} = V_2 Y_{(\uparrow 1)} V_3^*, \quad Y'_R = V_2 Y_R \bar{V}_2^*.$$

There is a projection map $\pi : \mathcal{C}_1 \to \mathcal{C}_3$, to a space isomorphic to $\mathcal{C}_3$, where the projection forgets the term Y_R. The fiber of this projection consists of symmetric matrices modulo the scaling relation $Y_R \mapsto \lambda^2 Y_R$, which is of dimension $11 = 12 - 1$, so that the dimension of the full moduli space is $\dim_{\mathbb{R}}(\mathcal{C}_3 \times \mathcal{C}_1) = 31$.

This classification of Dirac operators further clarifies the extension of the minimal standard model that one obtains in this way. Indeed, the parameters of this moduli space correspond to the masses and mixing angles of all the fermions in the model. In particular, this shows that one gets a particle physics model, which we refer to as νMSM which, in addition to the particle content of the minimal standard model, also has right handed neutrinos with Majorana mass terms. More precisely, we have the following interpretation for the parameters:

$$Y_{(\uparrow 3)} = \delta_{(\uparrow 3)} \qquad Y_{(\downarrow 3)} = U_{CKM}\, \delta_{(\downarrow 3)}\, U_{CKM}^*$$

$$Y_{(\uparrow 1)} = U_{PMNS}^*\, \delta_{(\uparrow 1)}\, U_{PMNS} \qquad Y_{(\downarrow 1)} = \delta_{(\downarrow 1)}$$

where the diagonal matrices $\delta_\uparrow$, $\delta_\downarrow$ consist of the Dirac masses and the matrices U are written in the form

$$U = \begin{pmatrix} c_1 & -s_1 c_3 & -s_1 s_3 \\ s_1 c_2 & c_1 c_2 c_3 - s_2 s_3 e_\delta & c_1 c_2 s_3 + s_2 c_3 e_\delta \\ s_1 s_2 & c_1 s_2 c_3 + c_2 s_3 e_\delta & c_1 s_2 s_3 - c_2 c_3 e_\delta \end{pmatrix}$$

with angles and phase $c_i = \cos\theta_i$, $s_i = \sin\theta_i$, $e_\delta = \exp(i\delta)$. Thus, we identify the matrix U_{CKM} with the Cabibbo–Kobayashi–Maskawa mixing matrix for the quarks and U_{PMNS} with the Pontecorvo–Maki–Nakagawa–Sakata matrix of neutrino mixing. The matrix Y_R gives the Majorana mass terms for the right-handed neutrinos.

There are very strict experimental and theoretical constraints and exclusion curves on the parameters of the mixing matrices, such as the unitarity triangle for the CKM matrix, illustrated in Figure 1.1, see [Burgess, Moore (2007)], [Schwartz (2013)]. These constraints can be viewed as identifying certain geometric submanifolds of the moduli space of Dirac operators. We will return to this point of view in our discussion of renormalization group flows in the next chapter.

In the model of [Chamseddine, Connes, Marcolli (2007)], this commutation of D_F with (1.20), which ensures that the photon is massless, is imposed as an extra condition and not derived from the initial input of the model. A recent reformulation of the finite geometry axioms, developed in [Farnsworth, Boyle (2015)], provides a framework where this condition follows from the axiomatic properties of the finite geometry and is not imposed as an additional constraint, see §1.3.7 below.

1.3.4 *Bosons and inner fluctuations*

While the fermion content of the model is completely specified by the representation space $\mathcal{H}_F$ of the finite geometry F, the bosons arise as inner fluctuations $D \mapsto D_A$ of the Dirac operator of the product geometry $X \times F$.

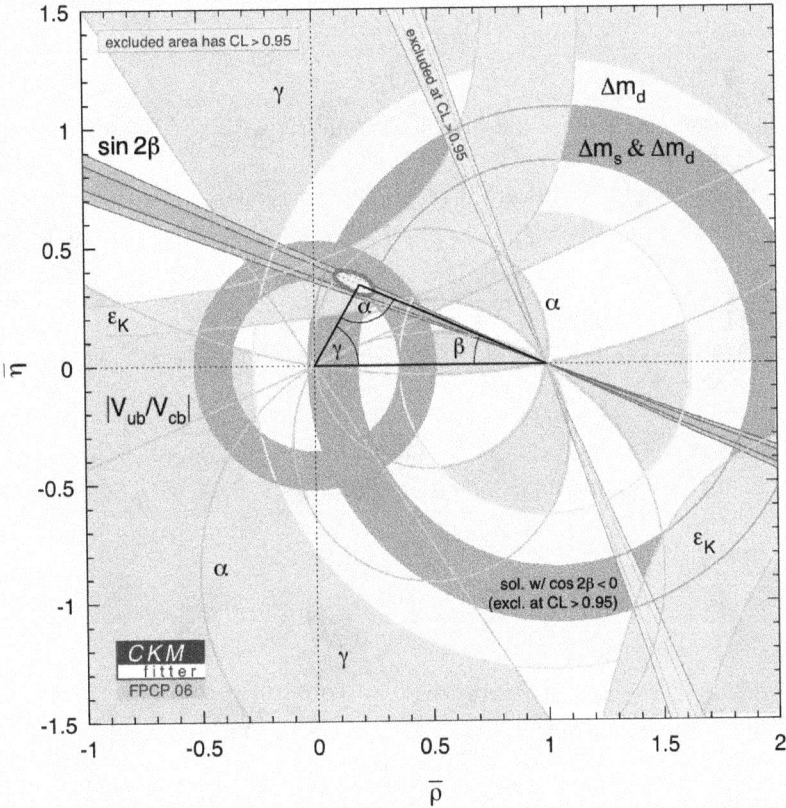

Fig. 1.1 Constraints on the CKM matrix. [CKMfitter Group (J. Charles et al.), Eur. Phys. J. C41, 1-131 (2005) [hep-ph/0406184], updated results and plots available at: http://ckmfitter.in2p3.fr]

In particular, there are inner fluctuations in the manifold directions X and inner fluctuation in the noncommutative direction F. These correspond, respectively, to the gauge bosons and the Higgs field. We recall the derivation of the bosons from [Chamseddine, Connes, Marcolli (2007)].

As we discussed before, when we introduced the spectral action (1.4), the inner fluctuations of the Dirac operator are given by

$$D \mapsto D_A = D + A + \varepsilon' \, J \, A \, J^{-1}$$

with $A = A^*$ a self-adjoint operator of the form

$$A = \sum a_j [D, b_j], \quad \text{with } a_j, b_j \in \mathcal{A}.$$

Here $\mathcal{A} = C^\infty(X) \otimes \mathcal{A}_F = C^\infty(X, \mathcal{A}_F)$ is the algebra of the product geometry and

$$D = \not{D}_X \otimes 1 + \gamma_5 \otimes D_F$$

is the Dirac operator of the product geometry, with the finite D_F as discussed above.

The inner fluctuations in the manifold direction are of the form $A^{(1,0)} = \sum_i a_i[\not{D}_X \otimes 1, a'_i]$ with $a_i = (\lambda_i, q_i, m_i)$, $a'_i = (\lambda'_i, q'_i, m'_i)$ in $\mathcal{A} = C^\infty(X, \mathcal{A}_F)$. These give:

- a $U(1)$ gauge field $\Lambda = \sum_i \lambda_i \, d\lambda'_i = \sum_i \lambda_i[\not{D}_X \otimes 1, \lambda'_i]$;
- an $SU(2)$ gauge field $Q = \sum_i q_i \, dq'_i$, with $q = f_0 + \sum_\alpha if_\alpha \sigma^\alpha$ and $Q = \sum_\alpha f_\alpha[\not{D}_X \otimes 1, if'_\alpha \sigma^\alpha]$
- a $U(3)$ gauge field $V' = \sum_i m_i \, dm'_i = \sum_i m_i[\not{D}_X \otimes 1, m'_i]$
- a reduction of the gauge field V' to $SU(3)$ obtained by passing to the unimodular subgroup $SU(\mathcal{A}_F)$ with a unimodular gauge potential with $\mathrm{Tr}(A) = 0$,

$$V' = -V - \frac{1}{3}\begin{pmatrix} \Lambda & 0 & 0 \\ 0 & \Lambda & 0 \\ 0 & 0 & \Lambda \end{pmatrix} = -V - \frac{1}{3}\Lambda 1_3.$$

The gauge bosons also correspond to the correct hypercharges. The $(1,0)$ part of $A + JAJ^{-1}$ acts on quarks and leptons, respectively, by

$$\begin{pmatrix} \frac{4}{3}\Lambda + V & 0 & 0 & 0 \\ 0 & -\frac{2}{3}\Lambda + V & 0 & 0 \\ 0 & 0 & Q_{11} + \frac{1}{3}\Lambda + V & Q_{12} \\ 0 & 0 & Q_{21} & Q_{22} + \frac{1}{3}\Lambda + V \end{pmatrix}$$

$$\begin{pmatrix} 0 & 0 & 0 & 0 \\ 0 & -2\Lambda & 0 & 0 \\ 0 & 0 & Q_{11} - \Lambda & Q_{12} \\ 0 & 0 & Q_{21} & Q_{22} - \Lambda \end{pmatrix}$$

The inner fluctuations in the noncommutative F direction are given by

$$\sum_i a_i[\gamma_5 \otimes D_F, a'_i] = \gamma_5 \otimes (A_q^{(0,1)} + A_\ell^{(0,1)}) \qquad (1.21)$$

with $a_i(x) = (\lambda_i, q_i, m_i)$ and $a'_i(x) = (\lambda'_i, q'_i, m'_i)$, and where $A_q^{(0,1)}$ and $A_\ell^{(0,1)}$ are given by

$$A_q^{(0,1)} = \begin{pmatrix} 0 & X \\ X' & 0 \end{pmatrix} \otimes 1_3 \qquad A_\ell^{(0,1)} = \begin{pmatrix} 0 & Y \\ Y' & 0 \end{pmatrix}$$

$$X = \begin{pmatrix} \Upsilon_u^* \varphi_1 & \Upsilon_u^* \varphi_2 \\ -\Upsilon_d^* \bar{\varphi}_2 & \Upsilon_d^* \bar{\varphi}_1 \end{pmatrix} \quad \text{and} \quad X' = \begin{pmatrix} \Upsilon_u \varphi_1' & \Upsilon_d \varphi_2' \\ -\Upsilon_u \bar{\varphi}_2' & \Upsilon_d \bar{\varphi}_1' \end{pmatrix}$$

$$Y = \begin{pmatrix} \Upsilon_\nu^* \varphi_1 & \Upsilon_\nu^* \varphi_2 \\ -\Upsilon_e^* \bar{\varphi}_2 & \Upsilon_e^* \bar{\varphi}_1 \end{pmatrix} \quad \text{and} \quad Y' = \begin{pmatrix} \Upsilon_\nu \varphi_1' & \Upsilon_e \varphi_2' \\ -\Upsilon_\nu \bar{\varphi}_2' & \Upsilon_e \bar{\varphi}_1' \end{pmatrix}$$

with the fields φ_i and φ_i' given by

$$\varphi_1 = \sum \lambda_i(\alpha_i' - \lambda_i'), \quad \varphi_2 = \sum \lambda_i \beta_i'$$

$$\varphi_1' = \sum \alpha_i(\lambda_i' - \alpha_i') + \beta_i \bar{\beta}_i', \quad \varphi_2' = \sum(-\alpha_i \beta_i' + \beta_i(\bar{\lambda}_i' - \bar{\alpha}_i')),$$

where we write the q_i and q_i' in the form

$$q = \begin{pmatrix} \alpha & \beta \\ -\bar{\beta} & \bar{\alpha} \end{pmatrix}.$$

Thus, the inner fluctuations in the F direction define a quaternion valued function $H = \varphi_1 + \varphi_2 j$, which we also write simply as $\varphi = (\varphi_1, \varphi_2)$. When one computes explicitly the asymptotic expansion of the spectral action, one finds that this field φ is the Higgs field. We refer the reader to [Chamseddine, Connes, Marcolli (2007)] and to Chapter 1 of [Connes and Marcolli (2008)] for the details of this derivation.

Field equations for the internal space metric fluctuations for the spectral triple of the standard model were considered in [Barrett, Dawe Martins (2006)], [Dawe Martins (2006)].

1.3.5 *Asymptotic expansion of the spectral action*

In the case of the product geometry $X \times F$, it is shown in [Chamseddine, Connes, Marcolli (2007)] that the asymptotic expansion of the spectral action (1.4) has terms of the following form

$$S_\Lambda[D, A] \sim \frac{1}{2\kappa_0^2(\Lambda)} \int R \sqrt{g}\, d^4x + \gamma_0(\Lambda) \int \sqrt{g}\, d^4x$$

$$+ \alpha_0(\Lambda) \int C_{\mu\nu\rho\sigma} C^{\mu\nu\rho\sigma} \sqrt{g}\, d^4x + \tau_0(\Lambda) \int R^* R^* \sqrt{g}\, d^4x$$

$$+ \frac{1}{2} \int |DH|^2 \sqrt{g}\, d^4x - \mu_0^2(\Lambda) \int |H|^2 \sqrt{g}\, d^4x$$

$$- \xi_0(\Lambda) \int R\,|H|^2 \sqrt{g}\, d^4x + \lambda_0(\Lambda) \int |H|^4 \sqrt{g}\, d^4x$$

$$+ \frac{1}{4} \int (G_{\mu\nu}^i G^{\mu\nu i} + F_{\mu\nu}^\alpha F^{\mu\nu\alpha} + B_{\mu\nu} B^{\mu\nu}) \sqrt{g}\, d^4x,$$

where a priori all the coefficients may be running with the energy scale Λ. The explicit form of the coefficients is a function of the momenta f_0, f_2, f_4, with $f_0 = f(0)$ and, for $k > 0$,

$$f_k = \int_0^\infty f(v) v^{k-1} dv,$$

which are free parameters of the model, and of the functions $\mathfrak{a}, \mathfrak{b}, \mathfrak{c}, \mathfrak{d}, \mathfrak{e}$ of the Yukawa parameters of the νMSM model (the coordinates on the moduli space of finite Dirac operators). These functions are explicitly of the form

$$\mathfrak{a} = \mathrm{Tr}(Y_\nu^\dagger Y_\nu + Y_e^\dagger Y_e + 3(Y_u^\dagger Y_u + Y_d^\dagger Y_d))$$

$$\mathfrak{b} = \mathrm{Tr}((Y_\nu^\dagger Y_\nu)^2 + (Y_e^\dagger Y_e)^2 + 3(Y_u^\dagger Y_u)^2 + 3(Y_d^\dagger Y_d)^2)$$

$$\mathfrak{c} = \mathrm{Tr}(MM^\dagger)$$

$$\mathfrak{d} = \mathrm{Tr}((MM^\dagger)^2)$$

$$\mathfrak{e} = \mathrm{Tr}(MM^\dagger Y_\nu^\dagger Y_\nu),$$

hence they are in principles functions of the energy scale Λ, through the corresponding dependence on Λ of the Yukawa parameters through the renormalization group flow of the model. We will discuss more in detail in the next chapter different forms of running for all these coefficients. For the moment we simply leave these indicated as functions $\mathfrak{a}(\Lambda), \mathfrak{b}(\Lambda), \mathfrak{c}(\Lambda), \mathfrak{d}(\Lambda), \mathfrak{e}(\Lambda)$, though, as we will see, in some versions of the model these are fixed at a particular energy scale, hence they do not always run. Then the coefficients in the asymptotic expansion of the spectral action have the explicit form:

$$\frac{1}{2\kappa_0^2(\Lambda)} = \frac{96 f_2 \Lambda^2 - f_0 \mathfrak{c}(\Lambda)}{24\pi^2}$$

$$\gamma_0(\Lambda) = \frac{1}{\pi^2}(48 f_4 \Lambda^4 - f_2 \Lambda^2 \mathfrak{c}(\Lambda) + \frac{f_0}{4}\mathfrak{d}(\Lambda))$$

$$\alpha_0 = -\frac{3 f_0}{10\pi^2}$$

$$\tau_0 = \frac{11 f_0}{60\pi^2}$$

$$\mu_0^2(\Lambda) = 2\frac{f_2 \Lambda^2}{f_0} - \frac{\mathfrak{e}(\Lambda)}{\mathfrak{a}(\Lambda)}$$

$$\xi_0 = \tfrac{1}{12}$$

$$\lambda_0(\Lambda) = \frac{\pi^2 \mathfrak{b}(\Lambda)}{2 f_0 \mathfrak{a}^2(\Lambda)}.$$

In the form given above of the asymptotic expansion of the spectral action, one uses the rescaled form of the Higgs field $H = \frac{\sqrt{a\,f_0}}{\pi}\varphi$, which has the effect of normalizing the kinetic term $\int \frac{1}{2}|D_\mu H|^2 \sqrt{g}\,d^4x$. Moreover, the Yang–Mills terms for the gauge bosons are all normalized. This implies that there is a preferred energy scale for the spectral action, which is the unification energy $\Lambda = \Lambda_{\text{unif}}$ where the three coupling constants (supposedly) unify. Away from unification energy the gauge boson terms would have the form

$$\frac{f_0}{2\,\pi^2} \int \left(g_3^2(\Lambda)\, G_{\mu\nu}^i\, G^{\mu\nu i} + g_2^2(\Lambda)\, F_{\mu\nu}^\alpha\, F^{\mu\nu\alpha} + \frac{5}{3}\, g_1^2(\Lambda)\, B_{\mu\nu}\, B^{\mu\nu}\right) \sqrt{g}\,d^4x,$$

where the coupling constants g_1, g_2, g_3 run with the energy scale Λ according to the renormalization group flow.

The expansion of the spectral action is obtained, as we mentioned before, using Gilkey's heat kernel expansion, [Gilkey (1995)], by showing that the square of the Dirac operator has the form $D_A^2 = \nabla^*\nabla - E$, for some bundle endomorphism. For more general differential operators $P = -(g^{\mu\nu}I\,\partial_\mu\partial_\nu + A^\mu\partial_\mu + B)$, with A, B bundle endomorphisms, and with $m = \dim M$, the heat kernel expansion is of the form

$$\operatorname{Tr} e^{-tP} \sim \sum_{n\geq 0} t^{\frac{n-m}{2}} \int_M a_n(x, P)\,dv(x).$$

For $P = \nabla^*\nabla - E$ and $E_{;\mu}{}^\mu := \nabla_\mu\nabla^\mu E$, with

$$\nabla_\mu = \partial_\mu + \omega'_\mu, \qquad \omega'_\mu = \frac{1}{2}\,g_{\mu\nu}(A^\nu + \Gamma^\nu \cdot \text{id})$$

$$E = B - g^{\mu\nu}(\partial_\mu\,\omega'_\nu + \omega'_\mu\,\omega'_\nu - \Gamma^\rho_{\mu\nu}\,\omega'_\rho)$$

$$\Omega_{\mu\nu} = \partial_\mu\,\omega'_\nu - \partial_\nu\,\omega'_\mu + [\omega'_\mu, \omega'_\nu],$$

the explicit form of the Seeley-DeWitt coefficients is given by

$$a_0(x, P) = (4\pi)^{-m/2}\operatorname{Tr}(\text{id})$$
$$a_2(x, P) = (4\pi)^{-m/2}\operatorname{Tr}\left(-\tfrac{R}{6}\,\text{id} + E\right)$$
$$a_4(x, P) = (4\pi)^{-m/2}\tfrac{1}{360}\operatorname{Tr}(-12R_{;\mu}{}^\mu + 5R^2 - 2R_{\mu\nu}\,R^{\mu\nu}$$
$$+ 2R_{\mu\nu\rho\sigma}\,R^{\mu\nu\rho\sigma} - 60\,R\,E + 180\,E^2 + 60\,E_{;\mu}{}^\mu$$
$$+ 30\,\Omega_{\mu\nu}\,\Omega^{\mu\nu}).$$

These are then used to obtain the terms listed above in the asymptotic expansion, see [Chamseddine, Connes, Marcolli (2007)].

It is also shown in [Chamseddine, Connes, Marcolli (2007)] that the fermionic term (1.17) for this geometry $X \times F$ delivers the correct fermionic terms, and fermion-boson interaction terms, of the Lagrangian of the νMSM extension of the minimal standard model. Here we focus only on the bosonic part, as that is the relevant part for cosmological applications.

1.3.6   *Coefficients of the gravitational terms*

We can then interpret the coefficients of the gravitational terms described above in the following way.

- Effective gravitational constant:

$$G_{\text{eff}}(\Lambda) = \frac{\kappa_0^2(\Lambda)}{8\pi} = \frac{3\pi}{192 f_2 \Lambda^2 - 2 f_0 \mathfrak{c}(\Lambda)}$$

- Effective cosmological constant

$$\gamma_0(\Lambda) = \frac{1}{4\pi^2}\left(192 f_4 \Lambda^4 - 4 f_2 \Lambda^2 \mathfrak{c}(\Lambda) + f_0 \mathfrak{d}(\Lambda)\right)$$

- Conformal non-minimal coupling of Higgs and gravity

$$\frac{1}{16\pi G_{\text{eff}}(\Lambda)} \int R \sqrt{g} d^4 x - \frac{1}{12} \int R |H|^2 \sqrt{g} d^4 x$$

- Conformal gravity

$$\frac{-3 f_0}{10\pi^2} \int C_{\mu\nu\rho\sigma} C^{\mu\nu\rho\sigma} \sqrt{g} d^4 x$$

- Gauss–Bonnet gravity

$$\frac{11 f_0}{60\pi^2} \int R^* R^* \sqrt{g} \, d^4 x.$$

We will discuss in the next chapter some cosmological consequences, for Early Universe models, of various possibilities for the running of these coefficients, and the relative magnitude of these terms.

It is important to keep in mind that the specific form of the coefficients of the gravitational terms discussed in this section depend on the specific choice of the finite geometry F used in the model of [Chamseddine, Connes, Marcolli (2007)]. Different choices of the almost commutative geometry $X \times F$, with other finite geometries F, would lead to a different form of the coefficients. The general ideas we discuss here and in the next chapter would apply to any choice of F. However, specific discussions of effects on cosmology of a modified gravity model based on the spectral action and its asymptotic expansion, depend on using the same finite geometry F used in [Chamseddine, Connes, Marcolli (2007)] and may be different for other choices of F.

1.3.7 *Fused algebra approach*

In [Boyle, Farnsworth (2014)] and [Farnsworth, Boyle (2015)], a reformulation of the noncommutative geometry of the standard model is given, where several of the properties of the spectral triple described above are combined together in a single object, referred to as the *fused algebra*. The main advantages of this formulation are that the order zero and order one conditions arise directly from the associativity property, and the additional *massless photon condition* is also implied by the associativity of the fused algebra and no longer has to be imposed by hand as in [Chamseddine, Connes, Marcolli (2007)]. Additionally, the symmetries of the fused algebra involve an additional $U(1)$, so that a scalar field σ of the type used in [Chamseddine, Connes (2012)] to correct the Higgs boson mass is also incorporated naturally in the model.

We summarize quickly the setting of this fused algebra approach, and refer the reader to [Boyle, Farnsworth (2014)] and [Farnsworth, Boyle (2015)] for a more detailed discussion.

The original data $(\mathcal{A}, \mathcal{H}, D, J)$ of a spectral triple with a real structure are modified as follows.

- The algebra $\mathcal{A}$ is not, a priori, required to be associative.
- The algebra $\mathcal{A}$ over $\mathbb{F}$ is extended to an algebra $\Omega\mathcal{A}$ generated by $\mathcal{A}$ and formal symbol da, for $a \in \mathcal{A}$, satisfying $d(ca) = c\,da$ for $c \in \mathbb{F}$, $d(a+a') = da + da'$, and $d(aa') = da\,a' + a\,da'$. If $\mathcal{A}$ is associative, then $\Omega\mathcal{A}$ should be as well, hence the relation $(a\,db)c = a(db\,c)$ should also hold.
- The map $d : \mathcal{A} \to \Omega\mathcal{A}$ induces $d : \Omega^m\mathcal{A} \to \Omega^{m+1}\mathcal{A}$, with $d^2 = 0$, where $\Omega^m\mathcal{A}$ is spanned linearly by monomials containing m differentials da_i, with $d(\omega_n\omega_m) = d(\omega_n)\omega_m + (-1)^n\omega_n\,d\omega_m$.
- The bimodule $\mathcal{H}$ is incorporated into a *fused algebra*

$$B_0 = \mathcal{A} \oplus \mathcal{H}$$

with product structure, for $b = a + h$ and $b' = a' + h'$, given by

$$b_0 b_0' = aa' + ah' + ha'.$$

This formulation makes it possible to define representations and bimodules also in the case where $\mathcal{A}$ may be non-associative.

- Associativity of B_0 implies: associativity of $\mathcal{A}$, left and right actions of $\mathcal{A}$ on $\mathcal{H}$, and the order-zero condition.

- The real structure is incorporated, together with the $\star$-structure on the algebra $\mathcal{A}$, as the anti-linear automorphism of B_0 given by

$$b_0 = a + h \mapsto b_0^* = a^* + Jh.$$

This has period 2 for KO dimension $0, 1, 6, 7 \bmod 8$ and period 4 for KO dimension $2, 3, 4, 5 \bmod 8$.

- The algebra $B_0 = \mathcal{A} \oplus \mathcal{H}$ is extended to an algebra $B = \Omega \mathcal{A} \oplus \mathcal{H}$, which incorporates the action of forms on $\mathcal{H}$ (modulo junk forms) through the Dirac operator, with $da\, h = [D, a]\, h$ and $h\, da = J[D, a]^* J^* h$

- The associativity condition for B implies the order-zero and order-one conditions for the Dirac operator D and the real structure J, and it also implies an additional constraint that, when applied to the almost-commutative geometry of the standard model, gives the massless photon condition.

- In the case of an almost-commutative geometry, one can consider a fused algebra B_F associated to the finite spectral tripe $(A_F, \mathcal{H}_F, D_F, J_F)$ and another fused algebra B_c for the continuous (manifold) part of the almost commutative geometry. For the finite spectral triple of the standard model, the symmetries of the B_F, obtained by classifying the derivations of this fused algebra, give a gauge group of the form

$$SU(3) \times SU(2) \times U(1) \times U(1)_{B-L},$$

where $SU(3) \times SU(2) \times U(1)$ is the same gauge group of the standard model obtained directly from the finite spectral triple, and the additional circle $U(1)_{B-L}$ leads to an extension of the νMSM model by an additional $U(1)$ gauge boson C_μ and a complex scalar field σ that is a singlet under $SU(3) \times SU(2) \times U(1)$ and has $B - L = 2$.

This type of extension of the νMSM model by the scalar field σ and the $U(1)$-gauge boson C_μ was considered in [Iso, Okada, Orikasa (2009)] and [Boyle, Farnsworth, Fitzgerald, Schade (2011)] in relation to cosmological models of inflation, and is compatible with the approach of [Chamseddine, Connes (2012)] for the correction of the Higgs mass value. For further developments of this fused algebra viewpoint see also [Brouder, Bizi, Besnard (2015)], [Boyle, Farnsworth (2016)], [Farnsworth (2016)].

1.3.8 *Supersymmetric theories*

Wim Beenakker, Thijs van den Broek, Walter van Suijlekom developed an approach to the Minimally Supersymmetric Standard Model (MSSM), within the framework of the noncommutative geometry approach, [Beenakker, van den Broek, van Suijlekom (2015)].

The main advantages provided by the MSSM as a particle physics model are: a greater stability of the Higgs mass under loop-corrections (where the contributions from superpartners compensate instabilities); since R-symmetry is preserved, the lightest $R = -1$ particle cannot decay and may provide a cold dark matter model; the renormalization group equations (RGE) for the coupling constants form a much smaller triangle, suggesting a better compatibility with unification.

They first extend the data of finite spectral triples to incorporate R-parity of supersymmetry,

$$R = (-1)^{2S+3B+L},$$

where S is spin, B is baryon number, and L is lepton number. All Standard Model particles (and an additional Higgs doublet) have $R = +1$, while all superpartners have $R = -1$. The data of a real even spectral triple $(\mathcal{A}, \mathcal{H}, D, J, \gamma)$ are the same as we discussed previously. Additionally, one requires the existence of a $\mathbb{Z}/2\mathbb{Z}$-grading $R : \mathcal{H} \to \mathcal{H}$. This gives a decomposition

$$\mathcal{H} = \mathcal{H}_{R=+1} \oplus \mathcal{H}_{R=-1}$$

with projectors $(1 \pm R)/2$. The compatibility of R with the algebra and the data J and γ is given by the requirement that, $\forall a \in \mathcal{A}$,

$$[R, \gamma] = [R, J] = [R, \pi(a)] = 0.$$

The Dirac operator splits as $D = D_+ + D_-$, with $\{D_-, R\} = [D_+, R] = 0$. In the Krajewski diagrams description of the finite spectral triples, one additionally colors vertices black/white for $R = \mp 1$.

In [Beenakker, van den Broek, van Suijlekom (2015)] they then proceed to identify several types of building blocks for such supersymmetric finite spectral triples, and investigate the corresponding terms in the spectral action that they respectively produce. The building blocks they consider are constructed in a sequence of steps, roughly summarized as:

- Adjoint representations
- Non-adjoint representations

- Extra interactions
- Majorana terms
- Other mass terms
- Interactions between blocks

Adjoint representations consist of a single component $\mathcal{A} = M_{N_j}(\mathbb{C})$ with a bimodule $\mathbb{C}^{N_j} \otimes \mathbb{C}^{N_j\,o} \simeq M_{N_j}(\mathbb{C})$, which they identify with the shorthand notation $\mathbf{N}_j \otimes \mathbf{N}_j^o$. Maintaining the requirement that the KO-dimension is 6, for compatibility with the Standard Model part, one requires $\{J, \gamma\} = 0$. This means having two copies with J exchanging them (and taking adjoints),

$$J(m, n) = (n^*, m^*),$$

so the grading γ has opposite values on the two copies. For $\mathcal{B}_j = M_{N_j,L}(\mathbb{C}) \oplus M_{N_j,R}(\mathbb{C})$ with additional terms in the adjoint representation in $End(\mathcal{H}_F)$, the R-parity gives $R|_{M_{N_j}(\mathbb{C})} = -1$.

These building blocks give rise to a supersymmetric Yang–Mills action. In order to match degrees of freedom for gauginos and gauge bosons, one needs to discard the trace part of the fermion, add a non-propagating auxiliary field G_j, which fixes the mismatch of degrees of freedom for off-shell theory, so that the on-shell agrees with the spectral action result. The action for this type of block involves

$$(\lambda'_{j,L}, \lambda'_{j,R}) \in L^2(\mathbb{S}_+ \otimes M_{N_j,L}(\mathbb{C})) \otimes L^2(\mathbb{S}_- \otimes M_{N_j,R}(\mathbb{C}))$$

and gives terms of the form

$$S(\lambda, \mathbb{A}) = \langle J_M \lambda'_{j,R}, \partial\!\!\!/_{\mathbb{A}} \lambda'_{j,L} \rangle - \frac{f(0)}{24\pi^2} \int_M \mathrm{Tr}(F^j_{\mu\nu} F^{j,\mu\nu}) + \mathcal{O}(\Lambda^{-2}),$$

with an additional term $-\frac{1}{2} \int_M \mathrm{Tr} G_j^2$, scaled to have normalized kinetic term $\lambda_j \mapsto \frac{1}{\sqrt{n_j}} \lambda_j$ with $n_j \delta_{ab} = \mathrm{Tr} T_j^a T_j^b$. The authors identify explicit transformation δA_j, $\delta \lambda_{j,L/R}$, δG_j under which this action is supersymmetric. Such blocks can occur only with multiplicity one, otherwise it would cause a breaking of supersymmetry, as one needs to have the same number of fermionic and bosonic degrees of freedom.

Non-adjoint representations account for non-gaugino fermions. One considers off-diagonal blocks $\mathbf{N}_i \otimes \mathbf{N}_j^o$, with opposite values of the γ-grading so that again the KO-dimension remains equal to 6. To have supersymmetry, we need a bosonic scalar superpartner. The interaction with gauge fields requires also gaugino degrees of freedom. These requirements can be achieved by combining a block $\mathbf{N}_i \otimes \mathbf{N}_j^o$ with two blocks $\mathcal{B}_i$ and $\mathcal{B}_j$, as in the previous case, giving

$$\mathbf{N}_i \otimes \mathbf{N}_j^o \oplus M_{N_i,L}(\mathbb{C}) \oplus M_{N_i,R}(\mathbb{C}) \oplus M_{N_j,L}(\mathbb{C}) \oplus M_{N_j,R}(\mathbb{C}) \oplus \mathbf{N}_j \otimes \mathbf{N}_i^o.$$

The general form of a Dirac operator D with $D^* = D$, $\{D, \gamma\} = 0$ and $[D, J] = 0$ is given by

$$\begin{pmatrix} 0 & 0 & A & 0 & B & 0 \\ 0 & 0 & M_i & 0 & 0 & JA^*J^* \\ A^* & M_i^* & 0 & 0 & 0 & 0 \\ 0 & 0 & 0 & 0 & M_j & JB^*J^* \\ B^* & 0 & 0 & M_j^* & 0 & 0 \\ 0 & JAJ^* & 0 & JBJ^* & 0 & 0 \end{pmatrix}$$

with $A : M_{N_i}(\mathbb{C})_R \to \mathbf{N}_i \otimes \mathbf{N}_j^o$ and $B : M_{N_j}(\mathbb{C})_R \to \mathbf{N}_i \otimes \mathbf{N}_j^o$. The terms M_i, M_j are supersymmetry breaking gaugino masses, hence one needs to set $M_i = M_j = 0$. The fermionic and bosonic contributions to the spectral action coming from this kind of blocks are also computed explicitly. In this case the trace parts remain and gives rise to $u(1)$ fields. Matching degrees of freedom requires identifying these $u(1)$ fields in pairs. Some off-shell fields G_j are added as before, and an explicit supersymmetry transformation of the action is computed.

In the case of a more general finite dimensional algebra $\mathcal{A}$, for example consisting of two blocks instead of one, $\mathcal{A} = M_{N_i}(\mathbb{C}) \oplus M_{N_j}(\mathbb{C})$, only certain combinations of blocks will have supersymmetry: for instance, two disjiont blocks $\mathcal{B}_i$ and $\mathcal{B}_j$ of first kind. The same is true in the case of an algebra $\mathcal{A} = M_{N_i}(\mathbb{F}) \oplus M_{N_j}(\mathbb{F})$, where $\mathbb{F}$ is either $\mathbb{R}, \mathbb{C}$, or $\mathbb{H}$. In the case of an algebra $\mathcal{A}$ with three blocks $M_{N_{ijk}}(\mathbb{C})$, having only building blocks of the second type is not compatible with supersymmetry. It is necessary to add a third type of building block $\mathcal{B}_{ijk}$ to restore supersymmetry: this accounts for the components of the Dirac operator of the form

$$D_{ij}{}^{kj} : \mathbf{N}_k \otimes \mathbf{N}_j^o \to \mathbf{N}_i \otimes \mathbf{N}_j^o$$

and similarly $D_{jk}{}^{ik}$ and $D_{ij}{}^{ik}$. These $D_{ij}{}^{kj}$ generate by inner fluctuations a scalar field $\tilde{\psi}_{ik}$, and need corresponding fermions to restore supersymmetry. The latter are provided by the blocks $\mathcal{B}_{ijk}$. Thus, this leads to combinations of blocks $\mathcal{B}_{ikl}, \mathcal{B}_{ikm}, \mathcal{B}_{jkl}, \mathcal{B}_{jkm}$ and (for reasons of supersymmetry) $\mathcal{B}_{ij}$ or $\mathcal{B}_{im}$. There is no need for building blocks with more than three indices because the data of finite spectral triples are either components of the algebra (one index N_j) and adjoint representations $\mathbf{N}_j \otimes \mathbf{N}_j^o$, or non-adjoint representations $\mathbf{N}_i \otimes \mathbf{N}_j^o$ (two indices), or components of Dirac operator with order-one condition $D_{ij}{}^{kj}$ (three indices), and everything else (such as multiple components) is expressible in terms of these basic building blocks.

In the SM case one obtains an extension νMSM of the minimal standard model that has right handed neutrinos and Majorana mass terms. Thus, one also expects these terms in further supersymmetric extensions. These are provided by representations $\mathbf{1} \otimes \mathbf{1}'^o \oplus \mathbf{1}' \otimes \mathbf{1}^o$ that are each other's antiparticles (that is, $(\mathbb{C} \oplus \mathbb{C})^{\oplus M}$ with multiplicity M), with the same component $\mathbb{C}$ of algebra acting on both, with Dirac components

$$D_{1'1}{}^{11'} : \mathbf{1} \otimes \mathbf{1}'^o \to \mathbf{1}' \otimes \mathbf{1}^o.$$

Other mass terms arise by considering two building blocks of second type with same indices and different grading. On these a Dirac operator that is self-adjoint, satisfies the order-one condition, and commutes with J gives a mass term in the action if it acts nontrivially only on generations, that is, on the M copies in

$$((\mathbf{N}_i \otimes \mathbf{N}_j^o)_L \oplus (\mathbf{N}_j \otimes \mathbf{N}_i^o)_R \oplus \mathbf{N}_i \otimes \mathbf{N}_j^o)_R \oplus (\mathbf{N}_j \otimes \mathbf{N}_i^o)_L)^{\oplus M}.$$

The construction of finite spectral triples with these building blocks ensures field content is supersymmetric. However, this does not automatically ensure that the action, obtained from the spectral action will also be supersymmetric. Possible sources of problems can be: a single building block of first type $\mathcal{B}_i$ with $N_i = 1$ has vanishing bosonic interactions; a single building block of second type $\mathcal{B}_{ij}$ with $R = -1$ has two interacting $u(1)$-fields (while gauginos do not); if there are at least three complex matrix components in $\mathcal{A}$, and building blocks $\mathcal{B}_{ij}$, $\mathcal{B}_{ik}$, then there are two interacting $u(1)$-fields, while gauginos do not. The approach followed in [Beenakker, van den Broek, van Suijlekom (2015)] to approach these possible problems consists of rewriting the four-scalar interactions generated by the spectral action as off-shell action via auxiliary fields. A general issue is having new contributions, for which one needs to obtain off-shell counterparts, so that the action remains supersymmetric.

A further aspect of the model of [Beenakker, van den Broek, van Suijlekom (2015)] is the supersymmetry breaking mechanism. If supersymmetry were exact in nature superpartners would be of equal mass as corresponding SM particles, which is obviously not the case. In the MSSM model, one should have a SUSY-breaking Higgs that gives mass to the SM particles and achieves electroweak symmetry breaking. There are two possible scenarios that are usually considered: spontaneous SUSY breaking (which is disfavored phenomenologically, as it would have one slepton/squark lighter than corresponding SM fermion), or a SUSY breaking Lagrangian. SUSY breaking terms in the Lagrangian should be *soft*: couplings of positive mass dimension, with no quadratically divergent loop corrections (the Higgs mass very sensitive to perturbative corrections that cause a stability problem). Soft SUSY breaking terms can be given by self-adjoint mass term for scalar bosons $\tilde{\psi}_\alpha$, a symmetric tensor $A_{\alpha\beta\gamma}$ of mass dimension 1, a matrix $B_{\alpha\beta}$ mass dimension 2, a gauge singlet linear coupling $C_\alpha \in \mathbb{C}$ of mass dimension 3, a gaugino mass terms $M \in \mathbb{C}$. In the Spectral Action, soft SUSY breaking terms arise from the Λ^2 terms. One finds scalar fields $\tilde{\psi}_{ij} \in \mathcal{C}^\infty(M, \mathbf{N}_i \otimes \mathbf{N}_j^o)$, fermions $\psi_{ij} \in \mathcal{C}^\infty(M, \mathbb{S} \otimes \mathbf{N}_i \otimes \mathbf{N}_j^o)$, and gauginos $\lambda_i \in L^2(M, \mathbb{S} \otimes M_{N_i}(\mathbb{C}))$ reduced to $su(N_i)$ after eliminating trace degrees of freedom. All the possible corresponding terms arising from the Spectral Action are: scalar masses, gaugino masses, linear couplings, bilinear couplings, and trilinear couplings. All these cases are analyzed in [Beenakker, van den Broek, van Suijlekom (2015)] in terms of the corresponding Krajewski diagrams.

The finite spectral triple of the MSSM is then constructed in terms of building blocks. The requirements are that it should contain the SM spectral triple and that the gauge group should be the same as for SM (up to finite groups), hence

$$\mathcal{A} = \mathcal{A}_{SM} = \mathbb{C} \oplus \mathbb{H} \oplus M_3(\mathbb{C}).$$

Inequivalent representations are indicated as in the SM case with $\mathbf{1}$, $\overline{\mathbf{1}}$, $\mathbf{2}$, $\mathbf{3}$, with $\overline{\mathbf{1}}$ a real representation with $\pi(\lambda)\nu = \bar{\lambda}\nu$. The gauge group (up to finite groups) is

$$G = U(1) \times SU(2) \times SU(3).$$

The superpartners with $R = -1$ are obtained from building blocks of the first type,

$$\mathcal{B}_1, \quad \mathcal{B}_{1_R}, \quad \mathcal{B}_{\overline{1}_R}, \quad \mathcal{B}_{2_L}, \quad \mathcal{B}_3.$$

In reducing the algebra from the left-right symmetric case $\mathbb{C} \oplus \mathbb{H}_L \oplus \mathbb{H}_R \oplus M_3(\mathbb{C})$ to $\mathbb{C} \oplus \mathbb{H} \oplus M_3(\mathbb{C})$, the $\mathbb{C} \hookrightarrow \mathbb{H}_R$ part acts as $(\lambda, \bar{\lambda})$ and $\mathbf{2}_R$ breaks to $1_R \oplus \overline{1}_R$. At this point in the construction there are too many fermionic degrees of freedom, which requires introducing some identifications. For each SM fermion one needs a building block of the second type with $R = +1$ (M copies to account for generations). The SM Higgs requires building blocks of the second type with $R = -1$ (one copy): these give Higgs/higgsino building blocks. These blocks determine the finite dimensional Hilbert space of the finite spectral triple

$$\mathcal{H}_F = \mathcal{H}_{F,R=+1} \oplus \mathcal{H}_{F,R=-1}.$$

There is an element of $\mathbb{C} \oplus \mathbb{H}_L \oplus \mathbb{H}_R \oplus M_3(\mathbb{C})$ implementing the R symmetry, given by

$$R = -(+,-,-,+) \otimes (+,-,-,+)^o.$$

This has $R = +1$ on all the SM fermions and $R = -1$ on the higgsino. The Yukawa couplings of fermions and Higgs require considering building blocks of the third type $\mathcal{B}_{ijk}$, which should contain the Higgs interaction of the SM part, but with different up and down Higgses. The analysis of these terms leads to four possible building blocks. The massless photon condition now requires for the D_+ part the condition $[D_+, \mathbb{C}_F] = 0$, with $\mathbb{C}_F = \{(\lambda, \bar{\lambda}, 0) \in \mathbb{C} \oplus \mathbb{H} \oplus M_3(\mathbb{C})\}$. The Majorana masses for the right handed neutrinos are obtained as in the νMSM case. In the MSSM model there is an up/down Higgs/higgsino interaction $\mu H_u \cdot H_d$, which requires additional building blocks. The counting of sparticles and hypercharges is given by: gauginos in $\mathbf{1} \otimes \mathbf{1}^o$, $\mathbf{2} \otimes \mathbf{2}^o$, $\mathbf{3} \otimes \mathbf{3}^o$ with zero hypercharges; higgsino in $\mathbf{1}_R \otimes \mathbf{2}_L^o$ and $\bar{\mathbf{1}}_R \otimes \mathbf{2}_L^o$ with hypercharges $+1$ and -1; correct hypercharges of sfermions: $\tilde{q}_L$ with $1/3$, $\tilde{u}_R$ with $4/3$, $\tilde{d}_R$ with $-2/3$, $\tilde{\ell}_L$ with -1, $\tilde{\nu}_R$ with 0, and $\tilde{e}_R$ with -2. This confirms the correct identification of MSSM sparticles. Unimodularity, which eliminates trace modes, is imposed for the correct matching of fermion/boson degrees of freedom.

The authors further derive a set of relations between parameters for supersymmetry of the action to hold, but the resulting constraints do not have solutions for any integer $M \in \mathbb{N}$, representing the number of generations. The conclusion that can be driven from this fact is that the spectral action

$$\text{Tr}(f(D_A/\mathbf{\Lambda})) + \frac{1}{2}\langle J\xi, D_A\xi\rangle$$

on an almost-commutative spectral triple $M \times F$, with F the finite spectral triple of the MSSM is itself *not* supersymmetric. Perhaps further extensions (beyond MSSM) may add more blocks and correct the problem. Or perhaps other possible approaches to incorporate supersymmetry in the NCG models of particle physics may be possible. We have given here only a very quick and sketchy account of the construction. We refer the reader to [Beenakker, van den Broek, van Suijlekom (2015)] for a precise and detailed account.

1.3.9 *Adinkras, SUSY algebras, and the spectral action*

This section is based on [Marcolli, Zolman (2016)]. It explores a possibly different approach to combine spectral triples with supersymmetry.

Instead of approaching directly a complicated supersymmetric model like the MSSM discussed in the previous section, we consider here a much simpler case, with a 1-dimensional spacetime (zero-dimensional space and a time direction), that is, the setting of supersymmetric quantum mechanics.

The off-shell supersymmetry algebras, or $(1|N)$-superalgebras, are generated by operators $Q_1, Q_2, \ldots, Q_N$ and ∂_t, with commutation relations

$$[Q_i, H] = 0 \tag{1.22}$$

$$\{Q_i, Q_j\} = 2\delta_{ij}H \tag{1.23}$$

where $H = i\partial_t$ is the Hamiltonian. Representations of these operators acting on bosonic and fermionic fields are given by

$$Q_k\phi_a = c\partial_t^\lambda \psi_b \tag{1.24}$$

$$Q_k\psi_b = \frac{i}{c}\partial_t^{1-\lambda}\phi_a \tag{1.25}$$

with $c \in \{-1, 1\}$ and $\lambda \in \{0, 1\}$, where $\{\phi_1, \ldots, \phi_m\}$ are real commuting bosonic fields and $\{\psi_1, \ldots, \psi_m\}$ are real anticommuting fermionic fields. The off-shell condition means that no other equation is satisfied except the commutation relations above. It was shown in [Faux, Gates (2005)] that there is a graphical way of classifying these supersymmetry algebras. The corresponding combinatorial objects are called Adinkras.

Adinkras consist of several data. The first set of data is referred to as an N-dimensional *chromotopology*. This is given by a finite connected graph A with the following properties:

- the graph is N-regular (all vertices have valence N) and bipartite;
- the edges $e \in E(A)$ are colored by colors in the set $\{1, 2, \ldots, N\}$;
- every vertex is incident to exactly one edge of each color;
- given two different colors $i \neq j$, the edges in $E(A)$ with colors i and j form a disjoint union of 4-cycles.

The bipartite structure on the graph corresponds to the subdivision into "bosons" and "fermions" (respectively, black/white colored vertices). In addition to this N-dimensional chromotopology structure, an Adinkra also has the following additional data:

- *ranking*: a function $h : V(A) \to \mathbb{Z}$ that defines a partial ordering. This can be represented by a height function that gives a vertical placement of vertices.

- *dashing*: a function $d : E(A) \to \mathbb{Z}/2\mathbb{Z}$ with values $0/1$, represented by drawing edges as solid or dashed.
- *odd-dashing*: every 4-cycle has an odd number of dashed edges.
- *well-dashed*: all 2-colored 4-cycles all have an odd-dashing.

An Adinkra is a well-dashed, N-chromotopology with a ranking on its bipartition such that bosons have even ranking and fermions have odd ranking. An N-chromotopology is *Adinkraizable* if it admits well-dashing and ranking as above. It was shown in [Faux, Gates (2005)] that Adinkras classify the $(1|N)$ superalgebras.

The N-cube is an Adinkra. Consider the 2^N vertices labelled with binary codewords of length N, and connect two vertices with an edge of color i if the Hamming distance between the corresponding binary strings is equal to 1 (that is, they differ in only one digit), and they differ exactly at the i-th digit. Define the ranking $h : V(A) \to \mathbb{Z}$ by setting $h(v)$ to be the number of digits in the binary string of the vertex v that are equal to 1. This determines a bipartion of vertices into bosons/fermions where even rank vertices are bosons, and odd rank ones are fermions. There are $2^{2^N - 1}$ possible choices of a dashing on the N-cube.

A more general construction of Adinkras was obtained in [Doran, Faux, Gates, Hübsch, Iga, Landweber, Miller (2011)], [Doran, Faux, Gates, Hübsch, Iga, Landweber (2008)] using coding theory. Let L be a linear binary code of dimension k, that is, a k-dimensional subspace of $\mathbb{Z}/2\mathbb{Z}^N$. For a codeword $c \in L$, the weight $wt(c)$ is the number of 1's in the word. The code L is even if every $c \in L$ has even weight. It is doubly-even if the weight of every codeword is divisible by 4. Consider the quotient space $\mathbb{Z}/2\mathbb{Z}^N / L$.

The chromotopology A_N on the N-cube determines a chromotopology on the graph $A = A_N / L$ with vertices the equivalence class of vertices and with an edge of color i between classes $[v]$ and $[w] \in V(A)$ whenever there is at least an edge of color i between representatives $v' \in [v]$ and a $w' \in [w]$. Then the graph A has the following properties:

- A has a loop iff L has a codeword of weight 1;
- A has a double edge iff L has a codeword of weight 2;
- A can be ranked iff A is bipartite iff L is an even code;
- A can be well-dashed iff L is a doubly-even code.

Coding theory a general source of constructions of Adinkras: it is shown in [Doran, Faux, Gates, Hübsch, Iga, Landweber, Miller (2011)] that all

Adinkraizable chromotopologies can be obtained from a cube Adinkras via linear codes.

Another very interesting recent development in the study of Adinkras and $(1|N)$-supersymmetry algebras was obtained in [Doran, Iga, Kostiuk, Landweber, Méndez-Diez (2015)], which uncovered a remarkable connection between these supersymmetry algebras and Grothendieck's theory of *dessins d'enfant*, [Schneps (1994)]. Grothendieck's dessins are a way of characterizing Riemann surfaces given by algebraic curves defined over number fields in terms of branched coverings of $\hat{\mathbb{C}} = \mathbb{P}^1(\mathbb{C})$, in such a way as to obtain a combinatorial and geometric description of the action of the absolute Galois group $\mathrm{Gal}(\bar{\mathbb{Q}}/\mathbb{Q})$. We will not consider here the Galois action, but we review briefly the main data of Belyi maps and dessins d'enfant. A Belyi maps is a meromorphic function $f : X \to \hat{\mathbb{C}}$, from a compact Riemann surface X to the Riemann sphere $\hat{\mathbb{C}} = \mathbb{P}^1(\mathbb{C})$, which is unramified outside of the points $\{0, 1, \infty\}$. A dessin is a bipartite graph Γ embedded on the surface X, with white vertices at points $f^{-1}(1)$, black vertices at points $f^{-1}(0)$, and edges along the preimage $f^{-1}(\mathcal{I})$ of the interval $\mathcal{I} = (0, 1)$. A Belyi pair (X, f) consists of a compact Riemann surface X with a Belyi map $f : X \to \hat{\mathbb{C}}$. Every Belyi pair defines a dessin and every dessin defines a Belyi pair. Those compact Riemann surfaces X that admit a Belyi map $f : X \to \hat{\mathbb{C}}$ are precisely the algebraic curves defined over a number field (Belyi's theorem [Belyi (1980)]). It is shown in [Doran, Iga, Kostiuk, Landweber, Méndez-Diez (2015)] that Adinkras determine dessins d'enfant. Namely, one first shows that it is possible to embed an Adinkra graph $A_{N,k}$ in a Riemann surface, by attaching 2-cells to consecutively colored 4-cycles, that is, to all 2-colored 4-cycles with color pairs $\{i, i + 1\}$ or $\{N, 1\}$. This determines an oriented compact Riemann surface $X_{N,k}$ of genus $g = 1 + 2^{N-k-3}(N - 4)$, if $N \geq 2$, and genus $g = 0$ otherwise. In [Marcolli, Zolman (2016)] it is shown how to obtain not only dessins d'enfant but also *origami curves* from Adinkras.

A result of [Cimasoni, Reshetikhin (2007)] shows that a *dimer configuration* on an embedded graph A on a Riemann surface X determines an isomorphism between *Kasteleyn orientations* on A (up to equivalence) and *spin structures* on X. A dimer configuration on a bipartite graph A is a *perfect matching*: a set of edges such that every vertex is incident to exactly one edge. As shown in [Doran, Iga, Kostiuk, Landweber, Méndez-Diez (2015)], taking edges of a fixed color on an Adinkra determines a dimer configuration. A Kasteleyn orientation of a graph embedded on a Riemann surface is an orientation of edges such that, when going around the boundary of a face counterclockwise, one goes against the orientation of an *odd* number of edges. An odd dashing of an Adinkra determines a Kasteleyn orientation. Two choices of dashings are equivalent if they can be obtained from one another by a sequence of "vertex changes", where the dash/solid drawing of each edge incident to the vertex is changed. Equivalent dashings give equivalent Kasteleyn orientations and determine the same spin structure on X.

A super-Riemann-surface M is locally modeled on $\mathbb{C}^{1|1}$ with local coordinates z (bosonic) and θ (fermionic), with a subbundle $\mathcal{D} \subset T\mathbb{C}^{1|1}$ defined by

$$D_\theta = \partial_\theta + \theta \partial_z$$

$$[D_\theta, D_\theta] = 2\partial_z$$

which gives $\mathcal{D} \otimes \mathcal{D} \simeq TM/\mathcal{D}$. For a general theory of super-Riemann-surfaces we refer the reader to [Manin (1991)], [Manin (1988)]. It is shown in [Doran, Iga, Kostiuk, Landweber, Méndez-Diez (2015)] that an odd dashing on an Adinkra determines a Super Riemann Surface structure on the Riemann surface $X_{N,k}$ of the dessin determined by the Adinkra.

The relation between Adinkras, dessins, and super-Riemann-surfaces shown in [Doran, Iga, Kostiuk, Landweber, Méndez-Diez (2015)] is used in [Marcolli, Zolman (2016)] to associate to a $(1|N)$-superalgebra a spectral geometry and a spectral action functional. First one associates to the $(1|N)$-superalgebra the corresponding Adinkra graph $A = A_{N,k}$. Then one proceeds as in [Doran, Iga, Kostiuk, Landweber, Méndez-Diez (2015)] and embeds A as a dessin on a compact Riemann surface X which has the structure of a super-Riemann-surface. The existence of a Belyi map $f : X \to \hat{\mathbb{C}}$ implies that there is a uniformization $\Phi : \mathbb{H} \to X = H\backslash\mathbb{H}$, where $\mathbb{H}$ is the hyperbolic upper half plane, ramified at $f^{-1}\{0, 1, \infty\}$, and where the uniformizing group is a finite index subgroup $H \subset \Delta$ of a Fuchsian triangle group $\Delta = \Delta_{p,q,r}$, with (p, q, r) integers such that $p^{-1} + q^{-1} + r^{-1} < 1$, with the Belyi map $f : X = H\backslash\mathbb{H} \to \widehat{\mathbb{C}} = \Delta_{p,q,r}\backslash\mathbb{H}$, and with p, q, r the ramification numbers at every point of $f^{-1}(0)$, $f^{-1}(1)$, $f^{-1}(\infty)$, respectively, see [Cohen, Itzykson, Wolfart (1994)]. In the case of the Riemann surfaces $X = X_{N,k}$, with $N > 4$, associated to the Adinkra graphs, the triangle groups are of the form $\Delta = \Delta_{N,N,2}$.

Given an Adinkra chromotopology $A_{N,k}$ and the associated Riemann surface $X_{N,k}$ with a Belyi map $f : X_{N,k} \to \hat{\mathbb{C}}$, we can then consider the spectral triple $(\mathcal{C}^\infty(X_{N,k}), L^2(X_{N,k}, \mathbb{S}_\mathfrak{s}), D)$ with D the Dirac operator $D = \slashed{D}_\mathfrak{s}$ and the spinor bundle $\mathbb{S}_\mathfrak{s}$ associated to the spin structure $\mathfrak{s}$

determined by the dashing of the Adinkra graph. The spectral action for such a geometry can be computed via the Selberg trace formula.

Lemma 1.3.1. *Let X be a compact Riemann surface of genus $g = g(X) \geq 2$, endowed with the hyperbolic metric of constant curvature -1. Let $\mathfrak{s}$ be a spin structure on X and $D = D_{\mathfrak{s}}$ the corresponding Dirac operator, acting on sections of the spinor bundle $\mathbb{S} = \mathbb{S}_{\mathfrak{s}}$ on X. Let $\chi : \bar{\Gamma} \to U(1)$ be the character determined by the spin structure $\mathfrak{s}$, as above. Let $h \in C^\infty(\mathbb{R})$ be an even compactly supported test function with support $\mathrm{supp}(h) = [-1, 1]$ and let $f \in \mathcal{S}(\mathbb{R})$ be the Fourier transform $f = \hat{h}$. The spectral action then satisfies*

$$\mathrm{Tr}(f(D/\Lambda)) = \Lambda^2 (g(X) - 1) \int_{\mathbb{R}} r f(r) \coth(\pi r) dr$$

$$+ \Lambda \sum_{P \in \mathcal{C}_{\bar{\Gamma}}} \sum_{\ell=1}^{\infty} \frac{\chi(P^\ell) \, \mathrm{arccosh}(\frac{t_P}{2}) \, h(\Lambda 2\ell \, \mathrm{arccosh}(\frac{t_P}{2}))}{\sinh(\ell \, \mathrm{arccosh}(\frac{t_P}{2}))} \quad (1.26)$$

where $\mathcal{C}_{\bar{\Gamma}}$ is the set of $\bar{\Gamma}$-conjugacy classes and $\Gamma = \bar{\Gamma}/\{\pm 1\}$.

Proof. The Selberg trace formula for the Dirac operator on a hyperbolic compact Riemann surface is obtained in [Bolte, Stiepan (2006)] by modifying the well known case of the Laplacian. One has

$$\sum_{j=0}^{\infty} f(\lambda_j) = (g(X) - 1) \int_{\mathbb{R}} r f(r) \coth(\pi r) \, dr$$

$$+ \sum_{P \in \mathcal{C}_{\bar{\Gamma}}} \mathrm{arccosh}(\frac{t_P}{2}) \sum_{\ell=1}^{\infty} \chi(P^\ell) \frac{h(2\ell \, \mathrm{arccosh}(\frac{t_P}{2}))}{\sinh(\ell \, \mathrm{arccosh}(\frac{t_P}{2}))}, \quad (1.27)$$

where $\{\lambda_j\} = \mathrm{Spec}(D_{\mathfrak{s}})$. This gives (1.26). $\qquad \square$

This version of the spectral geometry and the spectral action does not account for the super-Riemann-surface structure. In order to incorporate that part of the structure it is necessary to redefine the spectral action for a super-geometry and adapt the corresponding computation in terms of the Selberg trace formula.

A Selberg super trace formula for Super Riemann Surfaces, based on the Dirac Laplacian, was obtained in [Baranov, Manin, Frolov, Schwarz (1987)]. Additional results on the Selberg super zeta function were obtained in [Grosche (1990)], see also [Grosche (2013)]. We consider here the Dirac Laplacian $\Delta = 2Y D\bar{D}$ (the case $m = 0$ of the family of Dirac Laplacians considered in [Baranov, Manin, Frolov, Schwarz (1987)], [Grosche (1990)]), where $-(4Y^2)^{-1}$ is the superdeterminant of the metric tensor on the super upper half plane $S\mathbb{H}$, and $D = D_\theta = \theta\partial_z + \partial_\theta$ and $\bar{D} = D_{\bar\theta} = -\partial_{\bar\theta} + \bar\theta\partial_z$. Let $\{\lambda_j^B = ir_j^B + 1/2\}$ and $\{\lambda_j^F = ir_j^F + 1/2\}$ denote, respectively, the bosonic and fermionic spectra of Δ. The operator D satisfies $D^2 = \partial_z$, so it can be viewed as a square root of ∂_z. This is reflected in the structure of the spectrum, with respect to the Dirac spectrum on an ordinary Riemann surface.

Let f be a test function with the properties that $f(ix+1/2)$ is in $\mathcal{C}^\infty(\mathbb{R})$ with $f(ix+1/2) \sim O(x^{-2})$ for $x \to \pm\infty$, and with $f(iz+1/2)$ holomorphic for $|\Im(z)| \le 1 + \epsilon$. For a test function f as above. The supersymmetric spectral action of the Super Riemann Surface $SX = \Gamma \backslash S\mathbb{H}$ is given by

$$\mathcal{S}_{SX,\Delta,f}(\Lambda) = \mathrm{Tr}_s(f(\Delta/\Lambda)) = \sum_{j=0}^{\infty}(f(\frac{\lambda_j^B}{\Lambda}) - f(\frac{\lambda_j^F}{\Lambda})). \tag{1.28}$$

Let h denote the Fourier transform

$$h(t) = \frac{1}{2\pi}\int_{\mathbb{R}} e^{-itx} f(ix + 1/2)\, dx,$$

of a test function f chosen as above. Let $G(x, \chi)$ be the function

$$G(x, \chi) = h(x) + h(-x) - (\chi\, e^{-x/2}h(x) + \chi\, e^{x/2}h(-x)). \tag{1.29}$$

The supersymmetric spectral action can then be computed in terms of the Selberg supertrace formula.

Proposition 1.3.2. *Let f be a test function as above. The supersymmetric spectral action satisfies*

$$\mathcal{S}_{SX,\Delta,f}(\Lambda) = i\Lambda(g(X) - 1)\int_{\mathbb{R}} f(ir + 1/2)\tanh(\Lambda\pi r)\, dr$$

$$+ \sum_{\gamma\in\mathcal{C}(\Gamma)}\sum_{k=1}^{\infty}\frac{\lambda(\gamma)}{N_\gamma^{1/2} - N_\gamma^{-1/2}}G_\Lambda(\log N_\gamma, \chi(\gamma)), \tag{1.30}$$

where $G_\Lambda(x, \chi) = h_\Lambda(x) + h_\Lambda(-x) - (\chi\, e^{-x/2}h_\Lambda(x) + \chi\, e^{x/2}h_\Lambda(-x))$, with the function $h_\Lambda(t) = \Lambda e^{-\frac{t}{2}(\Lambda-1)}h(\Lambda t)$.

Proof. The Selberg supertrace formula is given by ([Baranov, Manin, Frolov, Schwarz (1987)], [Grosche (1990)])

$$\sum_{j=0}^{\infty}(f(\lambda_j^B) - f(\lambda_j^F)) = i(g-1)\int_R f(ir + 1/2)\tanh(\pi r)\, dr$$

$$+ \sum_{\gamma\in\mathcal{C}(\Gamma)}\sum_{k=1}^{\infty}\frac{\lambda(\gamma)}{N_\gamma^{1/2} - N_\gamma^{-1/2}}G(\log N_\gamma, \chi(\gamma)), \tag{1.31}$$

where the function $G(x, \chi)$ is given by (1.29). Here we identify the set $\mathcal{C}(\Gamma)$ of conjugacy classes of Γ with the oriented primitive closed geodesics and we write $\lambda(\gamma) = \ell(\gamma_0) = \log N_{\gamma_0}$ for the length of the unique element in the class of γ such that $\gamma = \gamma_0^m$ for some $m \in \mathbb{N}$. We write $N_\gamma = \exp(\ell(\gamma))$ for the exponentiated lengths. $\qquad\square$

1.3.10 *Grand Unified theories*

As we have seen in the previous sections, when constructing finite spectral triples with a real structure, very strong constraints on the possible Dirac operators arise from imposing the order-one condition

$$[[D, a], b^0] = 0$$

for all $a, b \in \mathcal{A}$, with $b^0 = Jb^*J^{-1}$. A natural question is what kind of models one can obtain if one relaxes the order-one condition. The interesting answer is that, in this way, one can obtain some GUT models like Pati–Salaam. This was shown in [Chamseddine, Connes, van Suijlekom (2013)] and [Chamseddine, Connes, van Suijlekom (2015)], which we summarize in this section.

It was shown in [Chamseddine, Connes, Marcolli (2007)] that imposing the order-one condition can be seen as a symmetry breaking mechanism, which breaks down the L/R symmetry of the algebra $\mathbb{C} \oplus \mathbb{H}_L \oplus \mathbb{H}_R \oplus M_3(\mathbb{C})$ to the SM algebra $\mathbb{C} \oplus \mathbb{H} \oplus M_3(\mathbb{C})$. In [Chamseddine, Connes (2008)] it is shown that the same argument applies with initial choice of algebra

$$\mathbb{H}_L \oplus \mathbb{H}_R \oplus M_4(\mathbb{C}),$$

which the order-one condition breaks down to $\mathbb{C} \oplus \mathbb{H} \oplus M_3(\mathbb{C})$.

If one does not impose the order-one condition, the general form of inner fluctuations of the Dirac operator differs from the usual case. Indeed, the usual argument for conjugation of fluctuated Dirac operator D_A by a unitary $U = hJuJ^{-1}$ as a gauge transformation

$$A \mapsto A^u = u[D, u^*] + uAu^*$$

only works if $[JuJ^{-1}, A] = 0$ for $A = \sum_j a_j[D, b_j]$, which requires the order-one condition. Without the order-one condition, the general form of inner fluctuations becomes

$$D' = D + A_{(1)} + \tilde{A}_{(1)} + A_{(2)},$$

with

$$A_{(1)} = \sum_j a_j[D, b_j]$$

$$\tilde{A}_{(1)} = \sum_j \hat{a}_j[D, \hat{b}_j], \quad \hat{a}_j = Ja_jJ^{-1}, \quad \hat{b}_j = Jb_jJ^{-1}$$

$$A_{(2)} = \sum_j \hat{a}_j[A_{(1)}, \hat{b}_j] = \sum_{j,k} \hat{a}_j a_k[[D, b_k], \hat{b}_j].$$

One can define a semigroup of inner perturbations

$$\text{Pert}(\mathcal{A}) = \{\sum_j a_j \otimes b_j^{op} \in \mathcal{A} \otimes \mathcal{A}^{op} : \sum_j a_j b_j = 1, \sum_j a_j \otimes b_j^{op} = \sum_j b_j^* \otimes a_j^{op}\},$$

acting on Dirac operators D by

$$\sum_j a_j \otimes b_j^{op} : D \mapsto \sum_j a_j D b_j.$$

The semigroup structure implies that inner fluctuations of inner fluctuations are still inner fluctuations, even without imposing the order-one condition.

When considering finite spectral triples without the order-one condition, one still assumes that the order-zero condition holds, $[a, b^0] = 0$ for all $a, b \in \mathcal{A}$, that is, that the Hilbert space $\mathcal{H}$ is an $\mathcal{A}$-bimodule. One also still assumes that the KO-dimension is equal to 6, for consistency with the Standard Model case. These two requirements imply that the center of the complexified algebra is

$$Z(\mathcal{A}_\mathbb{C}) = \mathbb{C} \oplus \mathbb{C}.$$

Further, the dimension of the Hilbert space is the square of an integer, hence $\mathcal{A}_\mathbb{C} = M_k(\mathbb{C}) \oplus M_k(\mathbb{C})$. Imposing a symplectic symmetry gives $k = 2a$ and an algebra of the form $M_a(\mathbb{H})$, where the chirality operator on $M_a(\mathbb{H})$ requires that a is even. A realistic physical assumption is that $k = 4$, which gives

$$\mathcal{A} = \mathbb{H}_R \oplus \mathbb{H}_L \oplus M_4(\mathbb{C}).$$

The inner automorphisms of the algebra $\mathcal{A}$ give a Pati–Salam type left-right model with gauge group

$$SU(2)_R \times SU(2)_L \times SU(4),$$

where the $SU(4)$ color group has the lepton number as 4-th color.

The dimension of the finite Hilbert space is

$$384 = 2^7 \times 3$$

where 3 is for generations, $\mathbf{2}_L$ and $\mathbf{2}_R$ correspond to $SU(2)_L$ and $SU(2)_R$, and further $16 = 1 + 15$ dimensions account for the $SU(4)$ representation, with further doubling for matter/antimatter.

Recall here that, for the Lie group $SU(N)$, the Lie algebra is $N^2 - 1$ dimensional, with a basis t_a. There are $N-1$ Casimir operators (the center of the universal enveloping algebra) that label irreducible representations. In the case of $SU(2)$, the Lie algebra is three dimensional, with basis σ_a given by the Pauli matrices, and with one Casimir operator $J^2 = \sigma_a\sigma_a$, with eigenvalues $j(j+1)$ where $j \in \frac{1}{2}\mathbb{Z}$. The irreducible representations are labelled by $p = 2j \in \mathbb{Z}_+$ and are of dimension $D^p = p + 1$. These are the usual angular momentum representations of quantum mechanics. In the case of $SU(3)$, the Lie algebra is eight dimensional, with basis λ_a given by the Gell-Mann matrices, and with two Casimir operators $\lambda_a\lambda_a$ and $f_{abc}\lambda_a\lambda_b\lambda_c$. The irreducible representations are labelled by these, with two quantum numbers $p, q \in \mathbb{Z}_+$ and have dimension

$$D^{pq} = \frac{1}{2}(p+1)(q+1)(p+q+2).$$

They are classified graphically as in the figure.

$$D^{pq}$$

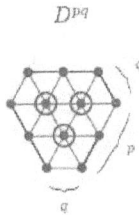

In the case of $SU(4)$, there are three Casimir operators and the irreducible representations are parameterized by three quantum numbers $p, q, r \in \mathbb{Z}_+$, with dimensions

$$D^{pqr} = \frac{1}{12}(p+1)(q+1)(p+q+2)(q+r+2)(p+q+r+3)$$

so that $D^{000} = 1$, $D^{100} = D^{001} = 4$, $D^{010} = 8$, $D^{200} = D^{002} = 10$, and $D^{101} = 15$.

Assuming that the unperturbed Dirac operator D satisfies the order-one condition when restricted to the SM subalgebra $\mathbb{C} \oplus \mathbb{H} \oplus M_3(\mathbb{C})$, as a condition of consistency with the standard model case, then it follows that the inner fluctuations in the vertical (noncommutative) direction, coming from the terms $A_{(2)}$, are composite, quadratic in those arising in the terms $A_{(1)}$. Then, with the order-one condition satisfied on the SM algebra, the Higgs has to be in the representations $(2_R, 2_L, 1)$, $(2_R, 1_L, 4)$ and $(1_R, 1_L, 1 + 15)$

(Marshak–Mohapatra model), otherwise one would have additional fundamental Higgs fields.

The Dirac operator D of the finite spectral triple is a 384×384 matrix, which the authors write in an explicit tensor notation. The inner fluctuations are also computed explicitly, again under the assumption that the order-one condition holds on the SM subalgebra. Taking a product geometry $M \times F$ of a 4-dimensional spacetime manifold and the finite geometry, the spectral action has an expansion

$$\mathrm{Tr}(f(D_A/\boldsymbol{\Lambda})) = \sum_{n=0}^{\infty} F_{4-n} \boldsymbol{\Lambda}^{4-n} a_n,$$

with a_n the Seeley deWitt coefficients of the heat kernel. The terms $F_k(u) = f(v)$, with $u = v^2$ give momenta $F_4 = 2f_4$, $F_2 = 2f_2$ with

$$F_4 = \int_0^\infty F(u)u\,du = 2 \int_0^\infty f(v)v^3\,dv, \quad F_2 = \int_0^\infty F(u)\,du = 2 \int_0^\infty f(v)v\,dv$$

and $F_0 = F(0) = f_0$, with remaining terms

$$F_{-2n} = (-1)^n F^{(n)}(0) = (-1)^n \left(\frac{1}{2v} \frac{d}{dv} \right)^n f|_{v=0}.$$

The a_2 coefficient contributes an Einstein–Hilbert term and Higgs terms; the a_0 coefficient contributes a volume term (the cosmological term); the a_4 coefficient contributes modified gravity terms, Yang–Mills terms, and Higgs terms. The Higgs terms coming from the a_2 coefficient give a Higgs potential with quartic terms and also a mass term, which are compatible with what expected in the Pati–Salaam model. We refer the reader to [Chamseddine, Connes, van Suijlekom (2013)] and [Chamseddine, Connes, van Suijlekom (2015)] for detailed expressions of these coefficients and the corresponding terms in the action.

Chapter 2

Renormalization Group Flows and Early Universe Models

In this chapter we look more closely at the dependence on the energy scale Λ of the modified gravity model derived from the asymptotic expansion of the spectral action, and how different possible assumptions about this dependence can lead to different scenarios for Early Universe models in cosmology. We begin by recalling some properties of the renormalization group flow in different particle physics models.

It is interesting to observe that, in all the current treatments of particle physics models based on noncommutative geometry, one makes an important assumption (and approximation) about the dependence on the energy scale. We have seen, from the form of the asymptotic expansion of the spectral action for the model of [Chamseddine, Connes, Marcolli (2007)], that there is a preferred energy scale associated to this action functional, which is the unification energy where the coupling constants of the three forces (should) meet. Thus, that is the scale at which it is natural to take initial conditions for the model and investigate geometric constraints on the choice of initial conditions (exclusion curves, relations between the parameters imposed by the geometry, etc.). One then obtains predictions at ordinary scales accessible by particle physics experiments by running the initial conditions down to lower energies using a renormalization group flow.

Ordinarily, the renormalization group equations should be determined by a beta function for the theory, which has one-loop (and higher-loop) contributions coming from the quantum field theory determined by the action functional of the model. However, the action functional in our case is the spectral action (1.4), with the added fermionic term (1.17). There is, at present, no good perturbative quantum field theory treatment of the spectral action (1.4) that would suffice to derive the beta functions for the renormalization group flow directly from it. Thus, what is customarily done is to use the renormalization group equation of the associated particle physics model obtained from the asymptotic expansion of the spectral action. However, doing so requires making certain choice and assumptions, in particular with respect to how one treats the presence of the gravitational terms, and the presence of particle physics content beyond the standard model. We will see in the coming sections that different choices of how to deal with these issues lead to different scenarios in the theory.

2.1   RGE flows

We review various facts about renormalization group flows in the minimal standard model and in the νMSM extension with right handed neutrinos with Majorana mass terms. The renormalization group equations (RGE) are a system of ordinary differential equations in the energy scale Λ, typically written in the form

$$\Lambda \frac{dF}{d\Lambda} = \beta_F(\Lambda),$$

where the beta function on the right hand side of the equation contains contributions at one-loop, two-loops, etc. coming from the Feynman diagrams of the asymptotic expansion of the quantum field theory.

2.1.1   *RGE flow from the Minimal Standard Model*

At one-loop order, the renormalization group equations in the minimal standard model have a property that greatly simplifies the analysis of the equations, namely the equations for the coupling constants decouple from the remaining equations. Thus, one can study separately the behavior of the equations for the coupling constants and then input the solutions as coefficient functions into the equations for the remaining parameters.

It is customary to write the equations for the coupling constants in terms of the variables

$$\alpha_i = \frac{g_i^2}{4\pi}.$$

The beta functions are given by

$$\beta_{g_i} = (4\pi)^{-2} \, b_i \, g_i^3, \quad \text{with} \quad b_i = (\frac{41}{6}, -\frac{19}{6}, -7),$$

and the renormalization group equations (at one-loop) for the coupling constants are then written as

$$\alpha_1^{-1}(\Lambda) = \alpha_1^{-1}(M_Z) - \frac{41}{12\pi} \log \frac{\Lambda}{M_Z}$$

$$\alpha_2^{-1}(\Lambda) = \alpha_2^{-1}(M_Z) + \frac{19}{12\pi} \log \frac{\Lambda}{M_Z}$$

$$\alpha_3^{-1}(\Lambda) = \alpha_3^{-1}(M_Z) + \frac{42}{12\pi} \log \frac{\Lambda}{M_Z}$$

where $M_Z \sim 91.188$ GeV is the mass of Z^0 boson, chosen to normalize the energy scale. It is well known that if one inputs the low energy values of the coupling constants and runs the RGE flow upwards towards higher energies, the coupling constants fail to unify exactly at a scale $\Lambda = \Lambda_{\text{unif}}$, with

$$g_3^2(\Lambda_{\text{unif}}) = g_2^2(\Lambda_{\text{unif}}) = \frac{5}{3} g_1^2(\Lambda_{\text{unif}}),$$

but meet instead in a triangle, as shown in Figure 2.1, plotted in the variable $t = \log(\Lambda/M_Z)$. The fact that one finds a triangle instead of having exact unification is regarded as a strong hint of the presence of new physics beyond the standard model, which would affect the RGE flow and lead to unification.

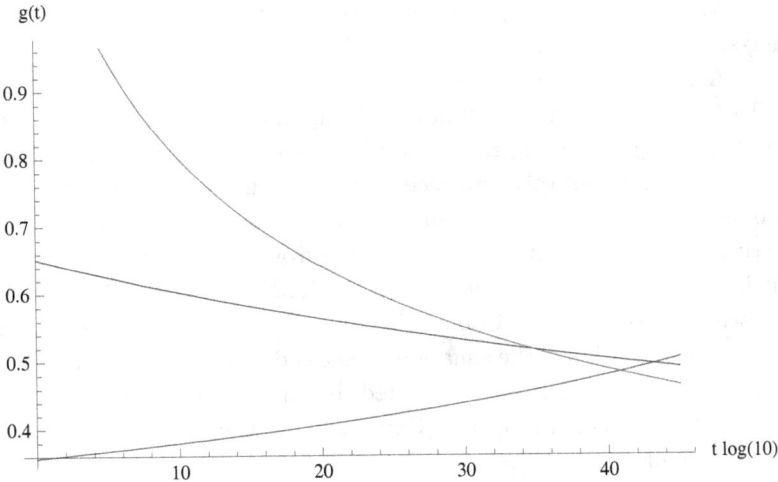

Fig. 2.1   1-loop RGE flow for the coupling constants, in the MSM, with "unification" at around 10^{17} GeV.

After separately solving the equations for the coupling constants, one can consider the remaining equations for the other parameters. These depend on the solutions for the coupling constants equations. For the complete expression of the renormalization group flow for the minimal standard model at one-loop and at two-loops, we refer the reader to [Arason et al. (1992)], [Donoghue et al. (1994)], [Sher (1989)]. We will return to see a more detailed form of the RGE flow in the case of the νMSM extension of the minimal standard model, where we will discuss the full set of equations more carefully. For the moment, we make a drastic simplification of the remaining equations, based on the customary assumption in the renormalization group analysis of the minimal standard model, that the Yukawa parameter of the top quark dominates over all other parameters. This reduces the complicated equations to just two simpler equations involving the Yukawa coupling y of the top quark, the Higgs quartic coupling λ, respectively given by the beta functions

$$\beta_y = \frac{1}{16\pi^2}\left(\frac{9}{2}y^3 - 8g_3^2 y - \frac{9}{4}g_2^2 y - \frac{17}{12}g_1^2 y\right). \tag{2.1}$$

$$\beta_\lambda = \frac{1}{16\pi^2}\left(24\lambda^2 + 12\lambda y^2 - 9\lambda(g_2^2 + \frac{1}{3}g_1^2) - 6y^4 + \frac{9}{8}g_2^4 + \frac{3}{8}g_1^4 + \frac{3}{4}g_2^2 g_1^2\right), \tag{2.2}$$

where g_1, g_2, g_3 are the solutions for the equations for the coupling constants. We will discuss in §2.1.3 the constraints that the almost commutative geometry imposes on the choice of the initial conditions at unification energy Λ_{unif} for these equations.

The choice of using the RGE flow of the minimal standard model in order to do the renormalization group analysis of the noncommutative geometry model clearly does not take into account the fact that the particle content of our model is an extension of the minimal standard model with right-handed neutrinos, where some of the terms coming from the Yukawa terms of the neutrinos may be large enough to affect the RGE flow. A simple correction was used in [Chamseddine, Connes, Marcolli (2007)], which keeps the same drastic approximation of the equations described here above, where only the equations (2.1) and (2.2) are considered, but introduces a correction term coming from the τ-neutrino, see [Chamseddine, Connes, Marcolli (2007)] for more details.

A different approach was taken in [Kolodrubetz, Marcolli (2010)] and [Marcolli, Pierpaoli (2010)], where the full renormalization group equations for the νMSM with right-handed neutrinos and Majorana mass terms are considered. The RGE analysis of the νMSM model requires additional care. The Majorana mass terms introduce non-renormalizable interaction terms, which can be dealt with by a sequence of effective field theory approximations that integrate out the heavier modes in coincidence with the see-saw scales for the neutrinos. We summarize this setting in §2.1.2.

2.1.2 *RGE flow from the νMSM*

The equations for the νMSM model were extensively studied by particle physicists, see for instance [Antusch et al. (2002)], [Antusch et al. (2005)] for detailed results. We summarize below the main ideas and the method used in these references.

At one-loop, the RGE equations for the νMSM model have beta functions:

$$16\pi^2 \, \beta_{g_i} = b_i \, g_i^3 \quad \text{with } (b_{SU(3)}, b_{SU(2)}, b_{U(1)}) = (-7, -\frac{19}{6}, \frac{41}{10})$$

$$16\pi^2 \, \beta_{Y_u} = Y_u(\frac{3}{2}Y_u^\dagger Y_u - \frac{3}{2}Y_d^\dagger Y_d + \mathfrak{a} - \frac{17}{20}g_1^2 - \frac{9}{4}g_2^2 - 8g_3^2)$$

$$16\pi^2 \, \beta_{Y_d} = Y_d(\frac{3}{2}Y_d^\dagger Y_d - \frac{3}{2}Y_u^\dagger Y_u + \mathfrak{a} - \frac{1}{4}g_1^2 - \frac{9}{4}g_2^2 - 8g_3^2)$$

$$16\pi^2 \, \beta_{Y_\nu} = Y_\nu(\frac{3}{2}Y_\nu^\dagger Y_\nu - \frac{3}{2}Y_e^\dagger Y_e + \mathfrak{a} - \frac{9}{20}g_1^2 - \frac{9}{4}g_2^2)$$

$$16\pi^2 \, \beta_{Y_e} = Y_e(\frac{3}{2}Y_e^\dagger Y_e - \frac{3}{2}Y_\nu^\dagger Y_\nu + \mathfrak{a} - \frac{9}{4}g_1^2 - \frac{9}{4}g_2^2)$$

$$16\pi^2 \, \beta_M = Y_\nu Y_\nu^\dagger M + M(Y_\nu Y_\nu^\dagger)^T$$

$$16\pi^2 \, \beta_\lambda = 6\lambda^2 - 3\lambda(3g_2^2 + \frac{3}{5}g_1^2) + 3g_2^4 + \frac{3}{2}(\frac{3}{5}g_1^2 + g_2^2)^2 + 4\lambda\mathfrak{a} - 8\mathfrak{b}.$$

We see from these equations that, as in the case of the minimal standard model, at one-loop the equations for the three coupling constants decouple from the remaining equations (and from each other). In fact, the equations for the coupling constants are exactly the same that we saw already for the minimal standard model, which can be solved explicitly in closed form. In the remaining equations, the functions $\mathfrak{a}, \mathfrak{b}$ are exactly as they appear in the expressions for the coefficients of the gravitational terms we described before, namely

$$\mathfrak{a} = \text{Tr}(Y_\nu^\dagger Y_\nu + Y_e^\dagger Y_e + 3(Y_u^\dagger Y_u + Y_d^\dagger Y_d))$$

$$\mathfrak{b} = \text{Tr}((Y_\nu^\dagger Y_\nu)^2 + (Y_e^\dagger Y_e)^2 + 3(Y_u^\dagger Y_u)^2 + 3(Y_d^\dagger Y_d)^2).$$

These equations differ from the full one-loop RGE equations for the minimal standard model in the coupled equations for the Dirac and the Majorana masses of neutrinos, because of the Majorana mass terms in

$$16\pi^2 \; \beta_M = Y_\nu Y_\nu^\dagger M + M(Y_\nu Y_\nu^\dagger)^T.$$

In the general case, the Majorana masses have a non-degenerate spectrum: the eigenvalues then correspond to three different see-saw scales. In order to deal with the fact that the Majorana mass terms introduce a non-renormalizable interaction, one considers different effective field theories in between these three see-saw scales, where each time one descends below a see-saw scale one also integrates out the heavier degrees of freedom. Thus, one can summarize the procedure of [Antusch et al. (2005)], to run down the renormalization group flow from unification energy, as the following sequence of steps:

- First run the RGE flow from unification energy Λ_{unif} down to the first see-saw scale, the largest eigenvalue of M, with a specified initial condition at unification energy.
- Then pass to an effective field theory, where the new matrix $Y_\nu^{(3)}$ is obtained by removing the last row of Y_ν in a basis in which M is diagonal, and with $M^{(3)}$ similarly obtained by removing the last row and column of M.
- Restart the new RGE flow, with these new matrices, and run it down until the second see-saw scale, with the boundary conditions at the first see-saw scale that match the value coming from the previous running.
- Introduce a second effective field theory with $Y_\nu^{(2)}$ and $M^{(2)}$ again obtained as in the previous step, and with matching boundary conditions.
- Run the induced RGE flow down to first see-saw scale.
- Introduce a new effective field theory with $Y_\nu^{(1)}$ and $M^{(1)}$ again obtained by the same procedure, and with matching boundary conditions.
- Run the induced RGE down to the electoweak energy scale Λ_{ew}.

Using the effective field theories $Y_\nu^{(N)}$ and $M^{(N)}$ between see-saw scales gives rise to a renormalization group flow that has singularities at the see-saw scales.

Figure 2.2 shows the running of the coefficients $\mathfrak{a}$ and $\mathfrak{b}$ near the top see-saw scale. One can see that there are singularities (first order phase transitions) at the see-saw scales, due to the fact of passing to the next effective field theory that integrates out the heavier modes after each see-saw scale.

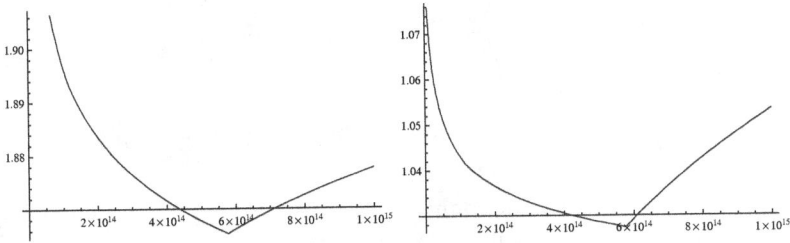

Fig. 2.2 Running of the coefficients $\mathfrak{a}$ and $\mathfrak{b}$ near the top see-saw scale.

2.1.3 Geometric constraints at unification

Geometrically, we can regard the one-loop RGE flow as a vector field on fibration over a three dimensional space, with fiber over each point given by a copy of the moduli space $\mathcal{C}_3 \times \mathcal{C}_1$ of Dirac operators of the finite geometry F. The coordinates in the base space specify a value of the three coupling constants and a point in the fiber $\mathcal{C}_3 \times \mathcal{C}_1$ specifies all the remaining parameters. Solving the uncoupled equations for the coupling constants determines a flow in the base space. Over each specified flowline in the base, one has a corresponding solution of the remaining equations, that is, a flow on the moduli space $\mathcal{C}_3 \times \mathcal{C}_1$. From this geometric viewpoint, relations between the parameters that exist independently of scale, correspond to submanifolds of the moduli space that are preserved by the flow.

The specific choice of the almost commutative geometry $X \times F$ of [Chamseddine, Connes, Marcolli (2007)] determines certain constraints and relations between the parameters at unification energy $\Lambda = \Lambda_{\text{unif}}$. These conditions restrict the possible choices of initial conditions at unification, even though these relations need not be preserved along the RGE flow. The main conditions imposed by the geometry are:

- λ parameter constraint:

$$\lambda(\Lambda_{unif}) = \frac{\pi^2}{2f_0} \frac{\mathfrak{b}(\Lambda_{unif})}{\mathfrak{a}(\Lambda_{unif})^2}$$

- Higgs vacuum constraint:

$$\frac{\sqrt{a}f_0}{\pi} = \frac{2M_W}{g}$$

- See-saw mechanism and $\mathfrak{c}$ constraint:

$$\frac{2f_2\Lambda_{unif}^2}{f_0} \leq \mathfrak{c}(\Lambda_{unif}) \leq \frac{6f_2\Lambda_{unif}^2}{f_0}$$

- Mass relation at unification:

$$\sum_{generations} (m_\nu^2 + m_e^2 + 3m_u^2 + 3m_d^2)|_{\Lambda=\Lambda_{unif}} = 8M_W^2|_{\Lambda=\Lambda_{unif}}$$

We refer the reader to [Chamseddine, Connes, Marcolli (2007)] and to Chapter 1 of [Connes and Marcolli (2008)] for a detailed discussion of how one obtains these conditions from the asymptotic expansion of the spectral action and the properties of the almost commutative geometry $X \times F$. Any use of RGE techniques within this model then requires the use of initial conditions at unification that are compatible with these constraints and low energy limits that are compatible with experimental data.

2.1.4   *Maximal mixing and initial condition at unification*

An explicit example of a choice of a set of initial conditions at unification, for the renormalization group flow of [Antusch et al. (2005)] for the νMSM model, which is compatible with all the geometric constraints listed above, and also compatible with experimental values at low energy (except for the Higgs mass, which requires a further correction term, see §2.4 below) was constructed in [Kolodrubetz, Marcolli (2010)], using the maximal mixing conditions

$$\zeta = \exp(2\pi i/3)$$

$$U_{PMNS}(\Lambda_{unif}) = \frac{1}{3}\begin{pmatrix} 1 & \zeta & \zeta^2 \\ \zeta & 1 & \zeta \\ \zeta^2 & \zeta & 1 \end{pmatrix}$$

and the initial values

$$\delta_{(\uparrow 1)} = \frac{1}{246}\begin{pmatrix} 12.2 \times 10^{-9} & 0 & 0 \\ 0 & 170 \times 10^{-6} & 0 \\ 0 & 0 & 15.5 \times 10^{-3} \end{pmatrix}$$

$$Y_\nu = U_{PMNS}^\dagger \delta_{(\uparrow 1)} U_{PMNS}.$$

2.1.5 *Sensitive dependence and fine tuning*

It was also shown in [Kolodrubetz, Marcolli (2010)] that the RGE flow of [Antusch et al. (2005)] presents a sensitive dependence on the initial conditions at unification. This was shown explicitly, by changing only one parameter in the diagonal matrix Y_ν and following its effect on the running of the top quark term. The graphs of Figure 2.3 show the behavior of the running of the top term for varying initial conditions. This presence of sensitive dependence on the initial conditions creates a possible fine tuning problem in the model.

2.2 Gravitational terms

The renormalization group equations discussed above, either for the minimal standard model or for its νMSM extension, apply to the running of the coupling constants and the Yukawa parameters, that is, to the particle physics content of the model. In particular, the chosen RGE flow determines the running of the expressions $\mathfrak{a}, \mathfrak{b}, \mathfrak{c}, \mathfrak{d}, \mathfrak{e}$, which are functions of the Yukawa parameters and the Majorana mass terms. One then needs to address separately the possible running of the gravitational terms in the asymptotic expansion of the spectral action.

There are two different possible approaches to this question. The first, followed in [Chamseddine, Connes, Marcolli (2007)], assumes that the relation between the coefficients $\kappa_0, \gamma_0, \alpha_0, \tau_0, \mu_0, \xi_0, \lambda_0$ in the gravitational terms and the parameters $\mathfrak{a}, \mathfrak{b}, \mathfrak{c}, \mathfrak{d}, \mathfrak{e}$ holds *only* at unification energy $\Lambda = \Lambda_{unif}$. In this sense, these relations have the same nature as the other geometric constraints at unification discussed above, and they only limit the possible choices of initial conditions at unification.

A second possible viewpoint, analyzed in [Marcolli, Pierpaoli (2010)], assumes that there is some interval of energies $\Lambda_{min} \leq \Lambda \leq \Lambda_{unif}$, whose precise size is unknown, but which may contain one or more of the see-saw scales, where the relation continues to hold, at least approximately, so that the running of the coefficients $\kappa_0, \gamma_0, \alpha_0, \tau_0, \mu_0, \xi_0, \lambda_0$ may be determined by the running of the expressions $\mathfrak{a}, \mathfrak{b}, \mathfrak{c}, \mathfrak{d}, \mathfrak{e}$. We discuss in §2.3 below some scenarios for very early universe models that can be derived in this setting.

The first approach requires independent information on the running of the gravitational parameters. For modified gravity models involving

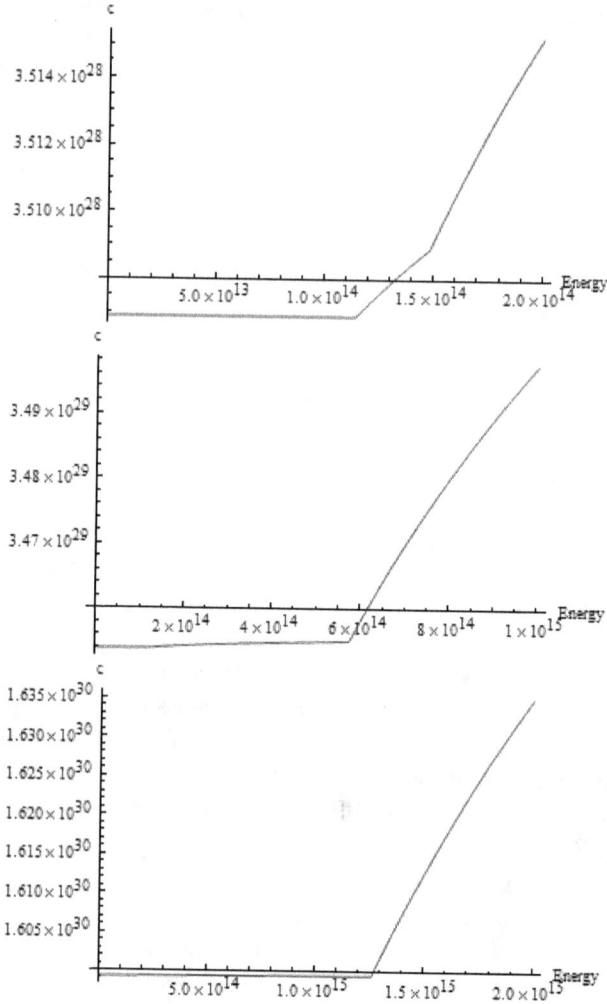

Fig. 2.3 Running of the top term, with singularities at the see-saw scales, for varying initial conditions in Y_ν.

conformal gravity (Weyl curvature) and a Gauss–Bonnet term,

$$\int \left(\frac{1}{2\eta} C_{\mu\nu\rho\sigma} C^{\mu\nu\rho\sigma} - \frac{\omega}{3\eta} R^2 + \frac{\theta}{\eta} R^* R^* \right) \sqrt{g} \, d^4 x,$$

models for the running of gravitational parameters were obtained in [Avramidi (1986)], [Codello, Percacci (2006)], [Donoghue (1994b)], with beta functions of the form

$$\beta_\eta = -\frac{1}{(4\pi)^2}\frac{133}{10}\eta^2$$

$$\beta_\omega = -\frac{1}{(4\pi)^2}\frac{25 + 1098\,\omega + 200\,\omega^2}{60}\eta$$

$$\beta_\theta = \frac{1}{(4\pi)^2}\frac{7(56 - 171\,\theta)}{90}\eta.$$

With initial conditions at unification compatible with the relations with the coefficients $\mathfrak{a}, \mathfrak{b}, \mathfrak{c}, \mathfrak{d}, \mathfrak{e}$, the running for the gravitational parameter then looks as in Figure 2.4. This gives plausible values (near fixed points) in low energy limit.

2.3 Early universe models

The form of the gravitational terms in the asymptotic expansion of the spectral action near unification energy $\Lambda = \Lambda_{\text{unif}}$ can be used as a model of cosmology in the very early universe. We report in this section some of the results of [Marcolli, Pierpaoli (2010)].

The relation between energy scales and the cosmological timeline can be summarized as follows (see [Guth (1982)]):

- Planck epoch: $t \leq 10^{-43}\,s$ after the Big Bang (unification of forces with gravity, quantum gravity)
- Grand Unification epoch: $10^{-43}\,s \leq t \leq 10^{-36}\,s$ (electroweak and strong forces unified; Higgs)
- Electroweak epoch: $10^{-36}\,s \leq t \leq 10^{-12}\,s$ (strong and electroweak forces separated)
- Inflationary epoch: possibly $10^{-36}\,s \leq t \leq 10^{-32}\,s$

Using the second approach described above for treating the running of the gravitational terms in the asymptotic expansion of the spectral action near unification scales, one obtains several possible scenarios in the very early universe (including a possible inflationary phase). These include several phenomena and gravity models that have been variously proposed and studied in the literature:

- Varying effective gravitational constant, [Jordan (1955)].
- Varying effective cosmological constant, [Overduin, Cooperstock (1998)].

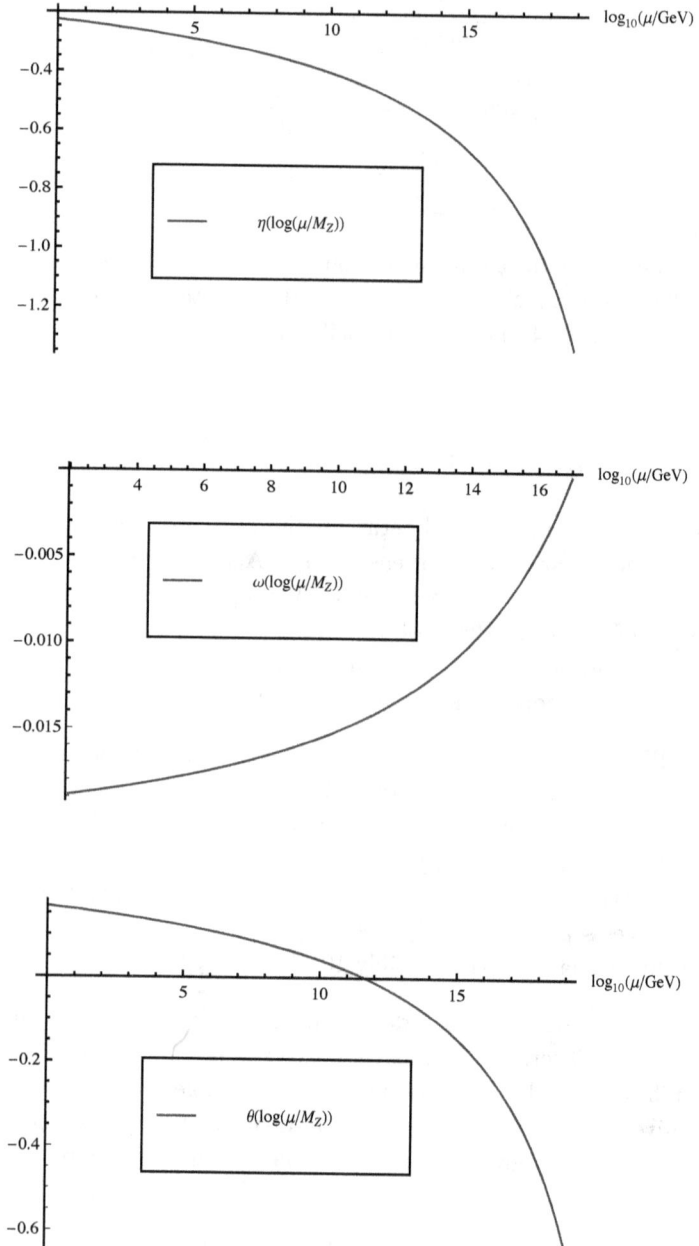

Fig. 2.4   Running of gravitational parameters from [Chamseddine, Connes, Marcolli (2007)].

- Antigravity in the early universe, [Linde (1980)].
- Gravity balls from the conformal coupling of the Higgs field to gravity, [Safonova, Lohiya (1998)].
- Primordial black holes and gravitational memory, [Carr (2001)], [Novikov et al. (1979)].
- Emergent Hoyle-Narlikar cosmologies, [Hoyle, Narlikar (1964)].
- Higgs or scalar field based slow-roll inflation, [De Simone, Hertzberg, Wilczek (2009)].

2.3.1 *Effective gravitational constant*

The effective gravitational constant in the asymptotic expansion of the spectral action has the form

$$G_{\text{eff}}(\Lambda) = \frac{\kappa_0(\Lambda)}{8\pi} = \frac{3\pi}{192\, f_2\, \Lambda^2 - 2\, f_0\, \mathfrak{c}(\Lambda)}.$$

If one follows the first approach to the running of the gravitational terms, this expression only fixes the value of $G_{\text{eff}}(\Lambda_{\text{unif}})$, while if one follows the second approach, one considers Λ in an interval near unification energy (possibly including some of the see-saw scales) and views this expression as a function of Λ, according to the running of the coefficient $\mathfrak{c}(\Lambda)$ and of the parameters f_2, f_0. As shown in [Marcolli, Pierpaoli (2010)], depending on the choice of the parameters f_0, f_2 that determine the initial condition for $G_{\text{eff}}(\Lambda)$ at $\Lambda = \Lambda_{\text{unif}}$, one can have sign changes leading to a phase of antigravity in the early universe, compatibly with restoring the desired value $G \sim (10^{19}\text{GeV})^{-2}$ of the Newton constant at lower energies (modern universe). In §2.3.3 we discuss more in detail mechanisms producing antigravity in the early universe. Notice, however, that the problem of sensitive dependence we mentioned in §2.1.5 implies that one gets very different behavior of $G_{\text{eff}}(\Lambda)$ according to the choice of the boundary conditions at unification.

A varying effective gravitational constant that runs with the energy scale Λ alters the form of the gravitational wave equations. The Einstein equations $R^{\mu\nu} - \frac{1}{2}g^{\mu\nu}R = \kappa_0^2 T^{\mu\nu}$, in the case of a cosmology of the form

$$g_{\mu\nu} = a(t)^2 \begin{pmatrix} -1 & 0 \\ 0 & \delta_{ij} + h_{ij}(x) \end{pmatrix},$$

after separating the trace and traceless part of h_{ij} give rise to the Friedmann equation

$$-3\left(\frac{\dot{a}}{a}\right)^2 + \frac{1}{2}\left(4\left(\frac{\dot{a}}{a}\right)\dot{h} + 2\ddot{h}\right) = \kappa_0^2(\mathbf{\Lambda})\,T_{00}.$$

The relation between the energy, the expansion factor $a(t)$ and time is given by the relation $\mathbf{\Lambda}(t) = 1/a(t)$. During the inflationary epoch $a(t) \sim e^{\alpha t}$ has exponential growth, while during the radiation dominated epoch it satisfies $a(t) \sim t^{1/2}$, and in the matter dominated era $a(t) \sim t^{2/3}$, see [Dodelson (2003)]. Thus, in the model based on the asymptotic expansion of the spectral action, assuming that $\mathbf{\Lambda}(t) = 1/a(t)$, and assuming that the term $192 f_2 \mathbf{\Lambda}^2$ in $G_{\text{eff}}(\mathbf{\Lambda})$ is dominant over $2 f_0 \mathfrak{c}(\mathbf{\Lambda})$, one obtains a modified equation. For example, if we consider the radiation dominated case, ordinary cosmology would give

$$h(t) = 2\pi G T_{00} t^2 + B + A\log(t) + \frac{3}{8}\log(t)^2$$

while in this type of spectral action model we would obtain a different behavior of the form

$$h(t) = \frac{4\pi^2 T_{00}}{288 f_2} t^3 + B + A\log(t) + \frac{3}{8}\log(t)^2.$$

2.3.2 *Effective cosmological constant*

The effective cosmological constant obtained from the asymptotic expansion of the spectral action is given by

$$\gamma_0(\mathbf{\Lambda}) = \frac{1}{4\pi^2}\left(192 f_4 \mathbf{\Lambda}^4 - 4 f_2 \mathbf{\Lambda}^2 \mathfrak{c}(\mathbf{\Lambda}) + f_0 \mathfrak{d}(\mathbf{\Lambda})\right).$$

Again, one can assume this expression purely as an initial condition at unification or as valid in an interval of energies, where it determines the running, as a function of the running of the terms $\mathfrak{c}(\mathbf{\Lambda})$ and $\mathfrak{d}(\mathbf{\Lambda})$. This expression depends on the additional parameter f_4, assuming that f_0 and f_2 have already been determined by the initial conditions at unification for the coupling constants of the three forces and for the effective gravitational constant. At a given (fixed) energy scale $\mathbf{\Lambda}$ where the expression above for the effective cosmological constant is valid, the vanishing of $\gamma_0(\mathbf{\Lambda})$ corresponds to the choice

$$f_4 = \frac{(4 f_2 \mathbf{\Lambda}^2 \mathfrak{c}(\mathbf{\Lambda}) - f_0 \mathfrak{d}(\mathbf{\Lambda}))}{192 \mathbf{\Lambda}^4}.$$

Again, because of sensitive dependence on the initial conditions, the behavior is very strongly influenced by the choice of the value at unification.

The example in Figure 2.5 shows that it is possible to find a running for $\gamma_0(\Lambda)$ that drastically changes the order of magnitude of the parameter, compatibly with it being very small in the modern universe, and at the initial condition at unification, but very large in an intermediate phase that corresponds in the cosmological timeline to an inflationary epoch.

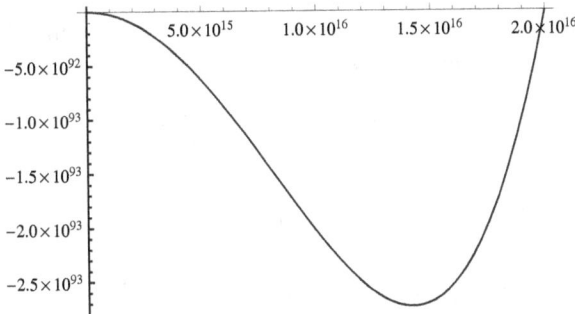

Fig. 2.5 Example of a possible running of $\gamma_0(\Lambda)$.

2.3.3 *Antigravity in the early universe*

Linde's hypothesis of antigravity in the early universe [Linde (1980)] was based on the presence of a conformal coupling of a field ϕ^2 to gravity,

$$\frac{1}{16\pi G} \int R \sqrt{g} d^4 x - \frac{1}{12} \int R \phi^2 \sqrt{g} d^4 x.$$

The presence of this field can be regarded as producing an effective gravitational constant of the form

$$G_{\text{eff},\phi}^{-1} = G^{-1} - \frac{4}{3}\pi \phi^2.$$

Unlike the original G, this G_{eff} now can vary both in time and in space (see §2.3.4 below). In particular, depending on the specific values of ϕ^2, this mechanism can produce phases of "antigravity" in the early universe.

While the model of [Linde (1980)] of antigravity in the early universe was criticized in [Pollock (1982)], similar scenarios were more recently considered within the setting of extra dimensions and brane world models, [Gregory at al. (2000)], [Guenther et al. (2005)], [Pollock (2000)].

In the spectral action model, we do have such a coupling, for the Higgs field, $\phi^2 = |H|^2$, and a conformal coupling

$$\frac{1}{16\pi G_{\text{eff}}(\Lambda)} \int R \sqrt{g} d^4 x - \frac{1}{12} \int R |H|^2 \sqrt{g} d^4 x.$$

Here, in addition to the change to the effective gravitational constant produced by the coupling to the field $|H|^2$, which recovers the same antigravity mechanism of [Linde (1980)], one also has the independent possibility of the running of $G_{\text{eff}}(\Lambda)$. The Linde mechanism of antigravity based on $|H|^2$ is present in our model, regardless of what approach one follows regarding the running of the gravitational parameters, while additional phases of antigravity, due only to the running of $G_{\text{eff}}(\Lambda)$ can be found if one follows the second approach described in §2.2 above. Thus, combining both effects, one has a variable effective gravitational constant of the form

$$G^{-1}_{\text{eff},H}(\Lambda) = G^{-1}_{\text{eff}}(\Lambda) - \frac{4}{3}\pi|H|^2.$$

This gives, for $G_{\text{eff}}(\Lambda) > 0$,

$$\begin{cases} G_{\text{eff},H}(\Lambda) < 0 \quad \text{when } |H|^2 > \dfrac{3}{4\pi G_{\text{eff}}(\Lambda)} \\[4mm] G_{\text{eff},H}(\Lambda) \geq 0 \quad \text{when } |H|^2 \leq \dfrac{3}{4\pi G_{\text{eff}}(\Lambda)}. \end{cases}$$

The conformal coupling of the Higgs field to gravity also affects the effective cosmological constant, through

$$\gamma_{0,H}(\Lambda) = \frac{\gamma_0(\Lambda)}{1 - 16\pi G_{\text{eff}}(\Lambda)\xi_0|H|^2} = \gamma_0(\Lambda)\frac{G_{\text{eff},H}(\Lambda)}{G_{\text{eff}}(\Lambda)}.$$

2.3.4 Gravity balls

In a range of energies close to unifications where the relations between parameters imposed by the asymptotic spectral action, a way to estimate the occurrence of such antigravity regimes can be obtained by first considering the Higgs self-coupling parameter λ_0. This satisfies

$$\lambda_0|_{\Lambda=\Lambda_{unif}} = \lambda(\Lambda_{unif})\frac{\pi^2\mathfrak{b}(\Lambda_{unif})}{f_0\mathfrak{a}^2(\Lambda_{unif})}.$$

In a range of energies near unification where this relation approximately holds, the equilibrium for H is given by

$$\ell_H(\Lambda,f_2) := \frac{\mu_0^2}{2\lambda_0}(\Lambda) = \frac{2\frac{f_2\Lambda^2}{f_0} - \frac{\mathfrak{c}(\Lambda)}{\mathfrak{a}(\Lambda)}}{\lambda(\Lambda)\frac{\pi^2\mathfrak{b}(\Lambda)}{f_0\mathfrak{a}^2(\Lambda)}} = \frac{(2f_2\Lambda^2\mathfrak{a}(\Lambda) - f_0\mathfrak{c}(\Lambda))\mathfrak{a}(\Lambda)}{\pi^2\lambda(\Lambda)\mathfrak{b}(\Lambda)},$$

where on both sides we emphasize the dependence on Λ and on the choice of the parameter f_2 (for fixed f_0). Negative gravity regimes then occur where

$$\ell_H(\Lambda,f_2) > \frac{3}{4\pi G_{\text{eff}}(\Lambda,f_2)}.$$

Gravity balls are space-time regions where negative gravity can occur, due to a change in the values of the field $|H|^2$, from an unstable to a stable equilibrium, see for instance [Safonova, Lohiya (1998)]. In our model, such gravity balls can occur, depending on the behavior of the effective gravitational constant $G_{\text{eff}}(\mathbf{\Lambda})$ and the estimates for the equilibria of $|H|^2$.

2.3.5 *Primordial black holes with gravitational memory*

Primordial black holes in the early universe were proposed in the 1960s by Zeldovich and Novikov, see [Novikov et al. (1979)]. Unlike stellar black holes, the promordial black holes predate the formation of stella objects and originate from the collapse of overdense spacetime regions. Other possible mechanisms that have been considered in relation to the formation of primordial black holes are cosmic loops and strings, and inflationary reheating. The behavior of primordial black hole is especially interesting in relation to cosmological models that allow a variable effective gravitational constant, see the discussion in [Barrow (1992)]. In fact, the mass loss rate due to evaporation via Hawking radiation and the Hawking temperature of primordial black holes depends on the value of the Newton constant so that the evolution of such black holes depends on the change in the Newton constant. The evaporation of primordial black holes has been proposed as a possible mechanism underlying γ-ray bursts, see [Belyanin et al. (1996)] and [Bugaev et al. (2009)]. Thus, investigating γ-ray bursts and their possible connection to primordial black holes may provide a window for an observational confirmation of a variable gravitational constant in the early universe. The evaporation law for primordial black hole and its dependence on the gravitational constant is related to the hypothesis of *gravitational memory*, see [Barrow (1992)], namely the question of whether the evolution of a primordial black hole follows the changing gravitational constant, or whether the primordial black hole retains a gravitational memory, of the value of the effective gravitational constant at the time of its formation, which may then be different from the one of surrounding space.

More precisely, in our model we have an effective gravitational constant of the form $G_{\text{eff},H}(\mathbf{\Lambda})$, which combines the effect of coupling of the Higgs field to gravity with the possible running of $G_{\text{eff}}(\mathbf{\Lambda})$ as discussed previously in this section. The evaporation law for black holes, under Hawking radiation is given by

$$\frac{d\mathcal{M}(t)}{dt} \sim -(G_{\text{eff},H}(t)\mathcal{M}(t))^{-2},$$

with Hawking temperature $T = (8\pi G_{\text{eff},H}(t)\mathcal{M}(t))^{-1}$. In terms of the energy scale one then has

$$\mathcal{M}^2 \, d\mathcal{M} = \frac{1}{\Lambda^2 G_{\text{eff},H}^2(\Lambda)} d\Lambda.$$

In the presence of gravity balls with $G_{\text{eff},H}(\Lambda)$ different from the value in surrounding regions, a primordial black holes can evolve according to a Hawking radiation law that carries the memory of the different value of the gravitational constant with respect to that of the surrounding spacetime.

2.3.6 *Emergent Hoyle–Narlikar cosmologies*

Near the phase transitions at the see-saw scales, the conformal gravity terms can become dominant and lead to an emergent Hoyle–Narlikar cosmology, described by the conformal gravity terms with the Higgs conformally coupled to gravity,

$$S_c = \alpha_0 \int C_{\mu\nu\rho\sigma} \, C^{\mu\nu\rho\sigma} \sqrt{g} \, d^4x + \frac{1}{2} \int |DH|^2 \sqrt{g} \, d^4x$$

$$- \xi_0 \int R \, |H|^2 \sqrt{g} \, d^4x + \lambda_0 \int |H|^4 \sqrt{g} \, d^4x.$$

In this approach of [Marcolli, Pierpaoli (2010)], based on running the coefficients of the gravitational terms, these conformal gravity terms are subdominant with respect to the usual Einstein–Hilbert action with cosmological term, except near the phase transitions in the early universe that correspond to the see-saw scales, where they can become dominant (again depending strongly on the boundary conditions at unification).

However, in other approaches to the spectral action model, where the gravitational terms do not significantly run in the early universe, the corrections to Einstein gravity due to the presence of the conformal gravity terms can be relevant regardless of scale. Under these assumptions, several effects due to the presence of these terms were investigated in [Nelson, Ochoa, Sakellariadou (2010)], [Nelson, Ochoa, Sakellariadou (2010b)], [Nelson, Sakellariadou (2010)], [Nelson, Sakellariadou (2009)].

2.3.7 *Slow-roll inflation*

In [De Simone, Hertzberg, Wilczek (2009)] a possible mechanism for slow-roll inflation based on non-minimal coupling of the Higgs field to gravity was proposed. The non-minimal coupling takes the form

$$\xi_0 \int R |H|^2 \sqrt{g} d^4 x,$$

as in our spectral action model, but it is non-conformally coupled, with $\xi_0 \neq 1/12$, and with the coefficient $\xi_0 = \xi_0(\Lambda)$ that runs with the energy scale. This determines an effective Higgs potential. Setting $\psi = \sqrt{\xi_0} \kappa_0 |H|$, one obtains a possible inflation scenario where an inflationary period corresponds to values $\psi >> 1$, the end of inflation occurs when $\psi \sim 1$, and the low energy regime corresponds to $\psi << 1$, see Figure 2.6.

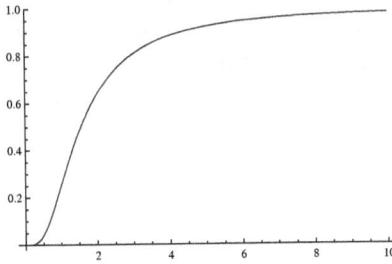

Fig. 2.6   A Higgs based slow-roll inflation scenario.

In the asymptotic expansion of the spectral action obtained in [Chamseddine, Connes, Marcolli (2007)] we have $\xi_0 = 1/12$ fixed at the conformal point, but we can still recover a similar Higgs based slow-roll inflation due to the fact that the parameter $\kappa_0 = \kappa_0(\Lambda)$ is running, see [Marcolli, Pierpaoli (2010)]. This gives

$$\psi(\Lambda) = \sqrt{\xi_0(\Lambda)} \kappa_0(\Lambda) |H| = \sqrt{\frac{\pi^2}{96 f_2 \Lambda^2 - f_0 \mathfrak{c}(\Lambda)}} \, |H|.$$

As in [De Simone, Hertzberg, Wilczek (2009)] one considers an effective Einstein metric $g_{\mu\nu}^E = f(H) g_{\mu\nu}$, for $f(H) = 1 + \xi_0 \kappa_0 |H|^2$, and the Higgs potential

$$V_E(H) = \frac{\lambda_0 |H|^4}{(1 + \xi_0 \kappa_0^2 |H|^2)^2}$$

For $\psi >> 1$ this approaches a constant value, as one expects in a slow–roll inflationary model, while it recovers the usual quartic potential for $\psi << 1$.

The feasibility of Higgs based slow–roll inflation models within the spectral action approach were investigated in [Buck, Fairbairn, Sakellariadou (2010)]. They found very strong constraints coming from the CMB data, which make an inflation scenario based on the conformal coupling of the Higgs to gravity in the spectral action expansion incompatible with the measured value of the top quark mass. We will see in the next chapter that better slow-roll inflation models in the spectral action setting are obtained by considering scalar perturbations of the Dirac operator and the non-perturbative form of the spectral action.

2.4 Higgs mass estimates

In the νMSM standard model extension obtained with the noncommutative geometry model of [Chamseddine, Connes, Marcolli (2007)], the low energy predictions obtained by RGE analysis involve an approximation, as discussed above, that only considers the Yukawa coupling for the top quark, the Higgs quartic coupling, and the three coupling constants, whose equations decouple from the remaining ones. Instead of using the RGE equations of the νMSM, as we discussed above, the presence of right handed neutrinos with Majorana masses is accounted for by a rough approximation that corrects the running of the top quark Yukawa parameter by a contribution from the τ neutrino, which is allowed to be comparably large by the see-saw mechanism present in the model. The resulting running for the Yukawa coupling $y = y(\Lambda)$ of the top quark is then used, as we discussed earlier in this chapter, to study the running of the Higgs self-coupling. Using the RGE of the minimal standard model (MSM), one has for the Higgs scattering parameter

$$\frac{f_0}{2\pi^2} \int \mathfrak{b}\, |\varphi|^4 \sqrt{g}\, d^4x = \frac{\pi^2}{2 f_0} \frac{\mathfrak{b}}{\mathfrak{a}^2} \int |\mathbf{H}|^4 \sqrt{g}\, d^4x$$

with a relation at unification

$$\tilde{\lambda}(\Lambda) = g_3^2 \frac{\mathfrak{b}}{\mathfrak{a}^2}.$$

The running of the Higgs scattering parameter is then given by

$$\frac{d\lambda}{dt} = \lambda\gamma + \frac{1}{8\pi^2}(12\lambda^2 + B)$$

$$\gamma = \frac{1}{16\pi^2}(12y^2 - 9g_2^2 - 3g_1^2) \quad B = \frac{3}{16}(3g_2^4 + 2g_1^2 g_2^2 + g_1^4) - 3y^4,$$

in the logarithmic variable $t = \log(\Lambda/M_Z)$. In this MSM approximation for the RGE flow one obtains

$$m_H^2 = 8\lambda \frac{M^2}{g^2}, \quad m_H = \sqrt{2\lambda} \frac{2M}{g}$$

which gives $\lambda(M_Z) \sim 0.241$ and an unrealistically large Higgs mass of up to ~ 170 GeV (or 168 GeV with the corrections from the see-saw mechanism), in clear contrast with the experimental value of around 125.5 GeV.

As we have seen, the NCG models of particle physics are very constrained on field content, so it is difficult to modify them to allow for additional fields, but it is natural to ask whether there is room for *other fields* that can alter the RGE flow and hence significantly lower the Higgs mass value. Another natural question is whether the RGE flow used in this argument is indeed the correct one. We have already seen the different between the RGE flow for the νMSM rather than the MSM model, where the effect of the Majorana mass terms can influence the equations and their solutions more significantly than in the simpler approximation described here. Another issue to consider is the presence of gravitational terms and their coupling to matter: usually the gravitational terms have a negligible effect, but it is possible that there may be acceptable boundary conditions at unification in the model that make these terms large enough at high energies to affect the RGE flow. Other issues can arise in connection to the possible existence of new physics (supersymmetry, GUTs). We discuss here two possible approaches, one based on an additional scalar field [Chamseddine, Connes (2012)], and one based on gravitational terms, [Estrada, Marcolli (2013b)].

2.4.1 *Scalar fields and the Higgs mass problem*

We recall here a possible solution to the Higgs mass problem proposed in [Chamseddine, Connes (2012)].

The action involving the Higgs and a coupled scalar field is given by

$$-\frac{2}{\pi} f_2 \Lambda^2 \int d^4x \sqrt{g} \left(\frac{1}{2} a \bar{H} H + \frac{1}{4} c\sigma^2\right)$$

$$+\frac{f_0}{2\pi^2} \int d^4x \sqrt{g} \left(\mathfrak{b}(\bar{H}H)^2 + \mathfrak{a}|\nabla_\mu H_a|^2 + 2\mathfrak{e}\bar{H}H\sigma^2 + \frac{1}{2}d\sigma^4 + \frac{1}{2}\mathfrak{c}(\partial_\mu\sigma)^2 \right),$$

where H is a Higgs dublet and σ is a scalar field.

The Higgs-singlet potential, after a rescaling of the fields, is given by

$$V = \frac{1}{4}(\lambda_h \bar{h}^4 + 2\lambda_{h\sigma}\bar{h}^2\bar{\sigma}^2 + \lambda_\sigma\bar{\sigma}^4) - \frac{2g^2}{\pi^2}f_2\Lambda^2(\bar{h}^2 + \bar{\sigma}^2),$$

with $\bar{h} = |k^u|h$, $\bar{\sigma} = |k^{\nu_R}|\sigma$ and $\bar{h} \mapsto \bar{h}g\sqrt{2/(n+3)}$, and $\bar{\sigma} \mapsto \bar{\sigma}2g$. Also, with $\lambda_\sigma = 8g^2$,

$$\lambda_h = \frac{n^2+3}{(n+3)^2}4g^2, \quad \lambda_{h\sigma} = \frac{2n}{n+3}4g^2,$$

where n here is a continuous parameter. Thus, there are two parameters in this model, a unification energy u and the parameter n, with

$$k_t(u) = g\sqrt{\frac{4}{n+3}}, \quad k_\nu = \sqrt{n}\,k_t.$$

The resulting RGE flow, involving the Yukawa parameters of the top quark k_t and the τ neutrino k_ν, and the Higgs and the singlet quartic couplings λ_h, λ_σ, and $\lambda_{h\sigma}$ is then given, in logarithmic variable $\mu = \log(\Lambda/M_Z)$, by

$$\frac{d}{d\mu}k_t = \frac{k_t}{32\pi^2}\left(-\left(\frac{17}{6}g_1^2 + \frac{9}{2}g_2^2 + 16g_3^2\right) + 9k_t{}^2 + 2k_\nu{}^2\right)$$

$$\frac{d}{d\mu}k_\nu = \frac{k_\nu}{32\pi^2}\left(-\left(\frac{3}{2}g_1^2 + \frac{9}{2}g_2^2\right) + 6k_t{}^2 + 5k_\nu{}^2\right)$$

$$\frac{d}{d\mu}\lambda_h = \frac{1}{16\pi^2}\left(\left(12k_t{}^2 + 4k_\nu{}^2 - (3g_1^2 + 9g_2^2)\right)\lambda_h\right.$$

$$\left. + 2\left(12\lambda_h^2 + \lambda_{h\sigma}^2 + \frac{3}{16}(g_1^4 + 2g_1^2g_2^2 + 3g_2^4) - 3k_t{}^4 - k_\nu{}^4\right)\right)$$

$$\frac{d}{d\mu}\lambda_{h\sigma} = \frac{\lambda_{h\sigma}}{16\pi^2}\left(\frac{1}{2}\left(12k_t{}^2 + 4k_\nu{}^2 - 3g_1^2 - 9g_2^2\right)\right.$$

$$\left. + 4\left(3\lambda_h + \frac{3}{2}\lambda_\sigma + 2\lambda_{h\sigma}\right)\right)$$

$$\frac{d}{d\mu}\lambda_\sigma = \frac{1}{16\pi^2}\left(8\lambda_{h\sigma}^2 + 18\lambda_\sigma^2\right).\,.$$

Again, as in the previous cases, one can separately run the equations for the coupling constants g_i, which are decoupled at one-loop, and then get the running of λ_h, $\lambda_{h\sigma}$, λ_σ. The resulting running obtained in [Chamseddine, Connes (2012)] is of the form shown in the figure, reproduced here from their paper.

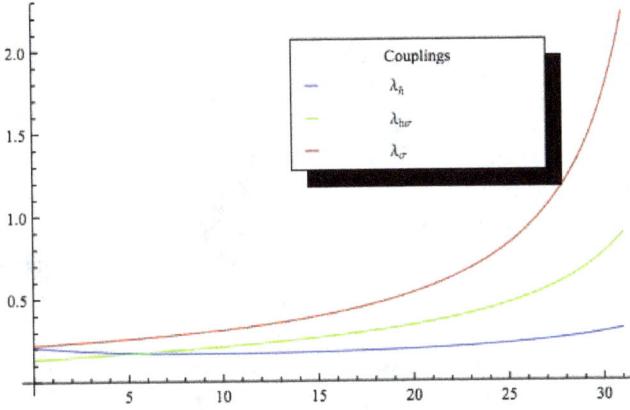

Expanding the scalar fields around the vacuum expectation value (vev) gives $\bar{h} = \bar{v} + \bar{\phi}$ and $\bar{\sigma} = \bar{w} + \bar{\tau}$, with potential

$$V \sim (-\frac{1}{4}\bar{v}^4\lambda_h - \frac{1}{2}\bar{v}^2\bar{w}^2\lambda_{h\sigma} - \frac{1}{4}\bar{w}^4\lambda_\sigma)$$

$$+ \bar{v}^2\bar{\phi}^2\lambda_h + 2\bar{v}\bar{w}\bar{\tau}\bar{\phi}\lambda_{h\sigma} + \bar{w}^2\bar{\tau}^2\lambda_\sigma$$

The expansion gives mass terms for $\bar{\phi}$ and $\bar{\tau}$,

$$\frac{1}{2}\left(\bar{\phi}\ \bar{\tau}\right) M^2 \begin{pmatrix} \bar{\phi} \\ \bar{\tau} \end{pmatrix}$$

$$M^2 = 2\begin{pmatrix} \lambda_h\bar{v}^2 & \lambda_{h\sigma}\bar{v}\bar{w} \\ \lambda_{h\sigma}\bar{v}\bar{w} & \lambda_\sigma\bar{w}^2 \end{pmatrix}$$

with eigenvalues

$$m_\pm^2 = \lambda_h\bar{v}^2 + \lambda_\sigma\bar{w}^2 \pm \sqrt{(\lambda_h\bar{v}^2 - \lambda_\sigma\bar{w}^2)^2 + 4\lambda_{h\sigma}^2\bar{v}^2\bar{w}^2}$$

One has an approximation

$$m_+^2 \sim 2\lambda_\sigma\bar{w}^2 + 2\frac{\lambda_{h\sigma}^2}{\lambda_\sigma}\bar{v}^2$$

$$m_-^2 \sim 2\lambda_h\bar{v}^2(1 - \frac{\lambda_{h\sigma}^2}{\lambda_h\lambda_\sigma}).$$

The Higgs mass is then reduced by a factor of $\sqrt{(1 - \frac{\lambda_{h\sigma}^2}{\lambda_h\lambda_\sigma})}$ which is around 0.78, so at low energy

$$m_t(0) = k_t(0)\frac{246}{\sqrt{2}}, \quad m_h(0) = 246\sqrt{2\lambda_h(0)(1 - \frac{\lambda_{h\sigma}^2(0)}{\lambda_h(0)\lambda_\sigma(0)})}$$

One obtains a stable Higgs mass for $\lambda_{h\sigma}^2 < \lambda_h \lambda_\sigma$. The estimate for the Higgs mass then depends on the parameters (u, n) and the authors show the curve in this parameter space that provides a correct Higgs mass of 125.5 GeV.

A scalar field σ in the NCG standard model can also be derived in a natural way from the fused algebra point of view of [Farnsworth, Boyle (2015)], where the symmetries of the fused algebra provide an extension of the symmetries of the standard model by an extra $U(1)_{B-L}$ gauge symmetry, and a single extra complex scalar field σ, which is a singlet under the SM gauge group $SU(3) \times SU(2) \times U(1)$, but has $B - L = 2$.

As we will discuss in the next chapter, scalar fields in the NCG spectral action model, arising from scalar perturbations of the Dirac operator, also have applications to cosmology as a source of slow-roll inflation potentials.

2.4.2   *Asymptotic safety and anomalous dimensions*

We discuss here an approach of [Estrada, Marcolli (2013b)] to the Higgs mass problem in the NCG standard model, based on asymptotic safety and anomalous dimensions. The asymptotic safety point of view is based on [Weinberg (1979)] and also [Reuter (1998)], [Shaposhnikov, Wetterich (2010)].

In this point of view, the gravitational terms introduce corrections to the RGE flow in the form of "anomalous dimensions" terms

$$\partial_t x_j = \beta_j^{\text{SM}} + \beta_j^{\text{grav}}$$

$$\beta_j^{\text{grav}} = \frac{a_j}{8\pi} \frac{\Lambda^2}{M_P^2(\Lambda)} x_j$$

where the a_j are the *anomalous dimensions*, and

$$M_P^2(\Lambda) = M_P^2 + 2\rho_0 \Lambda^2$$

is the scale dependence of the Newton constant estimated to $\rho_0 \sim 0.024$, as in [Shaposhnikov, Wetterich (2010)]. The goal here is to examine boundary conditions at unification compatible with the NCG model which may affect the Higgs running.

As we have discussed already, the coupling constants running (at one loop) without anomalous dimensions decouples to an ODE with exact solutions

$$u'(t) = A u(t)^3, \quad u(0) = B$$

$$u(t) = \pm \frac{1}{\sqrt{\frac{1-2AB^2 t}{B^2}}}$$

where A is determined by β function at one-loop and B by the values at $\Lambda = M_Z$,

$$g_1(0) = 0.3575, \qquad g_2(0) = 0.6514, \qquad g_3(0) = 1.221.$$

Solutions behave as plotted in the figure.

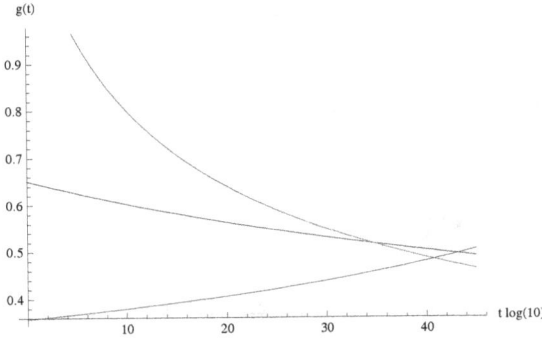

The coupling constants running with anomalous dimensions, taking $a_1 = a_2 = a_3 = a_g$, with $|a_g| \sim 1$ and negative sign, are given by

$$u'(t) = -a\,u(t) + A\,u(t)^3, \qquad u(0) = B$$

$$a = \frac{|a_g|}{16\pi\rho_0} \sim \frac{1}{16\pi\rho_0}$$

These also have exact solutions

$$u(t) = \frac{\pm\sqrt{a}}{\sqrt{A + \exp\left(2a\left(t + \frac{\log(-A + \frac{a}{B^2})}{2a}\right)\right)}}$$

The asymptotic safety effect is then shown in the figure.

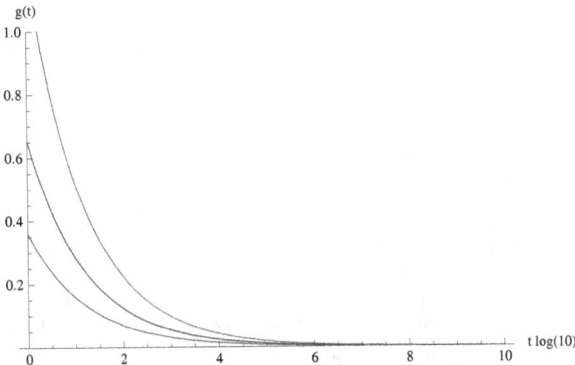

The running of the top Yukawa coupling without anomalous dimensions, as we have seen before, is given by

$$\beta_y = \frac{1}{16\pi^2}\left(\frac{9}{2}y^3 - 8g_3^2y - \frac{9}{4}g_2^2y - \frac{17}{12}g_1^2y\right)$$

in the MSM approximation as before, with solutions

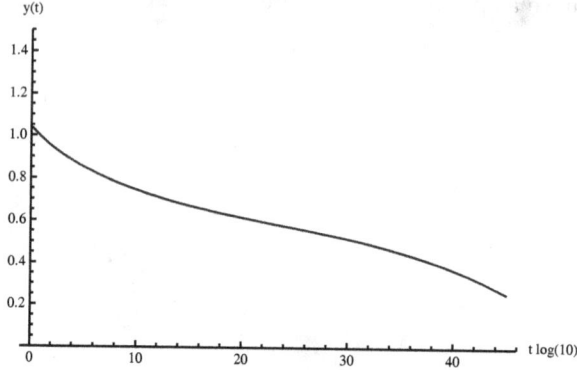

In the presence of anomalous dimensions one obtains

$$\partial_t y_t = -\frac{|a_y|}{16\pi\rho_0}y_t + \frac{1}{16\pi^2}\frac{9}{2}y_t^3.$$

This equation also has exact solutions of a form similar to $u(t)$ above but with parameters

$$a = \frac{|a_y|}{16\pi\rho_0}, \qquad A = \frac{9}{32\pi^2}.$$

The presence of anomalous dimensions also makes the Yukawa couplings asymptotically free.

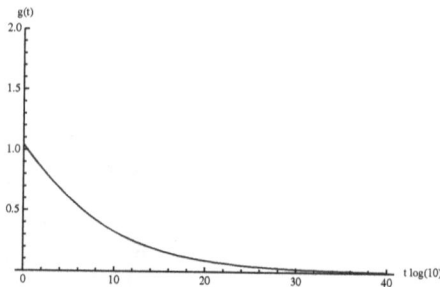

If the top Yukawa coupling contribution to the beta function of the Higgs self-coupling is dominant over the gauge contribution one can use a further simplification

$$\partial_t \lambda = \frac{a_\lambda}{16\pi\rho_0}\lambda + \frac{1}{16\pi^2}(24\lambda^2 + 12\lambda y^2 - 6y^4).$$

This gives a Riccati equation

$$\lambda' = q_0(t) + q_1(t)\lambda + q_2(t)\lambda^2$$

$$q_0(t) = \frac{-3y^4(t)}{8\pi^2}, \quad q_1(t) = \frac{a_\lambda}{16\pi\rho_0} + \frac{3y^2(t)}{4\pi^2}, \quad q_2(t) = \frac{3}{2\pi^2}.$$

After the standard change of variables one obtains

$$-\frac{u'}{u} = \lambda q_2$$

$$u'' - \left(q_1(t) + \frac{q_2'(t)}{q_2(t)}\right)u' + q_2(t)q_0(t)\, u = 0.$$

Using the general form of the solution for y_t,

$$y(t)^2 = \frac{a}{Ce^{2at} + A},$$

where the parameters a, A, and C are determined by the coefficients of the RGE and the initial condition, one gets a second order linear equation

$$\partial_t^2 u(t) - \left(\frac{a_\lambda}{16\pi\rho_0} + \frac{3}{4\pi^2}\frac{a}{Ce^{2at}+A}\right)\partial_t u - \frac{9}{16\pi^4}\left(\frac{a}{Ce^{2at}+A}\right)^2 u(t) = 0.$$

With another simple change of variables $x = e^{2at}$, $v(x) = v(e^{2at}) = u(t)$ this gives

$$(2ax)^2\partial_x^2 v(x) + \left(2a - \frac{a_\lambda}{16\pi\rho_0} - \frac{3}{4\pi^2}\frac{a}{Cx+A}\right)(2ax\partial_x v(x))$$

$$-\frac{9}{16\pi^4}\left(\frac{a}{Cx+A}\right)^2 v(x) = 0.$$

Thus, the general setting is an equation

$$y'' = \lambda_0 y' + s_0 y$$

$$\lambda_0 = \frac{-\frac{3a}{4\pi^2} + \left(2a - \frac{a_\lambda}{16\pi\rho_0}\right)(A + Cx)}{2ax(A + Cx)}$$

$$s_0 = -\frac{9}{64\pi^2 x^2 (A + Cx)^2}.$$

Consider an auxiliary functions $f_1(x) = (A + Cx)^{\alpha-3}$, so that

$$\theta = \frac{\sqrt{9 + 4\pi^2 \left(-9 + 4A \left(3 + 4A\pi^2\right)\right)}}{16A\pi^2}$$

$$\alpha = \frac{3}{2} + \frac{3}{16A\pi^2} + \theta$$

$$f_2(x) = x^\eta \left(\frac{Cx}{A}\right)^{1-\eta} \quad \text{with} \quad \eta = \frac{9}{64 A^2 \pi^2}$$

$$\beta = \alpha - \eta - 1 - \frac{3}{8A\pi^2}, \quad \gamma(x) = -\frac{Cx}{A}.$$

One obtains in this way a general solution of the equation $y'' = \lambda_0 y' + s_0 y$ above, which is of the form

$$v(x) = C_1 \, f_2(x) \, f_1(x) \, _2F_1\left(\alpha, \beta, 2 - \eta, \gamma(x)\right)$$

$$+ C_2 \, x^\eta \, f_1(x) \, _2F_1\left(\alpha - (2 + \tfrac{3}{8A\pi^2}), \beta + (2 + \tfrac{8A}{3})\eta, \eta\gamma(x)\right),$$

with $_2F_1(a, b, c, z)$ is the *Gauss hypergeometric function*. The corresponding solutions of the original Riccati equation that gives the RGE of the Higgs self coupling with anomalous dimensions is then

$$\lambda(t) = -\frac{2\pi^2}{3} \frac{u'(t)}{u(t)}, \quad \text{with} \quad u(t) = v(e^{2at}).$$

It is then possible to show that this allows for a choice of parameter $a_\lambda = 5.08$, with $A = 9/(32\pi^2)$, that is compatible with the top quark mass of $m_t \sim 171.3$ GeV and that gives a correct Higgs mass $m_H \sim 125.4$. The corresponding solution is plotted in the figure.

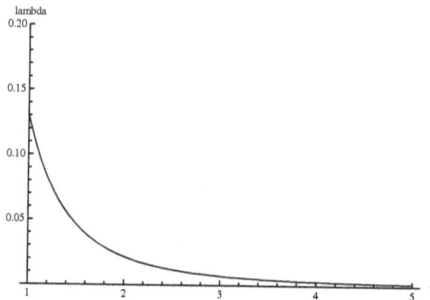

An inherent problem with this type of solutions of the Higgs mass problem in the NCG standard model is that they rely on a fine tuning of certain parameters. This can be a serious problem in the presence of sensitive dependence of RGE equations on initial conditions, which we have discussed earlier in this chapter.

Chapter 3

Cosmic Topology

The results in this chapter are based on [Marcolli, Pierpaoli, Teh (2011)], [Marcolli, Pierpaoli, Teh (2012)], and on [Ćaćić, Marcolli, Teh (2012)]. The main problem we consider is whether, and to what extent, the spectral action functional can distinguish between different cosmic topologies. The main idea behind this approach is the fact that, while the asymptotic expansion of the spectral action recovers the usual kind of local terms like the Einstein–Hilbert action for gravity and its higher derivative generalization (conformal, Gauss–Bonnet, etc.), the spectral action in its non-perturbative form also contains global information, coming from the fact that the Dirac operator on a manifold carries global information about the manifold. It is clear that the usual Einstein equations for gravity can only detect geometry but not distinguish between different topologies with the same local geometry (a 3-sphere from a spherical space form; a 3-torus from an infinite flat Euclidean 3-space, or from a Bieberbach manifold). However, a model of gravity derived from the spectral action may be able to provide information, which is in principle testable in an observational setting, which can distinguish the possible cosmic topologies.

While the Early Universe models discussed in the previous chapter depended on the existence of a preferred energy scale (unification energy) in the asymptotic expansion of the spectral actions, and on different possible assumptions on the running of the gravitational terms, the results we discuss in this section are completely independent on any such assumption. In fact, we will use the non-perturbative form (1.4) of the spectral action and obtain results directly from that form of our action functional for gravity. Some of the results we discuss here are independent also of the finite geometry F and only depend on the spectral action as a modified gravity model on an ordinary manifolds. We will find, towards the end of this chapter, some additional results that involve a dependence on F, but only in a relatively mild way, through the counting of the number of fermions in the associated particle physics model.

3.1 The problem of cosmic topology

The problem of cosmic topology can be stated as the question of identifying signatures of nontrivial (meaning non-simply connected) topologies in the spatial sections of space-time. Usually, what is meant by signatures is observable quantities, that can be read off data on the cosmic microwave background radiation (CMB), that would prefer certain candidate topologies over others.

It is known from [Kamionkowski et al. (1994)] that the geometry of the universe can be read in the CMB. Indeed, the anisotropies of the CMB depend upon the geometry of the universe (flat, positively or negatively curved) and this information can be detected through the location of the first Doppler peak which depends on the value of the curvature. This theoretical result made it possible to devise an observational test that could confirm the inflationary theory, which in particular predicts a flat or nearly flat geometry. The first experimental confirmation of the nearly flat geometry of the universe came in [de Bernardis et al. (2000)], through the Boomerang experiment (see also [Uzan et al. (2003)] for WMAP based results). Thus, the geometry of the universe leaves a measurable trace in the CMB. The observational result, moreover, shows that among the nearly flat cases, the slightly positively curve case is favored over the negatively curved one.

Thus, while the world of 3-dimensional manifolds is extremely complex, the available cosmological data already make a drastic selection among all the candidate topologies. Among all the constant curvature possibilities, this observational constraint rules out the largest and most complicated class: the negatively curved hyperbolic 3-manifolds. The remaining candidates are given by a much simpler list consisting of the positively curved spherical space forms, which are quotients of the round 3-sphere by finite groups of isometries, and the flat tori and Bieberbach manifolds, which are quotients of flat tori by finite groups of isometries. It is surprising that, while the mathematically "typical" case of 3-manifolds would be the hyperbolic one, our universe prefers the much smaller set of non-negatively curved candidates. It is then natural to ask whether there is a test for these

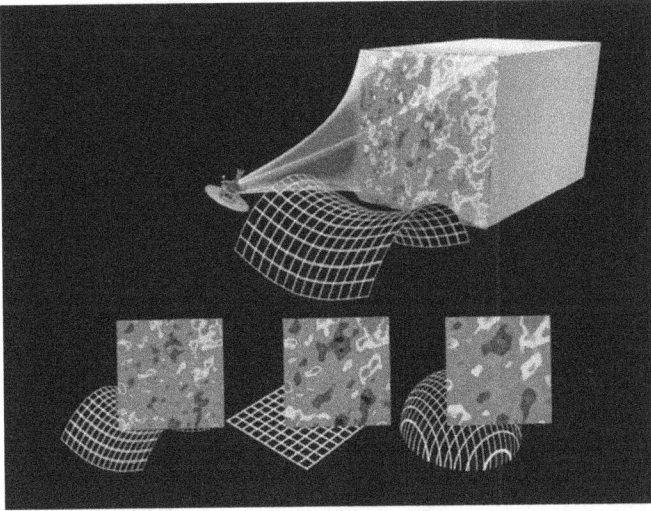

Fig. 3.1 Negatively curved, flat, and positively curved geometries and their effect on the CMB. Credit: NASA/WMAP Science Team.

possible candidate topologies that can also be read in the CMB. For a general introduction to the problem of cosmic topology, we refer the reader to [Lachièze-Rey, Luminet (1995)]. A first naive idea about detecting non-simply connected topologies relies on direct observation of periodicities in the catalog of observable astronomical objects. Objects that appear to the viewer as different objects would be just repeated images of the same object. However, extensive searches for correlation indicating the existence of such periodicities were inconclusive, leading to the belief that a fundamental domain for a closed topology must be at least as large as the last scattering surface of the CMB. Thus, the search for possible signatures of cosmic topology focused on the CMB.

The main methods used to approach the cosmic topology problem so far have focused on the following approaches:

- Statistical search for matching circles in the CMB sky: identify a nontrivial fundamental domain
- Anomalies of the CMB: quadrupole suppression, the small value of the two- point temperature correlation function at angles above 60 degrees, and the anomalous alignment of the quadrupole and octupole

- Residual gravity acceleration: gravitational effects from other fundamental domains
- Bayesian analysis of different models of CMB sky for different candidate topologies, [Niarchou, Jaffe (2007)].

A "circles in the skies" search, based on the first year of WMAP data, [Cornish et al. (2004)], failed to identify any non-simply connected topology. The "residual acceleration" approach of [Roukema, Rózański (2009)] addresses the problem of cosmic topology based on the prediction that a test particle that feels the gravitational influence of a nearby massive object should also feel a gravitational effect from the translates of the same massive objects in nearby fundamental domains of the group action. This would give rise to a gravitational effect qualitatively similar to dark energy. This effect vanishes at first order, but is non-trivial at higher order. It is shown in [Roukema, Rózański (2009)], based on this effect that cosmology dynamically prefers the most symmetric forms for a given topology. Cosmic topologies have been predicted also in brane-world scenarios [McInnes (2004)]. Despite the different methods employed to investigate this question, so far there have been no definitely conclusive results pointing to the presence of a specific non-simply connected topology. We present here a different approach based on the spectral action model.

3.2   The spectral action and cosmic topology

The approach of [Marcolli, Pierpaoli, Teh (2011)], [Marcolli, Pierpaoli, Teh (2012)], [Ćaćić, Marcolli, Teh (2012)] to the problem of cosmic topology starts from the spectral action as the action functional of our (modified) gravity model. The main ideas of this approach are summarized as follows:

- The nonperturbative form of the spectral action determines a scalar field with a slow-roll inflation potential.
- The underlying geometry (spherical or flat) affects the shape of the potential, so one obtains different inflation scenarios depending on the geometry and topology of the cosmos.
- The shape of the inflation potential is readable from cosmological data (CMB).

It was observed in [Chamseddine, Connes (2010)], in the case of S^3, that a perturbation of the Dirac operator by a scalar field $D^2 \mapsto D^2 + \phi^2$ gives a potential $V(\phi)$ for the scalar field coupled to gravity. For the spectral action on $S^3 \times S^1$, with the 3-sphere of radius a and the circle of size β, one finds

$$\text{Tr}(h((D^2 + \phi^2)/\Lambda^2))) = \pi \Lambda^4 \beta a^3 \int_0^\infty u h(u) du - \frac{\pi}{2} \Lambda^2 \beta a \int_0^\infty h(u) du$$

$$+ \pi \Lambda^4 \beta a^3 \, \mathcal{V}(\phi^2/\Lambda^2) + \frac{1}{2} \Lambda^2 \beta a \, \mathcal{W}(\phi^2/\Lambda^2)$$

where the potential functions are given by

$$\mathcal{V}(x) = \int_0^\infty u(h(u + x) - h(u)) du, \qquad \mathcal{W}(x) = \int_0^x h(u) du.$$

The parameter β, which can also be thought of as an inverse temperature, is related to the choice of a Euclidean compactification of spacetime, which is needed for the computation of the spectral action.

When Wick rotating back the geometry from Euclidean to the original Minkowskian, β loses this interpretation length, as the time direction is no longer compactified, but it can still be interpreted, as in [Chamseddine, Connes (2010)], as an inverse temperature. This interpretation allows for a time dependent β. In the cosmological setting, there are then proposals to interpret this inverse temperature β as a temperature of the cosmological horizon. In the setting of a Friedmann cosmology, the temperature of the photon fluid is proportional to $1/a(t)$, and the temperature of the cosmic horizon would certainly be time dependent, so that one would have to consider a function $\beta(t)$. The relation between the cosmic horizon and the scale factor $a(t)$ is discussed, for instance, in §7 of [Coles, Lucchin (1995)] and in §2.2 of [Dodelson (2003)]. At the start of the radiation dominated epoch, with the scale factor $a(t) \sim t^{1/2}$, the cosmic horizon behaves like $r_c \sim \frac{a_0}{\bar{a}(t)} \sim t^{1/2}$, with $T_c \sim (2\pi r_c)^{-1}$. The relation $\beta_c(t) \sim a(t)$ gives $\Lambda(t) \beta_c(t) \sim 1$. We will be assuming this type of behavior $\Lambda \beta \sim 1$. In addition to this estimate for the radiation dominated epoch, the choice is also justified geometrically: for $S^3_{a(t)} \times S^1_{\beta(t)}$, one obtains the same inflationary model regardless of the scale (time) at which one sets the initial condition, provided that $\Lambda(t) \sim 1/a(t)$, and $\Lambda(t) \sim 1/\beta(t)$. In the radiation dominated epoch this is consistent with the interpretation of β as inverse temperature of the cosmic horizon. With the assumption $\Lambda a \sim 1$, $\Lambda \beta \sim 1$, there are no more independently scaling parameters in the model.

3.2.1 *Slow-roll potential and slow-roll parameters*

The slow-roll models of inflation in the early universe are based on a potential $V(\phi)$ for a scalar field ϕ that runs the inflation. For a Minkowskian

Friedmann metric on a manifold $Y \times \mathbb{R}$,

$$ds^2 = -dt^2 + a(t)^2 ds_Y^2,$$

a phase of accelerated expansion corresponds to

$$\frac{\ddot{a}}{a} = H^2(1 - \epsilon),$$

where H is the Hubble parameter, where

$$H^2(\phi)\left(1 - \frac{1}{3}\epsilon(\phi)\right) = \frac{8\pi}{3m_{Pl}^2} V(\phi),$$

with m_{Pl} the Planck mass. The inflation phase corresponds to the values $\epsilon(\phi) < 1$.

Given a slow-roll potential, one can compute the associated slow-roll parameters,

$$\epsilon(\phi) = \frac{m_{Pl}^2}{16\pi}\left(\frac{V'(\phi)}{V(\phi)}\right)^2$$

$$\eta(\phi) = \frac{m_{Pl}^2}{8\pi}\frac{V''(\phi)}{V(\phi)}$$

$$\xi(\phi) = \frac{m_{Pl}^4}{64\pi^2}\frac{V'(\phi)V'''(\phi)}{V^2(\phi)}$$

These are quantities that are constrained and in principle can be determined on the basis of observation. In particular, important measurable quantities in cosmology include the spectral index n_s and the tensor-to-scalar ratio r

$$n_s \simeq 1 - 6\epsilon + 2\eta, \quad n_t \simeq -2\epsilon, \quad r = 16\epsilon,$$

$$\alpha_s \simeq 16\epsilon\eta - 24\epsilon^2 - 2\xi, \quad \alpha_t \simeq 4\epsilon\eta - 8\epsilon^2.$$

For a Friedmann metric describing an expanding universe, one can separate tensor and scalar perturbations of the metric, corresponding to the traceless and trace part. One writes scalar and tensor perturbations in the form

$$ds^2 = -(1 + 2\Phi)dt^2 + 2a(t)dBdt + a(t)^2((1 - 2\Psi)g_{ij} + 2\Delta E + h_{ij})dx^i dx^j$$

with $dB = \partial_i D dx^i$, $\Delta E = \partial_i \partial_j E$ and the tensor part of the perturbation h_{ij} satisfying $\partial^i h_{ij} = 0$ and $h_i^i = 0$. The intrinsic curvature perturbation

$$\mathcal{R} = \Psi - \frac{H}{\dot{\phi}}\delta\phi$$

describes the spatial curvature of a comoving hypersurface with constant ϕ. One expands $\mathcal{R}$ in Fourier modes

$$\mathcal{R} = \int e^{ikx} \, \mathcal{R}_k \, \frac{d^3k}{(2\pi)^{3/2}}.$$

One obtains from the Fourier modes a power spectrum $\mathcal{P}_s(k)$ for the scalar perturbations of the metric, from the two-point correlation function

$$\langle \mathcal{R}_k \mathcal{R}_{k'} \rangle = (2\pi^2)^3 \, \mathcal{P}_s(k) \, \delta^3(k + k').$$

The Fourier integral above (which refers to the flat space case) is replaced by an expansion in spherical harmonics in the case of the spherical geometry and a Fourier series in the case of a flat compact geometry. A power spectrum $\mathcal{P}_t(k)$ for the tensor fluctuations is obtained in a similar way from the two-point correlation functions of the Fourier modesl h_k of the tensor fluctuations

$$\langle h_k h_{k'} \rangle = (2\pi^2)^3 \, \mathcal{P}_t(k) \, \delta^3(k + k').$$

The power spectra $\mathcal{P}_s(k)$ and $\mathcal{P}_t(k)$ satisfy power laws

$$\mathcal{P}_s(k) \sim \mathcal{P}_s(k_0) \left(\frac{k}{k_0} \right)^{1 - n_s + \frac{\alpha_s}{2} \log(k/k_0)}$$

$$\mathcal{P}_t(k) \sim \mathcal{P}_t(k_0) \left(\frac{k}{k_0} \right)^{n_t + \frac{\alpha_t}{2} \log(k/k_0)}$$

Both the amplitude and the exponents of these power laws are constrained by observational parameters and predicted by slow-roll models of inflation. Indeed, in a slow-roll inflation model, one has

$$\mathcal{P}_s(k_0) \sim \frac{V^3}{(V')^2}, \quad \mathcal{P}_t(k_0) \sim V,$$

with a proportionality constant that depends on a power of the Planck mass m_{Pl}. We refer the reader to [Lidsey et al. (1997)], [Smith, Kamionkowski, Cooray (2006)], [Stewart, Lyth (1993)] for a more detailed account of the reconstruction of the potential $V(\phi)$ from the power spectra, and vice versa, and the observational constraint on the latter.

Thus, in the case of our model based on the spectral action, determining the potential $V(\phi)$ leads to associated predictions for the power spectra. We will show in the rest of this chapter that the information one obtains about the potential, using the non-perturbative form of the spectral action, shows that the resulting power spectra distinguish between (almost all of) the possible non-trivial topologies. The conclusion is that a model of gravity based on the spectral action detects cosmic topology in a way that is, in principle, testable in observations about the CMB.

3.2.2 *Spherical space forms*

An investigation of candidate cosmic topologies among the spherical space forms was conducted in [Luminet et al. (2003)]. Simulated CMB skies were studied for the various candidate topologies, see [Lehoucq et al. (2002)] and Figure 3.2. These initial results suggested that the Poincaré homology sphere (or dodecahedral space) could account for the missing large angle correlations of the two-point angular correlation function of the temperature spectrum of the CMB. However, the dodecahedral space fails to account for the quadrupole-octupole alignment, [Weeks, Gundermann (2007)]. The method of "cosmic crystallography" consists of analyzing the presence of spikes in the pair separation histogram for three dimensional catalogs of cosmic objects, and trying to deduce from them information about the topology. An analysis of how all the different possible spherical space forms may be detectable through such "crystallographic methods" was carried out in [Gausmann et al. (2001)]. See also another approach in [Dowker (2004)], in relation to the Casimir effect.

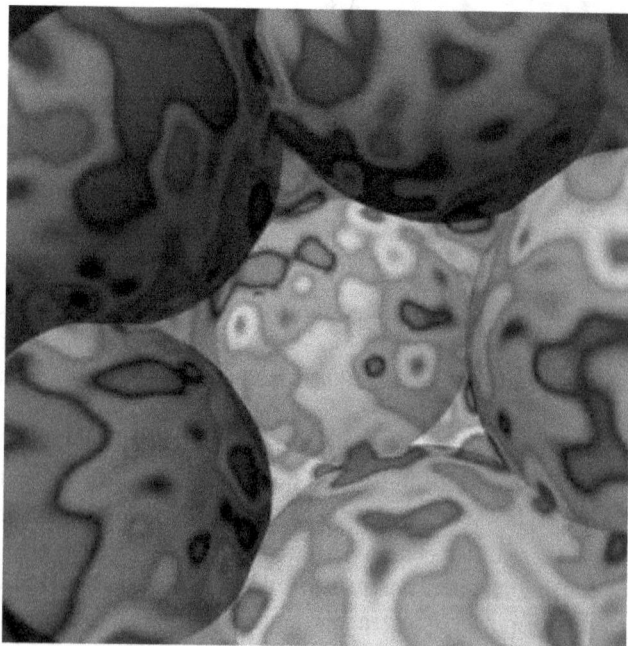

Fig. 3.2   CMB sky in a dodecahedral space, from [Luminet et al. (2003)].

The list of candidate cosmic topologies with underlying positively curved spherical geometry is given by the spherical space forms, which are quotients of the round 3-sphere by the action of a finite subgroup $\Gamma \subset SU(2)$ of isometries, see [Wolf (2011)]:

(1) *trivial group*: the 3-sphere S^3;
(2) *cyclic groups* of order N: the lens spaces $L_N = S^3/\Gamma_N$: quotient by a finite cyclic group $\Gamma_N \simeq \mathbb{Z}/N\mathbb{Z}$

$$\Gamma = \{ \begin{pmatrix} \omega & 0 \\ 0 & \omega^{-1} \end{pmatrix} : \omega^N = 1 \} \subset SU(2);$$

(3) *binary dihedral groups* (dicyclic groups) of order $4N$: quotients S^3/D_N by the dicyclic group generated by

$$B = \begin{pmatrix} e^{\frac{\pi i}{N}} & 0 \\ 0 & e^{\frac{-\pi i}{N}} \end{pmatrix}, \quad C = \begin{pmatrix} 0 & 1 \\ -1 & 0 \end{pmatrix}$$

(4) *binary tetrahedral group* of order 24: the quotient $S^3/2T$ of the sphere by the group of unit quaternions

$$2T = \{\pm 1, \pm i, \pm j, \pm k, \frac{1}{2}(\pm 1 \pm i \pm j \pm k)\}$$

(5) *binary octahedral group* of order 48: the quotient $S^3/2O$ of the sphere by the group $2O$ consisting of the same elements of $2T$ and the 24 additional elements obtained from

$$\frac{1}{\sqrt{2}}(\pm 1 \pm i)$$

by permuting the coordinates, with all sign combinations;
(6) *binary icosahedral group* of order 120: the Poincaré homology sphere (or dodecahedral space), quotient of S^3 by the group $2\mathcal{I}$ of unit quaternions consisting of the same 24 elements of $2T$ and the additional 96 elements obtained by applying all even permutations of coordinates to

$$\frac{1}{2}(\pm i \pm \phi^{-1} j \pm \phi k),$$

where ϕ is the golden ratio.

A particular case of the dicyclic group, for $N = 2$, is the quaternionic space $S^3/Q8$, the quotient by the group of quaternion units

$$Q8 = \{\pm 1, \pm i, \pm j, \pm k\}$$

whose possible occurrence as a candidate cosmic topology was considered promising according to the Bayesian approach of [Niarchou, Jaffe (2007)].

The Poincaré homology sphere is presently considered the most likely candidate for a possible spherical cosmic topology, and certainly the most widely studied in the cosmology literature on the subject.

In all of these cases, the spectrum of the Dirac operator (for the different possible spin structures) can be computed explicitly, along with the spectral multiplicities of the eigenvalues. We refer the reader to [Bär (1996)], [Bär (2000)], [Bär (1992)] (with a correction in [Marcolli, Pierpaoli, Teh (2011)]), [Ginoux (2008)] and the survey [Ginoux (2009)], for explicit results. An especially useful approach to obtain the spectral multiplicities is the method of generating functions of [Bär (1996)], which we summarize as follows:

- Spin structures on S^3/Γ are determined by homomorphisms

$$\epsilon : \Gamma \to \mathrm{Spin}(4) \cong SU(2) \times SU(2)$$

  lifting the inclusion $\Gamma \hookrightarrow SO(4)$ under the double cover $\mathrm{Spin}(4) \to SO(4)$, with $(A, B) \mapsto AB$.
- The Dirac spectrum on the quotient S^3/Γ is a subset of the spectrum on S^3;
- The multiplicities are given by a generating function: let ρ^+ and ρ^- be two half-spin irreducible representations and let $\chi^\pm$ be their characters, and let

$$F_+(z) = \frac{1}{|\Gamma|} \sum_{\gamma \in \Gamma} \frac{\chi^-(\epsilon(\gamma)) - z\chi^+(\epsilon(\gamma))}{\det(1 - z\gamma)}$$

$$F_-(z) = \frac{1}{|\Gamma|} \sum_{\gamma \in \Gamma} \frac{\chi^+(\epsilon(\gamma)) - z\chi^-(\epsilon(\gamma))}{\det(1 - z\gamma)}$$

Then $F_+(z)$ and $F_-(z)$ are generating functions of the spectral multiplicities:

$$F_+(z) = \sum_{k=0}^{\infty} m(\frac{3}{2} + k, D)z^k \qquad F_-(z) = \sum_{k=0}^{\infty} m(-(\frac{3}{2} + k), D)z^k.$$

It is important to remark that the Dirac spectrum with the multiplicities explicitly depends on the choice of the spin structure and can be different for different spin structures, see [Bär (2000)].

Using these methods and the explicit results on the Dirac spectra of spherical space forms, it was shown in [Teh (2013)], [Teh (2013b)] that the Dirac spectra can be decomposed into a finite union of arithmetic progressions and for each of these progressions there is a polynomial that interpolates the multiplicities, so that a Poisson summation formula argument analogous to that of [Chamseddine, Connes (2010)] can be applied.

For example, see [Marcolli, Pierpaoli, Teh (2011)], in the case of the quaternionic space $SU(2)/Q8$, the Dirac spectrum computed in [Ginoux (2008)] can be arranged into two arithmetic progressions:

$$\frac{3}{2} + 4k \quad \text{with multiplicity} \quad 2(k+1)(2k+1)$$

$$\frac{3}{2} + 4k + 2 \quad \text{with multiplicity} \quad 4k(k+1)$$

Polynomial interpolations of the spectral multiplicities for each of these progressions are given by

$$P_1(u) = \frac{1}{4}u^2 + \frac{3}{4}u + \frac{5}{16}$$

$$P_2(u) = \frac{1}{4}u^2 - \frac{3}{4}u - \frac{7}{16}$$

The spectral action can be computed by the Poisson summation formula applied to the rapidly decaying functions $g_i(u) = P_i(u)f(u/\Lambda)$ for each of these progressions. One obtains

$$\text{Tr}(f(D/\Lambda)) = \frac{1}{8}(\Lambda a)^3 \widehat{f}^{(2)}(0) - \frac{1}{32}(\Lambda a)\widehat{f}(0) + O(\Lambda^{-k}),$$

which is exactly $1/8$ of the action for S^3.

Another example considered in [Teh (2013)] is the dodecahedral space, for which the method described above gives generating functions $F_+(z)$ and $F_-(z)$ for the spectral multiplicities in the case of the trivial spin structure, with $F_+(z)$ of the form

$$-\frac{2(1 + 3z^2 + 4z^4 + 2z^6 - 2z^8 - 6z^{10} - 2z^{12} + 12z^{14} + 24z^{16} + 18z^{18} + 6z^{20})}{(-1 + z^2)^3(1 + 2z^2 + 2z^4 + z^6)^2(1 + z^2 + z^4 + z^6 + z^8)^2}$$

and $F_-(z)$ given by

$$-\frac{2z^{11}(6 + 18z^2 + 24z^4 + 12z^6 - 2z^8 - 6z^{10} - 2z^{12} + 2z^{14} + 4z^{16} + 3z^{18} + z^{20})}{(-1 + z^2)^3(1 + 2z^2 + 2z^4 + z^6)^2(1 + z^2 + z^4 + z^6 + z^8)^2}.$$

Again, it is shown in Proposition 11.1 of [Teh (2013)] that the spectrum can be separated into arithmetic progressions with 60 explicit polynomials interpolating the multiplicities,

$$P_k\left(\frac{3}{2} + k + 60j\right) = m_{\frac{3}{2}+k+60j}.$$

These polynomials satisfy

$$\sum_{k=0}^{59} P_k(u) = \frac{1}{2}u^2 - \frac{1}{8}.$$

Thus, when computing the non-perturbative spectral action via the Poisson summation formula one obtains

$$\mathrm{Tr}(f(D/\Lambda)) = \sum_{k=0}^{59} \sum_{j \in \mathbb{Z}} P_k(\frac{3}{2} + k + 60j) f(\frac{\frac{3}{2} + k + 60j}{\Lambda})$$

$$= \frac{1}{120}(\Lambda^3 \widehat{f}^{(2)}(0) - \frac{1}{4}\Lambda \widehat{f}(0)) + O(\Lambda^{-\infty}),$$

which is a fraction $1/120$ of the spectral action of the 3-sphere.

More generally, an explicit computation of the non-perturbative spectral action is then given in [Teh (2013)] for all the 3-dimensional spherical space forms. For $Y = S^3/\Gamma$, up to terms of order $O(\Lambda^{-\infty})$, one has

$$\mathrm{Tr}(f(D_Y/\Lambda)) = \frac{1}{\lambda_Y}\left(\Lambda^3 \widehat{f}^{(2)}(0) - \frac{1}{4}\Lambda \widehat{f}(0)\right) = \frac{1}{\lambda_Y}\mathrm{Tr}(f(D_{S^3}/\Lambda)),$$

where the factor $\lambda_Y = \#\Gamma$ is given by

Y spherical	λ_Y
sphere	1
lens N	$1/N$
binary dihedral $4N$	$1/(4N)$
binary tetrahedral	$1/24$
binary octahedral	$1/48$
binary icosahedral	$1/120$

Despite the fact that the spectra and the multiplicities depend on the spin structure, the spectral action computed in [Teh (2013)] via Poisson summation is independent of this choice. A more transparent explanation of why this is the case will be discussed in §3.2.4 below, in terms of the heat kernel.

One then shows in the same way that the spectral action on the 4-manifold $Y \times S^1$ is also a fraction $1/\lambda_Y$ of the spectral action on $S^3 \times S^1$,

$$\mathrm{Tr}(h(D_{Y \times S^1}/\Lambda)) = \frac{1}{\lambda_Y}\mathrm{Tr}(h(D_{S^3 \times S^1}/\Lambda)).$$

After introducing a scalar field perturbation of the Dirac operator $D_{Y \times S^1} \mapsto D_{Y \times S^1} + \phi^2$, the same computation of the spectral action gives a slow-roll potential

$$V_Y(\phi) = \frac{1}{\lambda_Y}V_{S^3}(\phi)$$

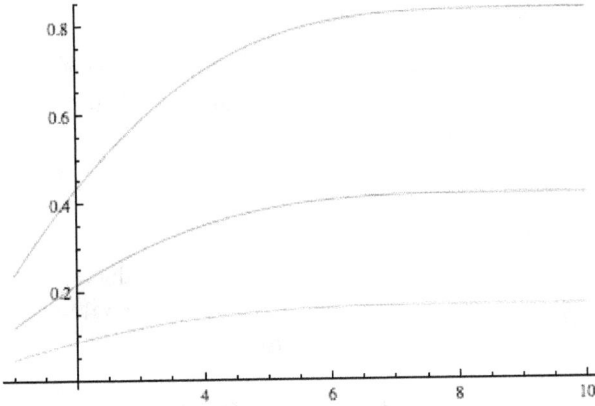

Fig. 3.3   Slow-roll potential for different spherical space forms.

where

$$V_{S^3}(\phi) = \pi\Lambda^4\beta a^3\mathcal{V}(\frac{\phi^2}{\Lambda^2}) + \frac{\pi}{2}\Lambda^2\beta a\mathcal{W}(\frac{\phi^2}{\Lambda^2})$$

is the slow-roll potential for $S^3 \times S^1$, see Figure 3.3.

The slow-roll parameters only depend on ratios V'/V and V''/V hence they are unaffected, when replacing the potential V by a constant multiple. They are the same as in the case of the sphere S^3 and are given by

$$\epsilon(x) = \frac{m_{Pl}^2}{16\pi}\left(\frac{h(x) - 2\pi(\Lambda a)^2\int_x^\infty h(u)du}{\int_0^x h(u)du + 2\pi(\Lambda a)^2\int_0^\infty u(h(u+x) - h(u))du}\right)^2$$

$$\eta(x) = \frac{m_{Pl}^2}{8\pi}\frac{h'(x) + 2\pi(\Lambda a)^2 h(x)}{\int_0^x h(u)du + 2\pi(\Lambda a)^2\int_0^\infty u(h(u+x) - h(u))du}.$$

The slow-roll parameters are independent of the compactification radius β and depend on Λ and a only through the quantity $\Lambda(t)a(t) \sim 1$.

Thus, the exponents of the power law for the power spectra $\mathcal{P}_s(k)$ and $\mathcal{P}_t(k)$ are also unaffected. However, the amplitudes of the power laws depend on the potential like $V^3/(V')^2$ or like V, hence these change by a factor of $1/\lambda_Y$. Notice that the factor $1/\lambda_Y$ is not sufficient to completely distinguish between all the spherical cases. For example a lens space with $N = 24$, a binary dihedral space with $N = 6$, and the binary tetrahedral case all have the same λ_Y, and other such ambiguities occur. However, it does distinguish most cases.

3.2.3 *Flat tori and Bieberbach manifolds*

The other set of possible candidates for cosmic topologies consists of the flat compact 3-manifolds. These include flat tori T^3 and the Bieberbach manifolds, which are quotients of tori by finite groups of isometries. The flat candidates for cosmic topologies were studied in [Gomero et al. (2000)], [Gomero et al. (2001)], [Riazuelo et al. (2004b)]. The quadrupole-octupole alignment problem appears to favor the flat geometries (flat tori) over the spherical ones [Aurich et al. (2007)]. Though the alignment obtained in this way is not strong enough to account for the observed anomaly, the flat cosmic topologies remain valid candidates.

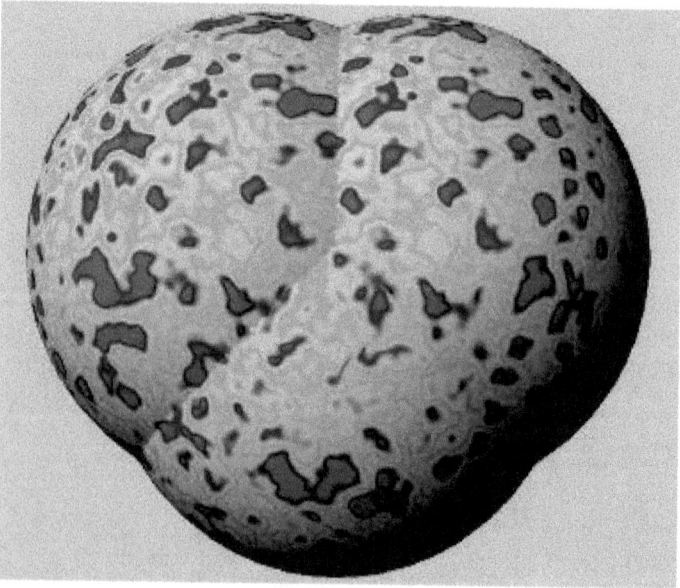

Fig. 3.4   CMB sky in a Bieberbach G6-cosmology from [Riazuelo et al. (2004b)].

The list of Bieberbach manifolds consists of the following cases:

(1) *flat torus*: quotient of $\mathbb{R}^3$ by the group generated by the translations t_i by a_i, with $a_1 = (0, 0, H)$, $a_2 = (L, 0, 0)$, and $a_3 = (0, S, 0)$, with $H, L, S \in \mathbb{R}_+^*$.

(2) *half turn space G2*: quotient of $\mathbb{R}^3$ by the group generated by α and translations t_i by a_i, with $a_1 = (0, 0, H)$, $a_2 = (L, 0, 0)$, and $a_3 =$

$(T, S, 0)$, with $H, L, S \in \mathbb{R}_+^*$ and $T \in \mathbb{R}$, with relations

$$\alpha^2 = t_1, \quad \alpha t_2 \alpha^{-1} = t_2^{-1}, \quad \alpha t_3 \alpha^{-1} = t_3^{-1}.$$

(3) *third turn space G3*: quotient of $\mathbb{R}^3$ by the group generated by α and translations t_i by a_i, with $a_1 = (0, 0, H)$, $a_2 = (L, 0, 0)$ and $a_3 = (-\frac{1}{2}L, \frac{\sqrt{3}}{2}L, 0)$, for H and L in $\mathbb{R}_+^*$ and relations

$$\alpha^3 = t_1, \quad \alpha t_2 \alpha^{-1} = t_3, \quad \alpha t_3 \alpha^{-1} = t_2^{-1} t_3^{-1}.$$

(4) *quarter turn space G4*: quotient of $\mathbb{R}^3$ by the group generated by α and translations t_i by a_i, with $a_1 = (0, 0, H)$, $a_2 = (L, 0, 0)$, and $a_3 = (0, L, 0)$, with $H, L > 0$, and relations

$$\alpha^4 = t_1, \quad \alpha t_2 \alpha^{-1} = t_3, \quad \alpha t_3 \alpha^{-1} = t_2^{-1}.$$

(5) *sixth turn space G5*: quotient of $\mathbb{R}^3$ by the group generated by α and translations t_i by a_i, with $a_1 = (0, 0, H)$, $a_2 = (L, 0, 0)$ and $a_3 = (\frac{1}{2}L, \frac{\sqrt{3}}{2}L, 0)$, $H, L > 0$, and relations

$$\alpha^6 = t_1, \quad \alpha t_2 \alpha^{-1} = t_3, \quad \alpha t_3 \alpha^{-1} = t_2^{-1} t_3.$$

(6) *Hantzsche–Wendt space G6*: quotient of $\mathbb{R}^3$ by the group generated by α, β, γ, and translations t_i by a_i, with $a_1 = (0, 0, H)$, $a_2 = (L, 0, 0)$, and $a_3 = (0, S, 0)$, with $H, L, S > 0$ and relations

$$\alpha^2 = t_1, \ \alpha t_2 \alpha^{-1} = t_2^{-1}, \ \alpha t_3 \alpha^{-1} = t_3^{-1},$$
$$\beta^2 = t_2, \ \beta t_1 \beta^{-1} = t_1^{-1}, \ \beta t_3 \beta^{-1} = t_3^{-1},$$
$$\gamma^2 = t_3, \ \gamma t_1 \gamma^{-1} = t_1^{-1}, \ \gamma t_2 \gamma^{-1} = t_2^{-1},$$
$$\gamma \beta \alpha = t_1 t_3.$$

Unlike the other cases, in the Hantzsche–Wendt space the fundamental domain is glued with a π-twist along each coordinate axis.

The different spin structures are specified by the choices of signs (corresponding to the two possible spin structures on a circle in each direction):

	δ_1	δ_2	δ_3
(a)	± 1	1	1
(b)	± 1	-1	1
(c)	± 1	1	-1
(d)	± 1	-1	-1

One correspondingly denotes the manifold with a given spin structure by writing $G2(a)$, $G2(b)$, $G2(c)$, $G2(d)$, etc.

We saw how, in the case of the spherical space forms, the spectral action of the 3-sphere serves as the model for all the other cases, up to a scale factor. In a similar way, the model for the spectral action of the Bieberbach manifolds is provided by the spectral action of the 3-dimensional flat tori T^3. This was computed in [Marcolli, Pierpaoli, Teh (2011)], using the Poisson summation formula method.

For a flat torus $T^2 = \mathbb{R}^3/\mathbb{Z}^3$, the spectrum is computed in [Bär (2000)] and it is given by

$$\pm 2\pi \, \|(m, n, p) + (m_0, n_0, p_0)\|, \quad (m, n, p) \in \mathbb{Z}^3$$

where each (m, n, p) occurs with multiplicity one. The constant vector (m_0, n_0, p_0) depends on the spin structure. The spectral action is then given by a sum

$$\text{Tr}(f(D^2_{T^3}/\Lambda^2)) = \sum_{(m,n,p) \in \mathbb{Z}^3} 2f\left(\frac{4\pi^2((m + m_0)^2 + (n + n_0)^2 + (p + p_0)^2)}{\Lambda^2}\right)$$

which can be computed using the Poisson summation formula over the 3-dimensional lattice $\mathbb{Z}^3$ and gives, up to terms of order $O(\Lambda^{-\infty})$,

$$\text{Tr}(f(D^2_{T^3}/\Lambda^2)) = \frac{\Lambda^3}{4\pi^3} \int_{\mathbb{R}^3} f(u^2 + v^2 + w^2) \, du \, dv \, dw.$$

On the 4-manifold $T^3 \times S^1$, with the flat torus T^3 of size ℓ and the circle of size β, we then have, up to $O(\Lambda^{-\infty})$ terms,

$$\text{Tr}(h(D^2_{T^3 \times S^1}/\Lambda)) = \frac{\Lambda^4 \beta \ell^3}{4\pi} \int_0^\infty u \, h(u) du.$$

This follows using

$$\sum_{(m,n,p,r) \in \mathbb{Z}^4} 2h\left(\frac{4\pi^2}{(\Lambda\ell)^2}((m+m_0)^2 + (n+n_0)^2 + (p+p_0)^2) + \frac{1}{(\Lambda\beta)^2}(r+\frac{1}{2})^2\right)$$

and applying the Poisson summation formula

$$\sum_{(m,n,p,r) \in \mathbb{Z}^4} g(m+m_0, n+n_0, p+p_0, r+\frac{1}{2}) = \sum_{(m,n,p,r) \in \mathbb{Z}^4} (-1)^r \, \widehat{g}(m, n, p, r)$$

to the function

$$g(u, v, w, y) = 2\,h\left(\frac{4\pi^2}{(\Lambda\ell)^2}(u^2 + v^2 + w^2) + \frac{y^2}{(\Lambda\beta)^2}\right).$$

It then follows that, fluctuating the Dirac operator by

$$D^2_{T^3 \times S^1} \mapsto D^2_{T^3 \times S^1} + \phi^2,$$

we obtain a slow-roll potential for ϕ of the form

$$V(\phi) = \frac{\Lambda^4 \beta \ell^3}{4\pi} \mathcal{V}(\phi^2/\Lambda^2) \tag{3.1}$$

$$\mathcal{V}(x) = \int_0^\infty u\left(h(u+x) - h(u)\right) du.$$

The case of a torus with three different lengths H, L, S is analogous, with the factor ℓ^3 replaced by the product HLS.

The resulting slow-roll parameters are different from those of the spherical cases, and are given by

$$\epsilon = \frac{m_{Pl}^2}{16\pi} \left(\frac{\int_x^\infty h(u) du}{\int_0^\infty u(h(u+x) - h(u)) du} \right)^2$$

$$\eta = \frac{m_{Pl}^2}{8\pi} \left(\frac{h(x)}{\int_0^\infty u(h(u+x) - h(u)) du} \right).$$

The case of Bieberbach manifolds was then computed in [Marcolli, Pierpaoli, Teh (2012)], using a lattice summation technique that separates out the spectrum into lattice sums suitable for the Poisson summation formula. For example, in the case of the $G3$ Bieberbach manifold, the spectrum $\lambda_{klm}^\pm$ has symmetries $R : l \mapsto -l, m \mapsto -m$, $S : l \mapsto m, m \mapsto l$, and $T : l \mapsto l - m, m \mapsto -m$ and one can decompose

$$\mathbb{Z}^3 = I \cup R(I) \cup S(I) \cup RS(I) \cup T(\tilde{I}) \cup RT(\tilde{I}) \cup \{l = m\},$$

with $I = \{(k, l, m) \in \mathbb{Z}^3 : l \geq 1, m = 0, \ldots, l-1\}$ and $\tilde{I} = \{(k, l, m) \in \mathbb{Z}^3 : l \geq 2, m = 1, \ldots, l-1\}$. This results in the regions of Figure 3.5, see [Teh (2013b)].

The general result for the Bieberbach manifold is then similar to that of the spherical cases. Namely, for Y any of the Bieberbach manifold, with an arbitrary choice of spin structure, the spectral action is given, up to terms $O(\Lambda^{-\infty})$ by

$$\mathrm{Tr}(f(D_Y^2/\Lambda^2)) = \frac{\lambda_Y \Lambda^3}{4\pi^3} \int_{\mathbb{R}^3} f(u^2 + v^2 + w^2) du\, dv\, dw,$$

where the proportionality factor λ_Y is of the form

$$\lambda_Y = \begin{cases} \frac{HSL}{2} & G2 \\[2mm] \frac{HL^2}{2\sqrt{3}} & G3 \\[2mm] \frac{HL^2}{4} & G4 \\[2mm] \frac{HL^2}{4\sqrt{3}} & G5 \\[2mm] \frac{HLS}{4} & G6 \end{cases}$$

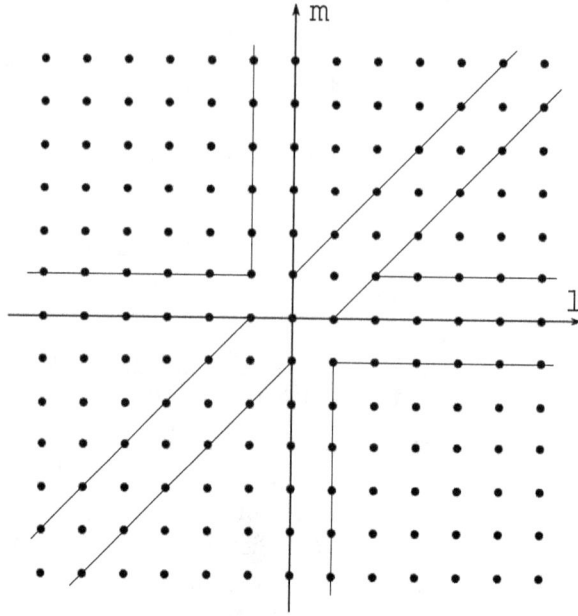

Fig. 3.5 Lattice summation for the Bieberbach manifold $G3$.

The rest of the argument then proceeds as in the spherical cases: one computes the spectral action on the 4-manifold $Y \times S^1$ and the slow-roll potential, which turns out to be of the form

$$V_Y(\phi) = \frac{\lambda_Y}{\pi^2 \ell^3} V_{T^3}(\phi),$$

with the potential for S^3 of the form (3.1) as above. It then follows that the slow-roll parameters are the same as in the case of the flat torus, while the amplitude of the potential changes according to the topology, see Figure 3.6. Computations of the spectral action for Bieberbach manifolds were also obtained in [Olczykowski, Sitarz (2011)].

3.2.4   *A heat kernel view*

The relation between the spectral action for the 3-sphere and for the other spherical space forms, and between that of the torus and of the Bieberbach manifolds, which we have seen explicitly in terms of Dirac spectra and multiplicities, can also be understood directly in terms of the heat kernel

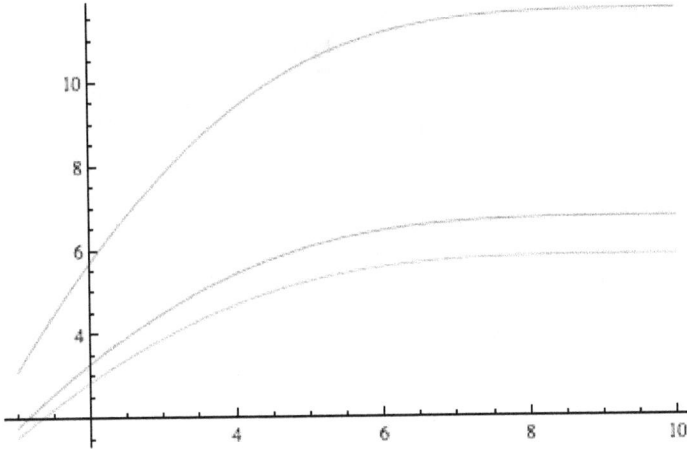

Fig. 3.6   Slow-roll potentials for different Bieberbach manifolds.

expansion, as discussed in [Ćаćić, Marcolli, Teh (2012)], [Ćаćić (2013b)]. We reproduce the argument here.

For a finite group Γ acting by unitary operators U_g on a Hilbert space $\mathcal{H}$, with $\mathcal{H}^\Gamma$ the subspace of Γ-invariant vectors, if A is a Γ-equivariant self-adjoint operator that is of trace-class, then its restriction to $\mathcal{H}^\Gamma$ is also trace class and the trace satisfies

$$Tr(A|_{\mathcal{H}^\Gamma}) = \frac{1}{\#\Gamma} \sum_{g \in \Gamma} \mathrm{Tr}(U_g A),$$

since the orthogonal projector onto $\mathcal{H}^\Gamma$ is given by

$$P_{\mathcal{H}^\Gamma} = \frac{1}{\#\Gamma} \sum_{g \in \Gamma} U_g.$$

For compact Riemannian manifolds $Y = X/\Gamma$, with the group Γ acting on X by isometries, let D_Y be the Dirac operator on $L^2(Y, S_Y) = L^2(X, S_X)^\Gamma$ obtained as restriction of the Dirac operator on X to the Γ-invariant square-integrable spinors. Then, using the previous formula, we see that, for all $t > 0$ the heat kernels are related by

$$\mathrm{Tr}(e^{-tD_Y^2}) = \frac{1}{\#\Gamma} \mathrm{Tr}(e^{-tD_X^2}) + \frac{1}{\#\Gamma} \sum_{g \in \Gamma \smallsetminus \{1\}} \mathrm{Tr}(U_g e^{-tD_X^2}),$$

where the terms in the sum are given by

$$U_g e^{-tD_X^2} \xi(x) = \int_X \rho(g)(xg^{-1}) \tilde{K}_X(t, xg^{-1}, x, \tilde{x}) \, \xi(\tilde{x}) \, dv(\tilde{x}),$$

with K_X the heat kernel of D_X and ρ the right action of Γ on the spinor bundle S_X. One then checks that all the traces

$$\mathrm{Tr}(U_g e^{-tD_X^2}) = \int_X \mathrm{Tr}(\rho(g)(xg^{-1})\tilde{K}_X(t, xg^{-1}, x, x))\, dv(x),$$

for $g \neq 1$, satisfy

$$\int_0^\infty \mathrm{Tr}(U_g e^{-\frac{s}{\Lambda^2}D_X^2})\varphi(s)\, ds = O(\Lambda^{-\infty}),$$

for a rapidly decaying function $\varphi \in \mathcal{S}(0,\infty)$. This estimate follows using [Kahle (2011)], from which one can show that, for all $n \in \mathbb{N}$, there is a $C_n > 0$, such that the heat kernel satisfies the estimate

$$\sup_{x \in X} \|K_X(t, xg^{-1}, x)\| \leq C_n t^n, \quad for\ all\ t > 0,$$

in the fiber-wise Hilbert–Schmidt norm. This leads to estimates

$$\left| \int_0^\infty \mathrm{Tr}(U_g e^{-\frac{s}{\Lambda^2}D_X^2})\varphi(s)\, ds \right| \leq \Lambda^{-2n} \cdot \mathrm{Vol}(X) \cdot C_n \cdot S_X \cdot \int_0^\infty s^n |\varphi(s)|\, ds,$$

with $S_X = \sup_{x \in X} \|\rho(g)(x)\|$.

For a test function that is a Laplace transform $f(u) = \int_0^\infty e^{-su^2} \varphi(s)\, ds$ of a rapidly decaying function $\varphi \in \mathcal{S}(0,\infty)$, we know the spectral action and the heat kernel are related by

$$\mathrm{Tr}(f(D_Y/\Lambda)) = \int_0^\infty \mathrm{Tr}(e^{-\frac{s}{\Lambda^2}D_Y^2})\varphi(s)\, ds,$$

hence, one obtains that the spectral actions on X and Y are related by

$$\mathrm{Tr}(f(D_Y/\Lambda)) = \frac{1}{\#\Gamma}\mathrm{Tr}(f(D_X^2/\Lambda)) + O(\Lambda^{-\infty}).$$

This recovers both the cases of the spherical space forms and the Bieberbach manifolds, respectively with X the sphere or the torus. The same argument applies to the spectral action for $Y \times S^1$ and the resulting slow–roll potential $V(\phi)$, which behave as seen in the previous subsections. In particular, this point of view make transparent the independence of the spectral action and the potential on the choice of the spin structure: in the spherical cases, one has the trivial spin structure on S^3, and in the torus case, the spectral action on T^3 manifestly does not depend on the spin structure on the torus.

3.2.5 *Gravity coupled to matter and slow-roll potential*

In the results discussed so far in this chapter, we have used the spectral action as an action functional for gravity on (Euclidean, compactified) spacetimes with spatial sections a candidate cosmic topology Y. However, we have not considered additional data of coupling of gravity to matter. As we discussed in the previous chapters, more realistic models are obtained when one considers an almost commutative geometry given (locally) by a product of the commutative spacetime directions with a finite noncommutative geometry F that determined the matter content of the model. It was shown in [Ćaćić, Marcolli, Teh (2012)] that replacing the spacetime manifold by an almost commutative geometry affects the shape of the inflation slow-roll potential $V(\phi)$ only by a constant multiplicative that counts the number of fermions in the model.

We discuss here briefly the results of [Ćaćić, Marcolli, Teh (2012)]. As in the previous cases, the problem can be approached either from the point of view of explicit Dirac spectra and multiplicities, or from the heat kernel perspective.

The almost commutative geometry is not assumed to be necessarily Cartesian. In the spherical case, let $\Gamma \subset SU(2)$ be a finite group acting by isometries on S^3, identified with the Lie group $SU(2)$ with the round metric. The spinor bundle on the spherical form S^3/Γ is given by $S^3 \times_\sigma \mathbb{C}^2 \to S^3/\Gamma$, where σ is the representation of Γ defined by the standard representation of $SU(2)$ on $\mathbb{C}^2$. We introduce the datum of an almost commutative geometry (in the sense of [Ćaćić (2012)]) by means of a unitary representation $\alpha :$ $\Gamma \to U(N)$. This defines a flat bundle $\mathcal{V}_\alpha = S^3 \times_\alpha \mathbb{C}^N$ endowed with a canonical flat connection. By twisting the Dirac operator with the flat bundle, one obtains an operator D_α^Γ on the spherical form S^3/Γ acting on the twisted spinors, which are the Γ-equivariant sections $C^\infty(S^3, \mathbb{C}^2 \otimes \mathbb{C}^N)^\Gamma$. Here the group Γ acts by isometries on S^3 and by $\sigma \otimes \alpha$ on $\mathbb{C}^2 \otimes \mathbb{C}^N$. The operator D_α^Γ is the restriction of the Dirac operator $D \otimes \mathrm{id}_{\mathbb{C}^N}$ to the subspace $C^\infty(S^3, \mathbb{C}^2 \otimes \mathbb{C}^N)^\Gamma \subset C^\infty(S^3, \mathbb{C}^2 \otimes \mathbb{C}^N)$. This operator D_α^Γ accounts for the pure gravity part of the model on the almost commutative geometry. The fiber $\mathbb{C}^N = \mathcal{H}_F$ determines the fermions in the model.

Explicit results on the spectra of twisted Dirac operators of the form D_α^Γ on the spherical space forms were obtained in [Cisneros-Molina (2001)]. The result is obtained using the decomposition, given by the Peter–Weyl theorem, $C^\infty(S^3, \mathbb{C}) = \oplus_k E_k \otimes E_k^*$, where E_k is the $k{+}1$-dimensional irreducible representation of $SU(2)$ on the space of homogeneous complex polynomials in two variables of degree k. Correspondingly, the Dirac operator $D \otimes \mathrm{id}_{\mathbb{C}^N}$ on $C^\infty(S^3, \mathbb{C}^2 \otimes \mathbb{C}^N) = \oplus_k E_k \otimes E_k^* \otimes \mathbb{C}^2 \otimes \mathbb{C}^N$ decomposes as a sum of pieces of the form $\mathrm{id}_{E_k} \otimes D_k \otimes \mathrm{id}_{\mathbb{C}^N}$, with $D_k : E_k^* \otimes \mathbb{C}^2 \to E_k^* \otimes \mathbb{C}^2$. This makes it possible to identify the multiplicities of the spectrum of the twisted Dirac operator D_α^Γ with the dimensions $\dim_{\mathbb{C}} \mathrm{Hom}_\Gamma(E_k, \mathbb{C}^2 \otimes \mathbb{C}^N)$. The latter are computed in terms of the pairing of the characters $\langle \chi_{E_k}, \chi_{\sigma \otimes \alpha} \rangle_\Gamma$ of the Γ-representation. The eigenvalues of the twisted Dirac operator D_α^Γ on S^3/Γ are then of the form

$$-\frac{1}{2} - (k+1) \text{ with multiplicity } \langle \chi_{E_{k+1}}, \chi_\alpha \rangle_\Gamma (k+1), \qquad \text{if } k \geq 0,$$

$$-\frac{1}{2} + (k+1) \text{ with multiplicity } \langle \chi_{E_{k-1}}, \chi_\alpha \rangle_\Gamma (k+1), \qquad \text{if } k \geq 1.$$

Let c_Γ denote the exponent of the group Γ, that is, the least common multiple of the orders of the elements of Γ, and let $k = c_\Gamma l + m$ with $0 \leq m < c_\Gamma$. The pairings of characters above satisfy the following recursive properties:

(1) if $-1 \in \Gamma$, then

$$\langle \chi_{E_k}, \chi_\alpha \rangle_\Gamma = \begin{cases} \frac{c_\Gamma l}{\#\Gamma}(\chi_\alpha(1) + \chi_\alpha(-1)) + \langle \chi_{E_m}, \chi_\alpha \rangle_\Gamma & \text{if } k \text{ is even,} \\ \frac{c_\Gamma l}{\#\Gamma}(\chi_\alpha(1) - \chi_\alpha(-1)) + \langle \chi_{E_m}, \chi_\alpha \rangle_\Gamma & \text{if } k \text{ is odd.} \end{cases}$$

(2) if $-1 \notin \Gamma$, then

$$\langle \chi_{E_k}, \chi_\alpha \rangle_\Gamma = \frac{N c_\Gamma l}{\#\Gamma} + \langle \chi_{E_m}, \chi_\alpha \rangle_\Gamma.$$

It is shown in [Ćaćić, Marcolli, Teh (2012)], [Teh (2013b)] that, for all the spherical space forms, there are polynomials P_m^+, and P_m^- that interpolate the multiplicities of the positive and negative eigenvalues of the twisted Dirac operator, with m ranging over residue classes mod c_Γ. For $k \equiv m \bmod c_\Gamma$, $k \geq 1$, and $\lambda = -1/2 \pm (k+1)$, the value $P_m^\pm(\lambda)$ gives the multiplicity of the eigenvalue λ. These polynomials satisfy $P_m^+(u) = P_{m'}^-(u)$, where, for a given m the index m' is given by

$$m' = \begin{cases} c_\Gamma - 2 - m, & \text{if } 1 \leq m \leq c_\Gamma - 2, \\ c_\Gamma - 1, & \text{if } m = c_\Gamma - 1, \\ c_\Gamma - 2 & \text{if } m = c_\Gamma. \end{cases}$$

One then computes the spectral action using the test functions $g_m(u) = P_m^+(u) f(u/\Lambda)$, to which one applies the Poisson summation formula. This gives

$$\text{Tr}(f(D_\alpha^\Gamma/\Lambda)) = \sum_m \sum_{l\in\mathbb{Z}} g_m(1/2 + c_\Gamma l + m + 1)$$

$$= \frac{N}{\#\Gamma} \sum_m \widehat{g_m}(0) + O(\Lambda^{-\infty})$$

$$= \frac{N}{\#\Gamma} \left(\int_\mathbb{R} u^2 f(u/\Lambda) - \frac{1}{4} \int_\mathbb{R} f(u/\Lambda) \right) + O(\Lambda^{-\infty})$$

$$= \frac{N}{\#\Gamma} \left(\Lambda^3 \widehat{f}^{(2)}(0) - \frac{1}{4}\Lambda\widehat{f}(0) \right) + O(\Lambda^{-\infty}).$$

The main step in the above consists in showing that the polynomials $P_m^\pm(u)$ satisfy the relation

$$\sum_{m=1}^{c_\Gamma} P_m^+(u) = \sum_{m=0}^{c_\Gamma-1} P_m^-(u) = \frac{Nc_\Gamma}{\#\Gamma} \left(u^2 - \frac{1}{4} \right). \qquad (3.2)$$

Thus, one again obtains that the spectral action for the twisted Dirac operator is a constant multiple of the spectral action on the 3-sphere

$$\text{Tr}(f(D_\alpha^\Gamma/\Lambda)) = \frac{N}{\#\Gamma} \text{Tr}(f(D_{S^3}/\Lambda)) + O(\Lambda^{-\infty}).$$

The cases of all the spherical space forms are given in completely explicit details in [Ćaćić, Marcolli, Teh (2012)] and [Teh (2013b)]. We describe here the example of the Poincaré dodecahedral space, to illustrate the technique. First one shows that the identity $P_m^+(u) = P_{m'}^-(u)$ recalled above is equivalent to $\beta_m^\alpha = \langle \chi_{E_{m-1}}, \chi_\alpha \rangle_\Gamma$ and $\gamma_m^\alpha = \langle \chi_{E_{m+1}}, \chi_\alpha \rangle_\Gamma$ satisfying the relation

$$\beta_m^\alpha + \gamma_{m'}^\alpha = \begin{cases} \chi_\alpha(1), & \text{if } 1 \le m \le c_\Gamma - 2 \\ 2\chi_\alpha(1), & \text{if } m = c_\Gamma - 1, c_\Gamma, \end{cases}$$

if $-1 \notin \Gamma$, while in the case where $-1 \in \Gamma$ it corresponds to the relation

$$\beta_m^\alpha + \gamma_{m'}^\alpha = \begin{cases} \chi_\alpha(1)(\chi_\alpha(1) + (-1)^{m+1}\chi_\alpha(-1)), & \text{if } 1 \le m \le c_\Gamma - 2, \\ 2\chi_\alpha(1)(\chi_\alpha(1) + \chi_\alpha(-1)), & \text{if } m = c_\Gamma - 1, \\ 2\chi_\alpha(1)(\chi_\alpha(1) - \chi_\alpha(-1)), & \text{if } m = c_\Gamma. \end{cases}$$

The Poincaré dodecahedral space is the quotient of S^3 by the binary icosahedral group $2\mathcal{I}$. This has order 120 and exponent 60. Its character table, with $\mu = \frac{\sqrt{5}+1}{2}$, and $\nu = \frac{\sqrt{5}-1}{2}$, is given by the following.

Class	1_+	1_-	30	20_+	20_-	12_{a+}	12_{b+}	12_{a-}	12_{b-}
Order	1	2	4	6	3	10	5	5	10
χ_1	1	1	1	1	1	1	1	1	1
χ_2	2	-2	0	1	-1	μ	ν	$-\mu$	$-\nu$
χ_3	2	-2	0	1	-1	$-\nu$	$-\mu$	ν	μ
χ_4	3	3	-1	0	0	$-\nu$	μ	$-\nu$	μ
χ_5	3	3	-1	0	0	μ	$-\nu$	μ	$-\nu$
χ_6	4	4	0	1	1	-1	-1	-1	-1
χ_7	4	-4	0	-1	1	1	-1	-1	1
χ_8	5	5	1	-1	-1	0	0	0	0
χ_9	6	-6	0	0	0	-1	1	1	-1

One then finds that (3.2) is satisfied, with

$$\sum_{m=1}^{c_\Gamma} \beta_m^\alpha = \begin{cases} 15, & \text{if } \chi_\alpha = \chi_1, \\ 31, & \text{if } \chi_\alpha \in \{\chi_2, \chi_3\}, \\ 45, & \text{if } \chi_\alpha \in \{\chi_4, \chi_5\}, \\ 60, & \text{if } \chi_\alpha = \chi_6, \\ 62, & \text{if } \chi_\alpha = \chi_7, \\ 75, & \text{if } \chi_\alpha = \chi_8, \\ 93, & \text{if } \chi_\alpha = \chi_9; \end{cases}$$

$$\sum_{m=0}^{c_\Gamma-1} \gamma_m^\alpha = \begin{cases} 16, & \text{if } \chi_\alpha = \chi_1, \\ 31, & \text{if } \chi_\alpha \in \{\chi_2, \chi_3\}, \\ 48, & \text{if } \chi_\alpha \in \{\chi_4, \chi_5\}, \\ 64, & \text{if } \chi_\alpha = \chi_6, \\ 62, & \text{if } \chi_\alpha = \chi_7, \\ 80, & \text{if } \chi_\alpha = \chi_8, \\ 93, & \text{if } \chi_\alpha = \chi_9. \end{cases}$$

The heat kernel point of view is also similar to the case discussed in §3.2.4. One again considers a test function f that is given by a Laplace transform of a function φ, so that the spectral action and the heat kernel are related by

$$\text{Tr}(f(D/\Lambda)) = \int_0^\infty \text{Tr}(e^{-\frac{s}{\Lambda^2}D^2})\, \varphi(s)\, ds.$$

Then the same argument discusses in §3.2.4 can be applied to the operator D_α^Γ and one obtains [Ćaćić, Marcolli, Teh (2012)], [Ćaćić (2013b)]

$$\mathrm{Tr}(f(D_\alpha^\Gamma/\Lambda)) = \frac{N}{\#\Gamma}\,\mathrm{Tr}(f(D_X/\Lambda)) + O(\Lambda^{-\infty}),$$

which now covers both the cases of the spherical and the flat geometries. The coefficient $N/\#\Gamma$ arises by identifying the operator $\tilde{D}_\alpha$ lifting D_α^Γ to the covering X with $D_X \otimes \mathrm{id}_N$, upon identifying $\mathcal{F}_X \otimes \mathcal{V}_X$ with $\mathcal{F}_X^{\oplus N}$, with the trivial bundle $\mathcal{F}_X = X \times \mathbb{C}^N$ and the Γ-equivariant $\mathcal{V}_X$ satisfying $\mathcal{V}_\alpha = \mathcal{V}_X/\Gamma$.

As we have seen in the previous cases, one similarly computes the spectral action on the product geometry $Y \times S^1$ and perturbs the Dirac operator in this product geometry by the scalar field ϕ^2 to obtain the associated slow–roll potential. This is then of the form

$$V(\phi) = \frac{N}{\#\Gamma}\,V_X(\phi),$$

where $V_X(\phi)$ is the potential for either the spherical case $X = S^3$ or the flat case $X = T^3$. This implies that the effect on the inflation potential of the presence of the finite noncommutative geometry is only seen in an overall factor that affects the amplitudes of the power laws for the scalar and tensor fluctuation, and which is equal to the number N of fermions in the model.

3.2.6 *Engineering inflation via Dirac spectra*

By allowing 3-manifolds with metric that no longer satisfy homogeneity and isotropy conditions, one can construct more general shapes of slow–roll potentials $V(\phi)$ using the spectral action. We introduced this method in [Marcolli, Pierpaoli, Teh (2011)] as geometric engineering of inflation potentials. The construction relies on results of [Dahl (2005)] and [Dahl (2003)], to modify the Dirac spectrum in such a way as to control easily the effect on the spectral action and on the slow-roll potential.

It is shown in [Dahl (2005)] that, for a given $L > 0$ and a given sequence of non-zero real numbers

$$-L < \lambda_1 < \lambda_2 < \lambda_3 < \cdots < \lambda_N < L, \tag{3.3}$$

it is possible to construct a Riemannian metric g on a given smooth compact spin 3-manifold M, so that, for D_M the resulting Dirac operator, the spectrum satisfies

$$(\mathrm{Spec}(D_M) \smallsetminus \{0\}) \cap (-L, L) = \{\lambda_j\}_{j=1,\dots,N},$$

and all these eigenvalues are simple. The construction is based on rescaling the metric and taking connected sum with a 3-sphere with a Berger metric, so that the topology is unaffected, but the Dirac spectrum changes in a controlled way, according to the surgery formula of [Dahl (2005)]. We assume that either our Dirac operators have trivial kernel or that the spectral action is modified to sum only over non-trivial eigenvalues. Then given an arbitrary progression $\{\lambda_n = \eta + n\lambda\}_{n\in\mathbb{Z}}$, with $\lambda > 0$ and $\eta \neq 0$, one applies the method described above to obtain a metric for which the part of the Dirac spectrum in $(-L, L)$ coincides with the elements of this progression contained in this interval. For a suitable class of test functions, we have

$$\mathrm{Tr}(f(D/\mathbf{\Lambda})) = \sum_{n\in\mathbb{Z}} f(\lambda_n/\mathbf{\Lambda}),$$

which can be computed via the Poisson summation formula

$$\sum_{n\in\mathbb{Z}} g(\eta + n\lambda) = \sum_{n\in\mathbb{Z}} \frac{1}{\lambda}\exp(\frac{2\pi in\eta}{\lambda})\widehat{g}(\frac{n}{\lambda})$$

applied to $g(u) = f(u/\Lambda)$, which leads to

$$\mathrm{Tr}(f(D/\mathbf{\Lambda})) = \frac{\mathbf{\Lambda}}{\lambda}\widehat{f}(0) + O(\mathbf{\Lambda}^{-\infty}).$$

Chapter 4

Algebro-geometric models in Cosmology

This chapter is based on the results of [Manin, Marcolli (2014)].

4.1 Spacetimes and complex geometry

The usefulness of complex geometry in general relativity has been observed since the development of twistor theory, see for instance [Penrose (1978)]. For an extensive and detailed mathematical treatment of twistor theory and complex geometry, we refer the reader to [Manin (1988)].

We recall here the basic geometric setting. Let T be a 4-dimensional complex vector space, and let $G(2, T)$ denote the Grassmannian of complex 2-planes in T, with $\mathcal{S}$ the tautological plane bundle over $G(2, T)$. We denote by T^* the dual of T and we consider

$$\tilde{\mathcal{S}} = \{(x, t) : x \in G(2, T), t \in \mathcal{S}_x^\perp \subset T^*\}.$$

The trivial vector bundle $\mathcal{T}$ on $G(2, T)$ with fiber T decomposes as

$$0 \to \mathcal{S} \to \mathcal{T} \to \tilde{\mathcal{S}}^* \to 0.$$

This gives an identification

$$\sigma : \Omega^1(G(2, T)) \overset{\simeq}{\to} \mathcal{S} \otimes \tilde{\mathcal{S}}.$$

The 1-forms in $\Omega^1(G(2, T))$ act on $\mathcal{S}^* \oplus \tilde{\mathcal{S}}^*$ mapping it to $\mathcal{S} \oplus \tilde{\mathcal{S}}$, via the maps $\Omega^1(G(2, T)) \otimes \mathcal{S}^* \to \tilde{\mathcal{S}}$ and $\Omega^1(G(2, T)) \otimes \tilde{\mathcal{S}}^* \to \mathcal{S}$ induced by σ. The matrices defining this action are the γ-matrices. One obtains a conformal metric on $G(2, T)$ given by $g = \epsilon \otimes \tilde{\epsilon}$, for $\epsilon \in H^0(G(2, T), \Lambda^2 \mathcal{S})$ and $\tilde{\epsilon} \in H^0(G(2, T), \Lambda^2 \tilde{\mathcal{S}})$.

4.1.1 *Complexified spacetimes and Grassmannians*

The choice of a fixed 2-plane $W \subset T$ determines a *big cell* in the Grassmannian $G(2,T)$

$$U_{\mathbb{C}} = U_{\mathbb{C}}(W) = \{x \in G(2,T) : S_x \cap W = \{0\}\}.$$

One identifies a big cell $U_{\mathbb{C}} = \mathcal{M}_{\mathbb{C}}^4$ with the *complexified Minkowski spacetime*, with the chosen plane $W \in G(2,T)$ providing the point ∞. A conformal metric on $G(2,T)$ induces a Minkowski metric on the real locus $U_{\mathbb{C}}(\mathbb{R}) = \mathcal{M}_{\mathbb{R}}^4$, which recovers the usual real 4-dimensional Minkowski spacetime.

One can then view the Grassmannian $G(2,T)$ as a compactification of the complexified Minkowski spacetime $\mathcal{M}_{\mathbb{C}}^4$: the boundary cells

$$G(2,T) \smallsetminus U_{\mathbb{C}} = \mathcal{C}(\infty)$$

determine the (complex) *light cone* $\mathcal{C}(\infty)$ with vertex at ∞ and base $L(\infty) = \mathbb{P}(S_{\infty}^*) \times \mathbb{P}(\tilde{S}_{\infty}^*)$. This type of complexification and compactification extends not only for the flat Minkowski spacetime, but also to a more general class of spacetimes: the *GS-manifolds* (endowed with Grassmann spinor structures), which are characterized by an isomorphism $\sigma : \Omega^1(M) \overset{\cong}{\rightarrow} S \otimes \tilde{S}$. These include the Friedmann-Robertson-Walker spacetimes. We refer the reader to [Manin (1988)] for more details.

4.1.2 *Twistor spaces*

One refers to T as the twistor space and to T^* as the dual twistor space. Penrose's twistor transform is given by the correspondence

$$\mathbb{P}^3 = G(1;T) \longleftarrow F(1,2;T) \longrightarrow G(2;T) \tag{4.1}$$

where $F(1,2;T)$ is the flag variety, with the arrows given by the two natural projection maps.

The Klein quadric $\mathcal{K}$ is the image $G(2;T) \hookrightarrow \mathbb{P}^5$ under the Plücker embedding. We can write the defining equation of the Plücker embedding as

$$\Lambda^{(12)}\Lambda^{(34)} - \Lambda^{(13)}\Lambda^{(24)} + \Lambda^{(23)}\Lambda^{(14)} = 0,$$

where the variables $\{\Lambda^{(ij)} : 1 \leq i < j \leq 4\}$ are labelled by the minors of a 2×4-matrix.

The quadric $\mathcal{K}$ has rulings by two families of planes, called α-planes and β-planes in the twistor literature. The α-planes are the images under the second projection of the fibers of the first projection in the Penrose transform diagram (4.1). In these planes every line is a light ray: the two $\mathbb{P}^1$'s in the base of a light cone $C(x)$ are the same as those in the product $\mathbb{P}(\mathcal{S}_\infty^*) \times \mathbb{P}(\tilde{\mathcal{S}}_\infty^*)$. The β-planes are similarly obtained from the dual Penrose diagram

$$G(3; T^*) \longleftarrow F(2, 3; T^*) \longrightarrow G(2; T).$$

4.2 Blowup models

Let $Y \subset X$ be a smooth subvariety Y of codimension $m + 1$ embedded inside a smooth projective variety X. The blowup $\mathrm{Bl}_Y(X)$ of X along Y is a smooth projective variety, endowed with a projection map $\pi : \mathrm{Bl}_Y(X) \to X$ that is an isomorphism on $X \smallsetminus Y$. This map is also denotes by bl_Y. The exceptional divisor of the blowup, that is the locus $E = \pi^{-1}(Y)$, is a projective bundle $E = \mathbb{P}(\mathcal{N})$ over Y with fiber $\mathbb{P}^m$.

4.2.1 *Gluing spacetimes*

Following [Manin, Marcolli (2014)], we are going to present a general setting for gluing (complexified) spacetimes across their conformal infinity, based on the blowup operation.

Consider two complex 4-dimensional manifolds $\mathcal{M}_-$ and $\mathcal{M}_+$ that are smooth, though not necessarily projective. Assume that $\mathcal{M}_-$ contains an embedded smooth projective 2-dimensional quadric $\mathbb{S}_- \subset \mathcal{M}_-$, with

$$\mathbb{S}_- \cong \mathbb{P}^1 \times \mathbb{P}^1.$$

Also suppose given a 3-dimensional complex space $\mathcal{L}$, which is isomorphic to a neighborhood of the vertex of the complex cone with base $\mathbb{P}^1 \times \mathbb{P}^1$, and which is embedded as a closed submanifold in $\mathcal{M}_+$. Assume we also have an explicit isomorphism between $\mathbb{S}_-$ in $\mathcal{M}_-$ and the "quadrics of null directions" $\mathbb{S}_+$ in $\mathcal{M}_+$.

We then proceed as follows. First perform a blowup of $\mathcal{M}_+$ at a point. The point is thought of here as a "big bang singularity" in the second aeon. The point is the vertex of the cone with base $\mathbb{P}^1 \times \mathbb{P}^1$. We obtain a new variety

$$\tilde{\mathcal{M}}_+ = \mathrm{Bl}_x(\mathcal{M}_+).$$

One then performs a gluing, via the isomorphism, of the quadric $\mathbb{S}_-$ with $\mathbb{S}_+$ inside the exceptional divisor of the blowup $\tilde{\mathcal{M}}_+$ and one obtains a new glued (complex) spacetime

$$\mathcal{M}_- \star_{\mathbb{S}} \tilde{\mathcal{M}}_+.$$

Specific cases include:

- the blowup $\mathrm{Bl}_x(Q)$ at a point of a quadric hypersurface $Q \subset \mathbb{P}^5$, with exceptional divisor $\simeq \mathbb{P}^3$;
- a projective space $\mathbb{P}^4$, seen as the compactification $\mathbb{P}^4 = \mathbb{A}^4 \cup \mathbb{P}^3$, blown up along a $\mathbb{P}^1$ inside the $\mathbb{P}^3$ at infinity, $\mathrm{Bl}_{\mathbb{P}^1}(\mathbb{P}^4)$, with exceptional divisor of the form $\mathbb{P}^1 \times \mathbb{P}^1$.

In order to identify the physical spacetimes inside their complexifications and compactifications, one also needs to keep track of the real structures in these blowup and gluing operations performed at the level of the complex varieties. The real involution on $\mathbb{S}$ comes from exchanging the two spinor bundles S and $\tilde{S}$, so that the set of real points (fixed points of the involution) is $\mathbb{S}(\mathbb{R}) = S^2$, the diagonal $\mathbb{P}^1(\mathbb{C})$ interpreted as the "sky". For $\mathcal{M}_- = \mathbb{P}^4$ the real locus is $\mathcal{M}_-(\mathbb{R}) = \mathbb{P}^4(\mathbb{R})$, a 4-dimensional real projective space. The boundary at infinity is a $\mathbb{P}^3(\mathbb{R})$, which contains an embedded 2-sphere S^2: the common base at infinity of all the light cones in the real Minkowski 4-space $\mathcal{M}_-(\mathbb{R}) \smallsetminus \mathbb{P}^3(\mathbb{R})$. For the Klein quadric, $\mathcal{M}_+(\mathbb{R}) = Q(\mathbb{R})$ is the set of real points of the quadric, $Q(\mathbb{R}) \subset \mathbb{P}^5(\mathbb{R})$. For any point $x \in Q(\mathbb{R})$, there is a 4-dimensional projective subspace $\mathbb{P}^4_x(\mathbb{R})$ of $\mathbb{P}^5(\mathbb{R})$ tangent to Q at x. Let $\overline{L}_x := Q(\mathbb{R}) \cap \mathbb{P}^4_x(\mathbb{R})$ denote the compactified light cone in $\mathcal{M}_+(\mathbb{R})$. In the complement $Q(\mathbb{R}) \setminus L_x$, through each point y one has an uncompactified light cone $L_y := \overline{L}_y \setminus (\overline{L}_y \cap \overline{L}_x)$. One can identify $Q(\mathbb{R}) \setminus L_x$ with an affine 4-dimensional Minkovski space, by projecting $Q(\mathbb{R}) \setminus L_x$ from x to a sufficiently general hyperplane $\mathbb{P}^4(\mathbb{R}) \subset \mathbb{P}^5(\mathbb{R})$.

The resulting blowup diagram, restricted to real loci, is given by

$$
\begin{array}{ccc}
C & \xrightarrow{\mathrm{bl}_x} & Q(\mathbb{R}) \\
\downarrow{\scriptstyle \mathrm{bl}_{S^2}} & & \\
\mathbb{P}^4(\mathbb{R}) & &
\end{array}
$$

The same C is obtained from $Q(\mathbb{R}) = \mathcal{M}_+(\mathbb{R})$, by blowing up the point $x \in Q(\mathbb{R})$, or from $\mathbb{P}^4(\mathbb{R}) = \mathcal{M}_-(\mathbb{R})$, by blowing up the base at infinity of all the light cones in $\mathbb{P}^3(\mathbb{R}) = \mathbb{P}^4(\mathbb{R}) \setminus \mathbb{A}^4(\mathbb{R})$.

When dealing with real structures, one also has to consider the issue of orientability and orientation. For $n \geq 2$ have $\pi_1(\mathbb{P}^n(\mathbb{R})) = \mathbb{Z}/2\mathbb{Z}$, with universal cover $S^n \to \mathbb{P}^n(\mathbb{R})$, where the sphere S^n parameterizes *oriented* lines in $\mathbb{R}^{n+1}$. Similarly, in the case of the Grassmannians, we have a real Grassmannian of *oriented* d-spaces, which is the universal cover (double cover) of the set of real points of the complex Grassmannian, which parameterizes unoriented spaces. The blowup diagram lifts to these double covering spaces, where one future and past boundaries. One can then use this diagram to glue future boundary of $\mathcal{M}_-$ to past boundary of $\mathcal{M}_+$.

In [Manin, Marcolli (2014)], this blowup model for the gluing of successive aeons is applied in two different ways (two crossover models):

(1) Identify the 3–dimensional projective space $\mathrm{bl}_x^{-1}(x)$ with the 3-dimensional projective space at infinity of $\mathbb{P}^4(\mathbb{R})$, so that the sphere of null–directions in $\mathrm{bl}_x^{-1}(x)$ is identified with the common base at infinity of all the light cones.

(2) Identify the divisor $\mathrm{bl}_{S^2}^{-1}(\mathbb{P}^3(\mathbb{R})) \subset \mathrm{Bl}_{S^2}(\mathbb{P}^4(\mathbb{R}))$, which describes the infinity of the first aeon, with the divisor $\mathrm{bl}_x^{-1}(L_x) \subset \mathrm{Bl}_x(Q(\mathbb{R}))$, which represents the Big Bang of the second aeon.

The second construction provides a model akin to Penrose's idea of conformally cyclic cosmology, while the first model can be applied to generate eternal inflation type scenarios, as we discuss below.

4.2.2 *Conformally cyclic cosmological models*

The model of conformally cyclic cosmology was proposed in [Penrose (2010)], as an alternative to both inflationary theories and BKL type singularities (mixmaster cosmologies). In this model the future infinity of one spacetime aeon is conformally glued to the Big-Bang of the next aeon, via a conformal scaling of the metric at $\pm\infty$, so that the conformal classes are matched instead of the metrics. For more details, we refer the reader to the very readable presentation in [Penrose (2010)]. This proposed model generated some debate, after the claim of [Gurzadyan, Penrose (2013)] of detection of patterns of circles in the CMB sky predicted by the conformally cyclic model. For a criticism of these findings, and for more details on the ensuing discussion, see for instance [Moss et al. (2012)], [Tod (2013)], [Wehus, Eriksen (2011)].

4.2.3   *Eternal inflation via trees of projective spaces*

The model we describe here may be thought of as an algebro-geometric counterpart to the discretized model of eternal inflation (the "Eternal Symmetree") developed in [Harlow, Shenker, Stanford, Susskind (2012)], see also [Susskind (2012)], to which we will return in §8.2 below, where the branching off of universes happens according to a tree structure, determined by the 2-adic Bruhat–Tits tree, with pruning that determines terminal vacua.

Here we also consider an underlying tree structure, which defines the underlying combinatorial pattern of branching, while at each branching the gluing is performed according to a blowup diagram similar to the one discussed above. More precisely, the data we consider are the following:

- a finite rooted tree τ with one outgoing flag (half–edge) at the root vertex and a number of incoming half–edges (leaves);
- a projective space $X_v = \mathbb{P}^d$ (in our case $d = 4$) for each vertex v of τ, with a choice of a hyperplane $H_v \subset X_v$, and a point $p_{v,f} \in X_v$ for each incoming flag f at v, with the property that $p_{v,f} \neq p_{v,f'}$ for $f \neq f'$ and $p_{v,f} \notin H_v$ for all f;
- for each edge e in τ, we perform the blowup of $X_{t(e)}$ at the point $p_{t(e),f_e}$ then glue the exceptional divisor $E_{t(e)}$ of the blowup $\mathrm{Bl}_{p_{t(e),f_e}}(X_{t(e)})$ to the hyperplane $H_{s(e)} \subset X_{s(e)}$.

This creates a tree of projective spaces, with a projective space $\mathbb{P}^4$ (thought of as a complexified and compactified spacetime) placed at each vertex and different projective spaces glued together according to the edges of the trees, where the gluing happens by attaching the boundary divisor of one spacetime to the blowup of a point in another, as in the first crossover model discussed above.

The resulting trees of projective spaces are a generalization of the trees of $\mathbb{P}^1$'s with marked points that appear in the moduli spaces of rational curves $\bar{\mathcal{M}}_{0,n}$. In fact there is a moduli space $T_{d,n}$ of stable deformations n–pointed rooted trees of projective spaces $\mathbb{P}^d$, which is a direct generalization of the moduli space of rational cuves with marked points, and is also closely related to the Fulton–MacPherson compactifications of configuration spaces. It was introduced and studied in [Chen, Gibney, Krashen (2009)]. The existence of a nice moduli space appears useful in terms of an associated multiverse picture, with multiverse fields interpreted as living on moduli spaces $T_{4,n}$.

One may also consider a possible dual construction, performed at the twistor space level. Consider $\mathbb{P}^3(\mathbb{C})$ as twistor transforms of a complexified and compactified spacetimes given by a Grassmannian $G(2, T)$. Consider then (stable) trees of $\mathbb{P}^3$'s parameterized by the associated moduli space $T_{3,n}$, using the same construction described above. The complex variety obtained as the result of the blowups and gluings of the various $\mathbb{P}^3(\mathbb{C})$ according to the underlying tree τ is a priori not necessarily the twistor space of another smooth 4–dimensional spacetime. However, the kind of technique developed in [Donaldson, Friedman (1989)] shows that the resulting space may have an algebro-geometric deformation that is a twistor space.

Notice that the construction illustrated here only provides a geometric setting for a possible inflationary model, but not an actual underlying physical mechanism (a field) that drives the inflation.

4.3 Time and elliptic curves

In the algebro-geometric models described above, for gluing together complexified and compactified spacetimes along their conformal infinity, a natural question that arises is how one should obtain a good cosmological time coordinate across the gluing. The approach of [Manin, Marcolli (2014)], [Manin, Marcolli (2015)] suggests that in a local model near the singularity the cosmological time can be interpreted as a motion along a geodesic on a modular curve.

This suggestion is motivated by the fact that one can express global time in an FRW cosmology in terms of a period integral for a family of elliptic curves. In fact, in a FRW cosmology

$$ds^2 := dt^2 - R(t)^2 \left(\frac{dr^2}{1 - kr^2} + r^2(d\theta^2 + \sin^2\theta \, d\phi^2) \right),$$

one writes the Friedmann equation

$$\dot{R}^2 = -k + \kappa(\rho_M + \rho_\gamma)R^2 + \frac{1}{3}\Lambda R^2$$

in terms of $\kappa = 8\pi G/3$, with ρ_M and ρ_γ the matter and radiation densities. The perfect fluid conservation equation

$$3\frac{\dot{R}}{R} = -\frac{\dot{\rho}}{(\rho + p)}$$

gives $\rho_M \sim R^{-3}$ and $\rho_\gamma \sim R^{-4}$. Solutions of the Friedmann equation are expressible in terms of an *elliptic curve* of equation $Y^2 = R^4 + aR + b$, with global time given by a period integral on this elliptic curve

$$\tau \cong \int_0^{R(t)} \frac{dR}{Y},$$

see [Newman (2013)], [Tod (2013)], and [Manin, Marcolli (2014)]. Thus, in a FRW universe time evolution is described by a real curve on an algebraic surface given by a family of elliptic curves, with variable coefficients a, b related to matter and energy densities.

The idea then is that the resetting of cosmological time in the passage between successive aeons happens through a real curve in complexified time. In $\mathcal{M}_\pm(\mathbb{R})$ compactifications of real spacetime, any time-like line is a $\mathbb{P}^1(\mathbb{R})$, which lives inside a $\mathbb{P}^1(\mathbb{C})$ of complexified time. Using an FRW model, near the past boundary of $\mathcal{M}_+(\mathbb{R})$, one can view time as a real curve on the modular curve $\mathbb{H}/\mathrm{PSL}_2(\mathbb{Z}) \simeq \mathbb{P}^1(\mathbb{C})$, corresponding to a family of elliptic curves as discussed above. In the gluing of aeons one then identifies the modular $\mathbb{P}^1(\mathbb{C})$ where near-singularity time of the second aeon lives with the complexified time $\mathbb{P}^1(\mathbb{C})$ of the previous aeon.

Another model in which one sees cosmological time as a parameter on a (geodesic) curve on $\mathbb{H}/\mathrm{PSL}_2(\mathbb{Z})$ is the near-singularity behavior of a BKL (mixmaster) cosmology, where the geodesic distances between successive intersection points of the geodesic with a Farey triangulation of $\mathbb{H}$ are proportional to logarithmic times of the Kasner epochs, [Manin, Marcolli (2015)]. We will discuss these mixmaster models in the next chapter.

4.4   Noncommutativity and gluing of spacetimes

Another way in which noncommutative geometry may play a role in cosmological model is through the hypothesis that, near the initial singularity Big-Bang, where quantum effect begin to play a role, one of the possibilities that can arise is some amount of noncommutativity in the spacetime coordinates. This type of noncommutativity is very different from the one we have seen in spectral action models of gravity coupled to matter, where typically spacetime coordinates are assumed to remain commutative, while extra dimensions, in the form of a finite geometry, are noncommutative. The spectral action models we discussed in the previous chapter have a typical energy scale around unification energy, while it is only around the Planck scale that one expects further noncommutativity, involving spacetime itself, to play a role. The question of noncommutativity in spacetime coordinates has been investigated from an observational point of view, in trying to identify possible detectable signatures in the background radiation, see [Nautiyal (2014)], [Shiraishi et al. (2014)].

In [Manin, Marcolli (2014)] it was shown that the two crossover models of gluing successive aeons spacetimes is compatible with the possibility of noncommutativity in spacetime coordinates. More precisely, natural noncommutative deformations exist for all the varieties involved, which are compatible with the blowups and the gluing operations used in the crossover models. The method used to obtain these noncommutative deformations is based on a far reaching generalization of the Connes–Landi theta-deformations described in §1.1.6 above, developed in [Cirio, Landi, Szabo (2011)], [Cirio, Landi, Szabo (2013)], which replaces the real noncommutative tori acting on Riemannian manifolds of the original theta-deformations with algebraic tori $\mathbb{G}_m^n$ acting on toric algebraic varieties (and on some more general torified varieties). In particular, this method provides natural noncommutative deformations of projective spaces and of Grassmannians, the latter obtained via a deformation of their Plücker embeddings in projective spaces.

Deformations of projective spaces (as toric varieties) are obtained in [Cirio, Landi, Szabo (2012)] by considering the homogeneous coordinate algebra with generators $\{w_i\}_{i=1,\dots,n+1}$ and relations $w_i w_j = q_{ij}^2 w_j w_i$, for $i, j = 1, \dots, n$, with deformation parameters $q_{ab} = \exp(\frac{i}{2}\theta_{ab})$, and with $w_{n+1} w_i = w_i w_{n+1}$, for $i = 1, \dots, n$. Noncommutative Grassmannian $G_\theta(d, n)$ have generators Λ^J, labelled by the minors $J = (j_1, \dots, j_d)$, with relations

$$\Lambda^J \Lambda^{J'} = \left(\prod_{\alpha,\beta=1}^{d} q_{j_\alpha, j'_\beta}^2 \right) \Lambda^{J'} \Lambda^{J}.$$

The compatibility with Plücker embedding $G_\theta(d, n) \hookrightarrow \mathbb{P}_\Theta(\wedge^d \mathbb{C}^n)$ is obtained by requiring that the skew-symmetric matrix Θ is related to θ by

$$\Theta^{JJ'} = \sum_{\alpha,\beta=1}^{d} \theta^{j_\alpha j'_\beta}.$$

For our purposes, we are interested in the cases of complexified and compactified spacetimes that are either $\mathbb{P}^4$ or the image under the Plücker embedding $G(2, 4) \subset \mathbb{P}^5$. The noncommutative Plücker embedding of the deformed $G_\theta(2, T)$ in $\mathbb{P}_\Theta^5$ is defined

$$q_{31} q_{32} q_{34} \Lambda^{(12)} \Lambda^{(34)} - q_{21} q_{23} q_{24} \Lambda^{(13)} \Lambda^{(24)} + q_{12} q_{13} q_{14} \Lambda^{(23)} \Lambda^{(14)} = 0,$$

where the q_{ij} are as discussed above.

To see that this noncommutative deformation is compatible with the crossover model, first observe that the decomposition of the Grassmannian into Schubert cells is compatible. More precisely, the Grassmannian $G(2, T)$, with T a complex 4-dimensional vector space, has a decomposition into six Schubert cells $C_{(j_1, j_2)}$, with

$$(j_1, j_2) \in \{(1, 2), (1, 3), (1, 4), (2, 3), (2, 4), (3, 4)\},$$

which are, respectively, of complex dimensions 0, 1, 2, 2, 3, 4, and correspond to 2×4–matrices in row echelon form. In terms of the Plücker embedding $G(2, T) \hookrightarrow \mathbb{P}^5$, if we write the defining equation for $G(2, T)$ in $\mathbb{P}^5$ as above, in the form

$$\Lambda^{(12)} \Lambda^{(34)} - \Lambda^{(13)} \Lambda^{(24)} + \Lambda^{(23)} \Lambda^{(14)} = 0,$$

then the Schubert varieties $X_{(j_1, j_2)}$ given by the closures of the Schubert cells $C_{(j_1, j_2)}$, are given by

$$X_{(1,2)} = \{V \in G(2, T) \mid \Lambda^{(13)} = \Lambda^{(14)} = \Lambda^{(23)} = \Lambda^{(24)} = \Lambda^{(34)} = 0\}$$

$$X_{(1,3)} = \{V \in G(2, T) \mid \Lambda^{(14)} = \Lambda^{(23)} = \Lambda^{(24)} = \Lambda^{(34)} = 0\}$$

$$X_{(1,4)} = \{V \in G(2, T) \mid \Lambda^{(23)} = \Lambda^{(24)} = \Lambda^{(34)} = 0\}$$

$$X_{(2,3)} = \{V \in G(2, T) \mid \Lambda^{(14)} = \Lambda^{(24)} = \Lambda^{(34)} = 0\}$$

$$X_{(2,4)} = \{V \in G(2, T) \mid \Lambda^{(34)} = 0\}$$

with $X_{(3,4)} = G(2, T)$. The deformation described above then induces compatible noncommutative deformations on all the Schubert cells. In particular, the closure $X_{(2,4)}$ of the 3-dimensional cell is the boundary at infinity $C(\infty)$ of the complexified spacetime $U = C_{(3,4)}$, identified with the big cell. This means that, in order to make the crossover models compatible with the noncommutative deformations, it suffices to use a θ deformation of the exceptional divisor $\mathbb{P}^3$ of the blowup of Q^4, where the deformation parameters match the deformation parameters of the $\mathbb{P}^3$ at infinity of $\mathbb{P}^4$. In the case of the Klein quadric Q^4 in $\mathbb{P}^5$, we use the quantization described above, and for the case of $\mathbb{P}^4(\mathbb{C})$ we consider a ruled surface defined by the equation $x_0 x_3 + x_1 x_2 = 0$ inside the $\mathbb{P}^3$ defined by setting $x_4 = 0$. This can be identified with the intersection of the Klein quadric with the $\mathbb{P}^3(\mathbb{C})$ given by $x_4 = 0$ and $x_5 = 0$ in $\mathbb{P}^5$. Thus, we can compatibly quantize the $\mathbb{P}^4(\mathbb{C})$ and the hyperplane $\mathbb{P}^3(\mathbb{C})$ of equation $x_4 = 0$, as well as the locus $x_0 x_3 + x_1 x_2 = 0$, using the same quantization parameters used for $\mathbb{P}^5(\mathbb{C})$ with the embedded Klein quadric Q^4.

Chapter 5

Mixmaster Cosmologies

The content of this chapter is based on [Manin, Marcolli (2015)] and on the earlier work [Manin, Marcolli (2002)] and [Marcolli (2007)]. Section 5.3 is based on [Estrada, Marcolli (2013)].

5.1 Kasner metrics and mixmaster universe models

Consider the real circle in $\mathbb{R}^3$ defined by the equations

$$p_a + p_b + p_c = 1, \qquad p_a^2 + p_b^2 + p_c^2 = 1.$$

Each point on this circle determines a Kasner metric with exponents (p_a, p_b, p_c). This is a metric with Minkowskian (or Euclidean) signature

$$\pm dt^2 + a(t)^2 dx^2 + b(t)^2 dy^2 + c(t)^2 dz^2$$

with three anisotropic scaling factors a, b, c of the form

$$a(t) = t^{p_a}, \qquad b(t) = t^{p_b}, \qquad c(t) = t^{p_c}, t > 0.$$

It is customary to parameterize the points (p_a, p_b, p_c) on the circle by a coordinate $u \in [1, \infty]$, by setting

$$p_1^{(u)} := -\frac{u}{1 + u + u^2} \in [-1/3, 0]$$

$$p_2^{(u)} := \frac{1 + u}{1 + u + u^2} \quad \in [0, 2/3] \tag{5.1}$$

$$p_3^{(u)} := \frac{u(1 + u)}{1 + u + u^2} \quad \in [2/3, 1].$$

Up to a permutation of the three coordinate axes, one can arrange the exponents in increasing order $p_1^{(u)} \le p_2^{(u)} \le p_3^{(u)}$.

The Kasner metrics provide the local model for a very interesting family of anisotropic cosmologies, known as the mixmaster universe models (or BKL cosmologies), first introduced in [Belinskii, Khalatnikov, Lifshitz (1970)]. The behavior of these spacetimes, near the initial singularity, is modeled by a sequence of Kasner metrics, organized into Kasner epochs, which are in turn subdivided into Kasner cycles. Within each epoch one direction dominates the expansion, while the other two undergo a series of oscillation during the Kasner cycles. At the end of each epoch a bounce occurs and a possibly different direction becomes responsible for expansion.

The succession of Kasner epochs and Kasner cycles and the exponents of the model Kasner metric are determined by an associated discrete dynamical system, which can be best formulated in terms of the shift of the continued fraction expansion, see [Khalatnikov et al. (1985)], [Mayer (1987)].

5.1.1   *The shift of the continued fraction expansion*

Irrational numbers $x \in \mathbb{R}_+ \smallsetminus \mathbb{Q}$, have a continued fraction expansion

$$x = k_0 + \cfrac{1}{k_1 + \cfrac{1}{k_2 + \dots}} =: [k_0; k_1, k_2, \dots], \quad k_s \in \mathbb{Z}_+$$

with $k_0 := [x]$, $k_1 = [1/x]$, etc. One writes $x = [k_0; k_1, k_2, \dots]$ The one-sided shift of the continued fraction expansion is determined by the partially defined map

$$T : [0,1] \to [0,1] \qquad T : x \mapsto \frac{1}{x} - \left[\frac{1}{x} \right].$$

When applied to an irrational number in $[0,1]$ it acts on the digits of the continued fraction by

$$[0; k_1, k_2, k_3, \dots, k_n, \dots] \overset{T}{\mapsto} [0; k_2, k_3, \dots, k_{n+1}, \dots]$$

The double sides shift of the continued fraction expansion is similarly defined as

$$\widetilde{T} : [0,1]^2 \to [0,1]^2 \quad \widetilde{T} : (x,y) \mapsto \left(\frac{1}{x} - \left[\frac{1}{x} \right], \frac{1}{y + [1/x]} \right)$$

On pairs of points in $[0,1]^2 \cap (\mathbb{R}^2 \smallsetminus \mathbb{Q}^2)$ with continued fractions

$$x = [0; k_0, k_1, k_2, \dots], \qquad y = [0; k_{-1}, k_{-2}, \dots],$$

it acts as

$$\frac{1}{x} - \left[\frac{1}{x}\right] = [0; k_1, k_2, \dots], \qquad \frac{1}{y + [1/x]} = \frac{1}{k_0 + y} = [0; k_0, k_{-1}, k_{-2}, \dots].$$

Encoding a pair $(x, y) \in [0, 1]^2 \cap (\mathbb{R}^2 \setminus \mathbb{Q}^2)$ with a single doubly infinite sequence $(k) := [\dots k_{-2}, k_{-1}, k_0, k_1, k_2, \dots], k_i \in \mathbb{N}$, the invertible double-sided shift acts as $\widetilde{T}(k)_s = k_{s+1}$. There is a $\widetilde{T}$-invariant density on $[0, 1]^2$ given by

$$d\mu(x, y) = \frac{dx \, dy}{\log 2 \cdot (1 + xy)^2}$$

5.1.2 *Continued fractions and the mixmaster universe*

According to [Belinskii, Khalatnikov, Lifshitz (1970)], [Khalatnikov et al. (1985)], a typical solutions of Einstein equations Bianchi IX type with $SO(3)$–symmetry oscillates (near the initial singularity) close to a sequence of Kasner type solutions. The discretization of the dynamics is based on the following results:

- *Kasner epochs*: for logarithmic time $d\Omega := -\dfrac{dt}{abc} \to \infty$, there is an increasing sequence $\Omega_0 < \Omega_1 < \cdots < \Omega_n < \dots$, such that the interval $[\Omega_n, \Omega_{n+1})$ is n-th Kasner epoch.
- *Kasner exponents*: there is a sequence of $u_n \in (1, +\infty)$, $n \in \mathbb{N}$, such that, at the start Ω_n of the n-th Kasner epoch, the spacetime is modeled by a Kasner metric with exponents (5.1) with $u = u_n$.
- *Kasner cycles*: there are $k_n = [u_n]$ Kasner cycles within each Kasner epoch; within each Kasner cycle the spacetime is modeled by a Kasner metric with exponents (5.1) with u successively taking on the values $u_n - 1, u_n - 2, \dots, u_n - [u_n]$.
- *Epochs transition*: at the end of each Kasner epoch the parameter u undergoes a transformation to the next Kasner epoch of the form

$$u_{n+1} = \frac{1}{u_n - [u_n]}.$$

- *Permutation of spatial axes*: at the end of each Kasner epoch and of each Kasner cycle the three spatial directions are permuted, respectively according to the permutation (1)(23) and (12)(3).
- *Kasner times*: the sequence of Kasner times is determined by the recursion

$$\Omega_{n+1} = [1 + \delta_n k_n (u_n + 1/\{u_n\})]\Omega_n,$$

where $x_n = u_n - k_n$ and $\eta_n = (1 - \delta_n)/\delta_n$, with

$$\eta_{n+1}x_n = \frac{1}{k_n + \eta_n x_{n_1}}.$$

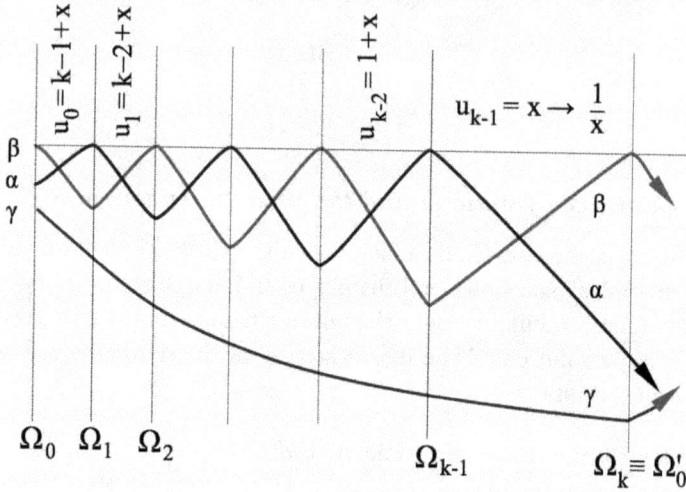

Fig. 5.1   The sequence of Kasner epochs and Kasner cycles in the mixmaster universe.

Figure 5.1 illustrates the mixmaster dynamics. This description of the discretized dynamics shows that trajectories of the mixmaster universe dynamics are parameterized by pairs $(x, y) \in [0,1]^2 \cap (\mathbb{R}^2 \smallsetminus \mathbb{Q}^2)$

$$x = [0; k_0, k_1, k_2, \ldots], \qquad y = [0; k_{-1}, k_{-2}, \ldots]$$

where the first coordinate x specifies the number of Kasner cycles in each Kasner epoch, through the successive digits of its continued fraction expansion, while the second digit y specifies the sequence of the Kasner logarithmic times. The transition from one Kasner epoch to the next is then given by the action of the double sided shift of the continued fraction

$$\tilde{T} : (x, y) \mapsto \left(\frac{1}{x} - \left[\frac{1}{x} \right], \frac{1}{y + [1/x]} \right).$$

We will discuss in Section 5.1.3 below, how the permutations of the spatial directions also fit in this coding of the dynamics.

5.1.3 *Continued fractions and modular curves*

On the upper half plane $\mathbb{H} = \{z = x + iy \in \mathbb{C} \,|\, y = \Im(z) > 0\}$ with the hyperbolic metric

$$ds^2 = \frac{dx^2 + dy^2}{y^2},$$

geodesics are vertical straight lines orthogonal to the x-axis and circular arcs (half-circles) orthogonal to the x-axis. The Lie group $\mathrm{PSL}_2(\mathbb{R})$ acts transitively and isometrically by fractional linear transformation

$$\gamma = \begin{pmatrix} a & b \\ c & d \end{pmatrix} : z \mapsto \frac{az + b}{cz + d}.$$

Let $\mathcal{M} = \mathbb{H}/\mathrm{PSL}_2(\mathbb{Z})$ be the quotient of $\mathbb{H}$ by the action of the discrete subgroup $\mathrm{PSL}_2(\mathbb{Z})$ of $\mathrm{PSL}_2(\mathbb{R})$. Similarly, we denote by $\mathcal{M}_\Gamma = \mathbb{H}/\Gamma$ the quotient by the action of some finite index subgroups $\Gamma \subset \mathrm{PSL}_2(\mathbb{Z})$. The complex curve (real surface) $\mathcal{M}$ is called the modular curve (modular surface). Similarly the quotients $\mathcal{M}_\Gamma$ are modular curves (of level Γ). The modular curves play a very important role because of their modular interpretation. Namely, let $E_\tau = \mathbb{C}/\Lambda$ be an elliptic curve obtained as the quotient of $\mathbb{C}$ by a rank 2 lattice $\Lambda = \mathbb{Z} + \tau\mathbb{Z}$ modulus $\tau \in \mathbb{H}$. Two elliptic curves are isomorphic, $E_\tau \simeq E_{\tau'}$ if and only if the moduli $\tau \sim \tau'$ are in the same orbit of the $\mathrm{PSL}_2(\mathbb{Z})$ action on $\mathbb{H}$. The fundamental domain of the $\mathrm{PSL}_2(\mathbb{Z})$ action on $\mathbb{H}$ is illustrated in Figure 5.2. Thus, the quotient

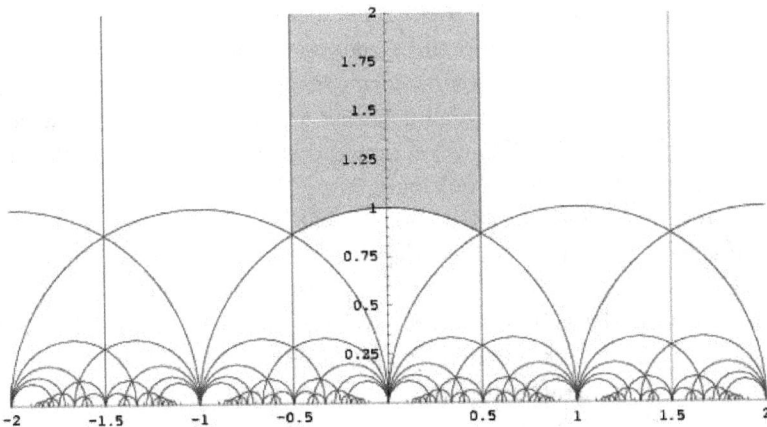

Fig. 5.2   The fundamental domain of the action of $\mathrm{PSL}_2(\mathbb{Z})$ on $\mathbb{H}$.

$\mathcal{M} = \mathbb{H}/\mathrm{PSL}_2(\mathbb{Z})$ parameterizes (isomorphism classes of) elliptic curves: it is the moduli space of elliptic curves. Similarly the other modular curves $\mathcal{M}_\Gamma$ are moduli spaces of elliptic curves enriched with certain kinds of level structures. The modular curves are compactified by adding cusps to the upper half plane, namely by replacing $\mathbb{H}$ with $\overline{\mathbb{H}} := \mathbb{H} \cup \{\mathbb{Q} \cup \{\infty\}\}$. Adding $\mathbb{P}^1(\mathbb{Q}) = \mathbb{Q} \cup \{\infty\}$ suffices for the compactification of modular curves, compatibly with their modular interpretation: the cusps in the moduli space correspond to degenerations of the elliptic curve E_τ to a cylinder given by the multiplicative group $\mathbb{G}_m(\mathbb{C}) = \mathbb{C}^*$. In [Manin, Marcolli (2002)] a different, non-algebro-geometric, compactification is obtained by adding to $\mathbb{H}$ the entire boundary $\mathbb{P}^1(\mathbb{R})$ and interpreting the resulting spaces $\mathbb{P}^1(\mathbb{R})/\Gamma$ as noncommutative spaces, and as moduli spaces of degenerations of elliptic curves to noncommutative tori.

The Farey tessellation of the hyperbolic plane is obtained by drawing all the vertical lines with $\Re(z) = n, n \in \mathbb{Z}$, and all the half circles connecting pairs of cusps $(p/q, p'/q')$ in $\mathbb{Q}$ with $pq' - p'q = \pm 1$. This subdivides $\mathbb{H}$ into a union of ideal triangles (with vertices at cusps): the resulting Farey tessellation is illustrated in Figure 5.3, represented in the Poincaré disk model of the hyperbolic plane. All the triangles in the tessellation are isometric, in the hyperbolic metric, to the ideal triangle Δ with vertices at $\{0, 1, \infty\}$ and sides the geodesics connecting them. The relation between the Farey tessellation and the fundamental domains of the $\mathrm{PSL}_2(\mathbb{Z})$ action can be seen as follows. The group S_6 acts by hyperbolic isometries of Δ, by permutating the vertices, with a unique fixed point in Δ given by $\rho = \exp(\pi i/3)$. The three geodesic arcs connecting ρ with the points $i, 1+i, \frac{1+i}{2}$, respectively, subdivide Δ into three geodesic quadrangles, each with one corner at a cusp. Each of these quadrangles is a copy of the fundamental domain for $\mathrm{PSL}_2(\mathbb{Z})$. The added geodesic arc form a tree in $\mathbb{H}$, which is the Cayley graph of the $\mathrm{PSL}_2(\mathbb{Z})$ action and is the dual graph of the Farey triangulation. This is shown in Figure 5.3.

The results of [Series (1985)] showed that, using the Farey tessellation, it is possible to obtain a coding of geodesics on the modular curve $\mathcal{M} = \mathbb{H}/\mathrm{PSL}_2(\mathbb{Z})$ in terms of the symbolic dynamics of the continued fraction expansion with the shift operator. We give a quick outline of how the coding is obtained. Let $\mathcal{B}$ denote the set of oriented geodesics β in $\mathbb{H}$ with endpoints that are irrational points $\beta_{-\infty}, \beta_\infty$ in the boundary $\mathbb{P}^1(\mathbb{R})$, with $\beta_{-\infty} \in (-1, 0)$, and $\beta_\infty \in (1, \infty)$. Expanding the endpoints according to their continued fraction expansion, we obtain

$$\beta_{-\infty} = -[0; k_0, k_{-1}, k_{-2}, \ldots], \qquad \beta_\infty = [k_1; k_2, k_3, \ldots].$$

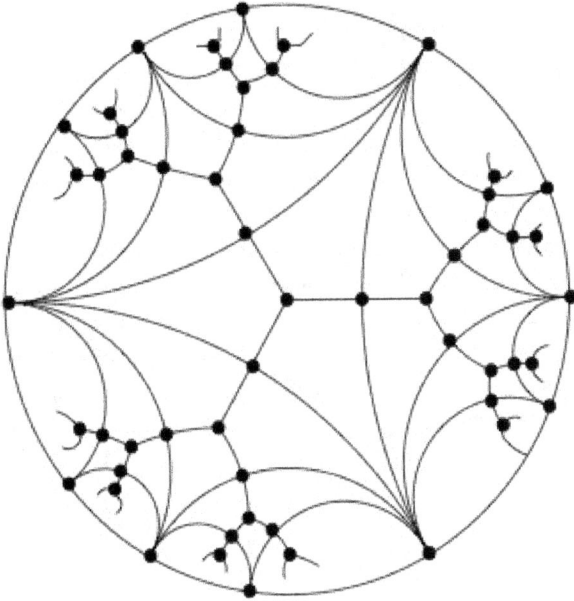

Fig. 5.3   Farey tessellation and fundamental domains of $\mathrm{PSL}_2(\mathbb{Z})$.

Since specifying the endpoints completely determines the geodesic, β is determined by the doubly infinite sequence of digits

$$[\ldots k_{-2}, k_{-1}, k_0, k_1, k_2, \ldots]$$

specifying the digits of the continued fraction expansion of the two endpoints. Consider then the intersection point $x = x(\beta)$ of β with the positive imaginary axis in $\mathbb{H}$. Starting at this point and moving along β (in either the positive or the negative direction) one crosses a sequence of triangles of the Farey tessellation, entering each triangle through one side and leaving through a different one. The ideal intersection point of the two sides remains either to the left or to the right of the oriented geodesic as it cuts across the triangle. One then obtains two sequences (for the two directions of motion) in the alphabet $\{L, R\}$,

$$\ldots L^{k_{-3}} R^{k_{-2}} L^{k_{-1}} R^{k_0} \qquad L^{k_1} R^{k_2} L^{k_3} R^{k_4} \ldots$$

It is shown in [Series (1985)] that the exponents of this sequence, which counts how many consecutive times the ideal vertex is to the left or to the right are indeed the same as the digits of the continued fraction expansion of the two endpoints. Acting on this sequence with the invertible double-sided shift of the continued fraction expansion corresponds to moving to another fundamental domain of the $\mathrm{PSL}_2(\mathbb{Z})$ action, see [Series (1985)], hence the resulting geodesics on the modular curve $\mathcal{M}$ correspond to the continued fraction expansions up to the action of the shift.

Equivalently, as discussed in [Manin, Marcolli (2015)], one can see the geodesic β transversing the triangles of the Farey tessellation as a billiard trajectory in a single triangle Δ, by folding the tessellation. A geodesic with irrational endpoints correspond to a billiard trajectory in Δ that never hits the corners. Thus, for example, a geodesic β with $x(\beta) = x_0$ in $(0, i\infty)$, for which the first digit in the continued fraction expansion of β_∞ is $k_0 = 1$, the billiard trajectory starting at x_0 reaches the opposite side $(1, i\infty)$ of Δ and then bounces off to the third side $(0, 1)$. Similarly, if $k_0 = 2$, after bouncing off the opposite side $(1, i\infty)$, the trajectory returns to the initial side $(0, i\infty)$, after which it gets reflected to $(0, 1)$. The billiard trajectory goes back and forth k_0 times between the sides $(0, i\infty)$ and $(1, i\infty)$, then is reflected to $(0, 1)$, from $(1, i\infty)$ if k_0 is odd, or from $(0, i\infty)$ if k_0 is even. By keeping track of these bounces in the billiard Δ and of the ideal vertices between the successive sides where the trajectory bounces, one can equivalently code the geodesic by a doubly infinite sequence in an alphabet of three letters $\{a, b, c\}$, labeling the three vertices $\{0, 1, i\infty\}$ at infinity of Δ. Note that the billiard Δ is not a flat surface, unlike other billiard models used in numerical studies of mixmaster dynamics: it is endowed with the constant negative curvature hyperbolic metric of $\mathbb{H}$.

A recent study of mixmaster dynamics in terms of billiards in the hyperbolic plane was done in [Lecian (2013)]. For numerical studies of mixmaster dynamics based on other types of (flat) billiard models, see for instance [Berger (2002)].

5.1.4   *Kasner times and geodesic lengths*

Using the same description of geodesics β in $\mathbb{H}$ as billiard trajectories in Δ, in addition to the discrete coding of the dynamics by the words $L^{k_1} R^{k_2} L^{k_3} \ldots$ and $\ldots R^{k-2} L^{k-1} R^{k_0}$, one can keep track of an additional (continuous) piece of information, namely the successive positions of the bouncing: the intersection points of β with sides of Farey triangles. This means considering a sequence of the following form, with $x_0 = x(\beta)$:

$$\ldots L^{k_{-1}} \, x_{-1} \, R^{k_0} \, x_0 \, L^{k_1} \, x_1 \, R^{k_2} \, x_2 \, L^{k_3} \, x_3 \, R^{k_4} \ldots$$

In [Manin, Marcolli (2015)] it is shown that the Kasner times are proportional to the geodesic distances between successive bounces, hence that one can identify the distance along geodesic with logarithmic cosmological time, and legitimately think of time as moving along the geodesic, see the previous discussion in §4.3, based on [Manin, Marcolli (2014)].

More precisely, the relation between Kasner times and geodesic lengths, when $n \to \infty$, $n \in \mathbb{N}$, is given by

$$\log \frac{\Omega_{2n}}{\Omega_0} \simeq 2 \sum_{r=0}^{n-1} \text{dist}\,(x_{2r}, x_{2r+1}), \tag{5.2}$$

where dist on the right-hand side denotes the hyperbolic distance between consecutive intersection points of the geodesic with sides of the Farey tesselation. Indeed it is known from the analysis of the mixmaster dynamics of [Khalatnikov et al. (1985)] that

$$\log \frac{\Omega_{2s}}{\Omega_0} \simeq - \sum_{p=1}^{2s} \log(x_p^+ x_p^-)$$

$$= \sum_{p=1}^{2s} \log([k_{p-1}, k_{p-2}, k_{p-3}, \ldots] \cdot [k_p, k_{p+1}, k_{p+2}, \ldots]),$$

where we write $x_n^+ = [0; k_n, k_{n+1}, k_{n+2}, \ldots]$ and $k_n^- = [0; k_{n-1}, k_{n-2}, \ldots]$, with $\delta_n = x_n^+/(x_n^+ + x_n^-)$ in the recursive formula for the Kasner times. On the other hand, from the coding of geodesics of [Series (1985)] it is also known that

$$\text{dist}(x_0, x_1) = \frac{1}{2} \log([k_0, k_{-1}, k_{-2}, \ldots] \cdot [k_1, k_2, \ldots] \cdot [k_1, k_0, k_{-1}, \ldots] \cdot [k_2, k_3, \ldots])$$

and, more generally, that

$$\text{dist}\,(x_{2r}, x_{2r+1}) = \frac{1}{2} \log([k_{2r}, k_{2r-1}, k_{2r-2}, \ldots] \cdot [k_{2r+1}, k_{2r+2}, \ldots] \cdot$$

$$\cdot [k_{2r+1}, k_{2r}, k_{2r-1}, \ldots] \cdot [k_{2r+2}, k_{2r+3}, \ldots]).$$

Combining these yields the estimate (5.2).

5.2 Modular curves, C^*-algebras, and mixmaster models

This section is based on [Marcolli (2007)] and on §1.2.2 of [Manin, Marcolli (2002)]. As discussed above, one can see trajectories of a discretized version of the mixmaster universe dynamics as describing geodesics in the modular curve $\mathbb{H}/\mathrm{PSL}_2(\mathbb{Z})$. In fact, as shown in §1.2.2 of [Manin, Marcolli (2002)], if one also takes into consideration, in the same picture, the alternation of the space axes responsible for expansion and oscillations at the end of each Kasner epoch and Kasner cycle, one obtains a geodesic on the modular curve $\mathbb{H}/\Gamma_0(2)$.

More precisely, we consider the following general setting. Let G be a finite index subgroup of a group Γ that will be either $\mathrm{PGL}_2(\mathbb{Z})$ or $\mathrm{PSL}_2(\mathbb{Z})$. Let X_G denote the quotient $X_G = G \backslash \mathbb{H}^2$, where $\mathbb{H}^2$ is, as above, the 2-dimensional real hyperbolic plane. Let $\mathbb{P}$ denote the coset space $\mathbb{P} = \Gamma/G$. We can write the quotient X_G equivalently as $X_G = \Gamma \backslash (\mathbb{H}^2 \times \mathbb{P})$. The quotient space X_G has the structure of a non-compact Riemann surface, which can be compactified by adding the set of cusp points at infinity,

$$\bar{X}_G = G \backslash (\mathbb{H}^2 \cup \mathbb{P}^1(\mathbb{Q})) \simeq \Gamma \backslash \left((\mathbb{H}^2 \cup \mathbb{P}^1(\mathbb{Q})) \times \mathbb{P} \right). \tag{5.3}$$

In particular, we consider the congruence subgroups $G = \Gamma_0(p)$, with p a prime, given by matrices

$$g = \begin{pmatrix} a & b \\ c & d \end{pmatrix}$$

satisfying $c \equiv 0 \mod p$, and in particular the case where $p = 2$. We consider also the noncommutative compactification, with boundary given by the quotient $\Gamma \backslash (\mathbb{P}^1(\mathbb{R}) \times \mathbb{P})$. Since this is a "bad quotient" when considered as an ordinary space (the action has dense orbits), it should be thought of as a noncommuative space, with the quotient replaces by a crossed product algebra. The action of Γ on $\mathbb{P}^1(\mathbb{R}) \times \mathbb{P}$ can be described, in a way similar to what we discussed already, in terms of a generalization of the Gauss shift of the continued fraction expansion, extended to act also on the discrete variable in the finite coset space $\mathbb{P}$. We have

$$T : [0,1] \times \mathbb{P} \to [0,1] \times \mathbb{P}$$

$$T(x,t) = \left(\frac{1}{x} - \left[\frac{1}{x} \right], \begin{pmatrix} -[1/x] & 1 \\ 1 & 0 \end{pmatrix} \cdot t \right). \tag{5.4}$$

The quotient space of $\mathbb{P}^1(\mathbb{R}) \times \mathbb{P}$ by the $\Gamma = \mathrm{PGL}_2(\mathbb{Z})$ action can be identified with the space of orbits of the dynamical system T on $[0,1] \times \mathbb{P}$.

Using, as above, the notation $x = [k_1, k_2, \ldots, k_n, \ldots]$ for the continued fraction expansion of the point $x \in [0,1]$, with convergents $p_n(x)/q_n(x)$, the Gauss shift T acts on the x variable in $[0,1]$ as the one-sided shift $T[k_1, k_2, \ldots, k_n, \ldots] = [k_2, k_3, \ldots, k_{n+1}, \ldots]$. The element $g_n(x)^{-1}$, with

$$g_n(x) = \begin{pmatrix} p_{k-1}(x) & p_k(x) \\ q_{k-1}(x) & q_k(x) \end{pmatrix} \in \Gamma,$$

acts on $[0,1] \times \mathbb{P}$ as T^n.

The Lyapunov exponent

$$\lambda(x) := \lim_{n \to \infty} \frac{1}{n} \log |(T^n)'(x)|$$

$$= 2 \lim_{n \to \infty} \frac{1}{n} \log q_n(x)$$

measures the exponential rate of divergence of nearby orbits, hence it provides a measure of how chaotic the dynamical system T is. The unit interval splits into a union of T-invariant level sets of the Lyapunov exponent λ, of varying Hausdorff dimension, plus an exceptional set where the limit defining λ does not converge. This Lyapunov spectrum was further studied in [Marcolli (2003)].

In the discretized version of the mixmaster universe dynamics discussed above, we have seen that, at the end of each Kasner cycle, a transformation $u \mapsto u - 1$ is accompanied by a permutation $(12)(3)$ of the space axes, while at the end of each Kasner epoch a transformation $u \to 1/u$ is accompanied by a permutation $(1)(23)$ of the three space axes.

It is shown in [Manin, Marcolli (2002)] that these two sets of transformations correspond, by setting $k_n = [u_n] = [1/x_n]$, to the action of the shift T as above, with $x_{n+1} = Tx_n \in [0,1]$ and with the transformation

$$\begin{pmatrix} -k_n & 1 \\ 1 & 0 \end{pmatrix}$$

acting on the discrete set $\mathbb{P}^1(\mathbb{F}_2) = \{0, 1, \infty\}$, identified with the finite coset space $\mathbb{P} = \Gamma/\Gamma_0(2)$. This acts as the permutation $0 \mapsto \infty$, $1 \mapsto 1$, $\infty \mapsto 0$, if k_n is even, and $0 \mapsto \infty$, $1 \mapsto 0$, $\infty \mapsto 1$ if k_n is odd. After an identification

$$\begin{aligned} 0 = [0:1] &\mapsto z \\ \infty = [1:0] &\mapsto y \\ 1 = [1:1] &\mapsto x. \end{aligned} \tag{5.5}$$

of the three points of the coset space with the three spatial axes, this action is indeed the same as the permutation $(1)(23)$ of the space axes (x, y, z), if k_n is even, or as the product of the permutations $(12)(3)$ and $(1)(23)$ if k_n is odd. This is precisely what is obtained in the mixmaster universe model by the repeated series of cycles within a Kasner era followed by the transition to the next era.

Consider then the modular curve $X_{\Gamma_0(2)} = \Gamma_0(2)\backslash\mathbb{H}$. Any geodesic on $X_{\Gamma_0(2)}$ not ending at cups can be coded in terms of data (ω^-, ω^+, s), where $(\omega^\pm, s)$ are the endpoints in $\mathbb{P}^1(\mathbb{R}) \times \{s\}$, $s \in \mathbb{P}$, with $\omega^- \in (-\infty, -1]$ and $\omega^+ \in [0, 1]$. In terms of the continued fraction expansion, we can write

$$\omega^+ = [k_0, k_1, \ldots k_r, k_{r+1}, \ldots]$$
$$\omega^- = [k_{-1}; k_{-2}, \ldots, k_{-n}, k_{-n-1}, \ldots].$$

The shift acts on these data by

$$T(\omega^+, s) = \left(\frac{1}{\omega^+} - \left[\frac{1}{\omega^+} \right], \begin{pmatrix} -[1/\omega^+] & 1 \\ 1 & 0 \end{pmatrix} \cdot s \right)$$

$$T(\omega^-, s) = \left(\frac{1}{\omega^- + [1/\omega^+]}, \begin{pmatrix} -[1/\omega^+] & 1 \\ 1 & 0 \end{pmatrix} \cdot s \right).$$

Thus, geodesics on $X_{\Gamma_0(2)}$ can be identified with orbits of T on the set of data (ω, s). This means that each infinite geodesic γ on the modular curve $X_{\Gamma_0(2)}$ not ending at cusps determines a mixmaster universe trajectory, and vice versa.

One can use this description to look for special classes of solutions of the mixmaster dynamics, associated to special classes of geodesics on the modular curve. For example, a closed geodesic would correspond to a periodic orbit, hence to a mixmaster dynamics that also exhibit periodicity. Another interesting class of solutions corresponds to geodesics on the modular curve $X_{\Gamma_0(2)}$ that wander only a finite distance into the cusps. In terms of the continued fraction expansion, these correspond to data (ω^+, s) with ω^+ in the set E_N given by all points of $[0, 1]$ with all digits in the continued fraction expansion bounded by N. These sets E_N are called the Hensley Cantor sets. Their properties were studied in [Hensley (1992)]. For the corresponding shift operator $T : E_N \times \mathbb{P} \to E_N \times \mathbb{P}$, there is an associated Perron-Frobenius operator of the form

$$(\mathcal{L}_{\beta,N} f)(x, s) = \sum_{k=1}^{N} \frac{1}{(x+k)^\beta} f\left(\frac{1}{x+k}, \begin{pmatrix} 0 & 1 \\ 1 & k \end{pmatrix} \cdot s \right). \qquad (5.6)$$

As shown in [Marcolli (2003)], this operator has a unique invariant measure μ_N, with density satisfying $\mathcal{L}_{2\dim_H(E_N),N} f = f$, where

$$\dim_H(E_N) = 1 - \frac{6}{\pi^2 N} - \frac{72 \log N}{\pi^4 N^2} + O(1/N^2)$$

is the Hausdorff dimension of the Hensley Cantor set E_N. The top eigen-value η_β of $\mathcal{L}_{\beta,N}$ is related to the Lyapunov exponent by

$$\lambda(x) = 2\frac{d}{d\beta}\eta_\beta|_{\beta=2\dim_H(E_N)},$$

for μ_N-almost all $x \in E_N$.

A consequence of this characterization of the time evolution in terms of the dynamical system (5.4) is that one can construct *moduli spaces* of mixmaster universe trajectories. For instance, in the case of the class of trajectories just described, the moduli space is the quotient of $E_N \times \mathbb{P}$ by the action of the shift T, since data (ω^+, s) that differ by the action of T determine the same solution, up to a shift in the choice of the initial time. This quotient is a "bad quotient" in the ordinary sense, but it is a "good quotient" and a "nice space" when regarded as a noncommutative space. To describe more explicitly this noncommutative space, we identify the dynamical system given by the Gauss shift (5.4) with the subshift of finite type described by the Markov partition

$$\mathcal{A}_N = \{((k,t),(\ell,s))|U_{k,t} \subset T(U_{\ell,s})\},$$

for $k, \ell \in \{1, \ldots, N\}$, and $s, t \in \mathbb{P}$, with sets $U_{k,t} = U_k \times \{t\}$, where $U_k \subset E_N$ are the clopen subsets where the local inverses of T are defined,

$$U_k = \left[\frac{1}{k+1}, \frac{1}{k}\right] \cap E_N.$$

This Markov partition determines a matrix A_N, with entries $(A_N)_{kt,\ell s} = 1$ if $U_{k,t} \subset T(U_{\ell,s})$ and zero otherwise. The 3×3 submatrices $A_{k\ell} = (A_{(k,t),(\ell,s)})_{s,t\in\mathbb{P}}$ of the matrix A_N are of the form

$$A_{k\ell} = \begin{cases} M_1 = \begin{pmatrix} 0 & 0 & 1 \\ 0 & 1 & 0 \\ 1 & 0 & 0 \end{pmatrix} & \ell = 2m \\[20pt] M_2 = \begin{pmatrix} 0 & 0 & 1 \\ 1 & 0 & 0 \\ 0 & 1 & 0 \end{pmatrix} & \ell = 2m+1 \end{cases}$$

Irreducibility of A_N means that the corresponding directed graph is strongly connected, namely any two vertices are connected by an oriented path of edges. Since the matrix A_N has the form

$$A_N = \begin{pmatrix} M_1 & M_1 & M_1 & \cdots & M_1 \\ M_2 & M_2 & M_2 & \cdots & M_2 \\ M_1 & M_1 & M_1 & \cdots & M_1 \\ \vdots & & & \cdots & \vdots \end{pmatrix},$$

irreducibility follows from the irreducibility of A_2, which corresponds to the directed graph illustrated in the figure.

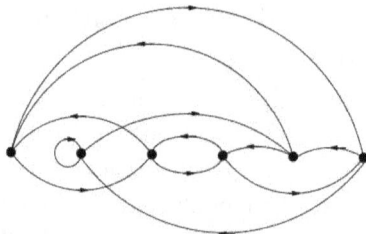

The non-commutative space determined by this Markov partition is the Cuntz–Krieger C^*-algebra $\mathcal{O}_{A_N}$, which is the universal C^*-algebra generated by partial isometries S_{kt} satisfying the relations

$$\sum_{(k,t)} S_{kt} S_{kt}^* = 1,$$

$$S_{\ell s}^* S_{\ell s} = \sum_{(k,t)} A_{(k,t),(\ell,s)} S_{kt} S_{kt}^*.$$

One can then reinterpret the Perron–Frobenius operator (5.6) in terms of the unique equilibrium KMS state on this C^*-algebra, with respect to a time evolution $\sigma_t^{u,h}(S_k) = \exp(it(u-h)) S_k$, for $u = \log \eta_\beta = P(\beta)$, the topological pressure (the top eigenvalue of (5.6)) and $h(x) = -\beta/2 \log |T'(x)|$. A state φ on a C^*-algebra satisfies the KMS condition at inverse temperature β if, for any a, b in a dense subalgebra of "analytic elements", the relation $\varphi(\sigma_t(a)\,b) = \varphi(b\,\sigma_{t+i\beta}(a))$ holds for all $t \in \mathbb{R}$. For the time evolution described above there is a unique KMS$_1$ state φ with

$$\sum_k \varphi(S_k^* e^h a\, S_k) = e^u\, \varphi(a),$$

for $a \in \mathcal{O}_{A_N}$. For $a = f \in C(E_N \times \mathbb{P})$ this satisfied $\sum_k S_k^* e^h f\, S_k = \mathcal{L}_h(f)$, where the Ruelle transfer operator

$$\mathcal{L}_h(f)(x,s) = \sum_{(y,r) \in T^{-1}(x,s)} \exp(h(y))\, f(y,r)$$

is the same as the Perron–Frobenius operator (5.6).

5.3 Noncommutative mixmaster cosmologies

This section is based on the noncommutative deformations of mixmaster cosmologies constructed in [Estrada, Marcolli (2013)].

In the noncommutative geometry models of matter coupled to gravity there is no noncommutativity in the spacetime coordinates, only in the extra dimensions. Moreover, these particle physics models live naturally at unification energy and there is, at this point, no clear picture of how the models should be extended to higher energies. It is expected, though, that closer to the Planck scale the model should incorporate more noncommutativity also involving spacetime itself. Thus, it is interesting to look for simple cosmological models where one can see how a noncommutative deformation in the spacetime directions can occur and how mechanisms can be introduced, related to inflation scenarios, that can cause the universe to exit an early noncommutative phase and transition to a modern universe with commutative spacetime coordinates. We discuss a simple model here, based on mixmaster universes with spatial sections that are tori, as in [Estrada, Marcolli (2013)], and we will discuss in §5.4 below a more elaborate model, based on the Bianchi IX gravitational instantons with $SU(2)$-symmetry, from [Manin, Marcolli (2015)].

5.3.1 *Mixmaster universes with torus sections*

Consider again the Kasner metrics (Lorentzian or Riemannian)

$$ds^2 = \mp dt^2 + t^{2p_1} dx^2 + t^{2p_2} dy^2 + t^{2p_3} dz^2, \qquad (5.7)$$

where the exponents p_1, p_2, p_3 satisfy the constraints $\sum_i p_i = 1$ and $\sum_i p_i^2 = 1$. As we have seen in the previous sections, mixmaster universe models are constructed by gluing together Kasner metrics according to a discrete dynamical system based on the Gauss shift of the continued fraction expansion, which determines the successing Kasner exponents p_1, p_2, p_3 of different cycles and epochs.

Observe that the metric (5.7) can be considered either on a spacetime whose spatial sections are topologically $\mathbb{R}^3$ or on spatial sections that are topologically tori T^3. We will focus on the latter possibility in order to obtain noncommutative deformations of these mixmaster cosmologies.

Let $T^3 = \mathbb{R}^3/\Lambda$ be a 3-dimensional torus, with Λ a lattice in $\mathbb{R}^3$. Let $\{\tau_1, \tau_2, \tau_3\}$ be a basis for Λ and let $\Lambda^\vee$ be the dual lattice with dual basis $\{w_1, w_2, w_3\}$. There are eight spin structures on T^3. They are classified by eight vectors $\{\mathfrak{s} = (\mathfrak{s}_1, \mathfrak{s}_2, \mathfrak{s}_3) \mid \mathfrak{s}_i \in \{0, 1\}\}$, where the value of each $\mathfrak{s}_i$ distinguishes whether the spin structure on each of the three circle directions is twisted or untwisted. On the circle S^1, the spinors for the two possible spin structures can be identified with

$$\Gamma(S^1, \mathbb{S}) = \{\psi : \mathbb{R} \to \mathbb{C} \mid \psi(t + 2\pi) = \pm\psi(t)\},$$

and the Dirac operator $-i\frac{d}{dt}$ has eigenfunctions $\psi_k(t) = \exp(2k\pi i t)$ for the trivial spin structure and $\psi_k(t) = \exp((2k + 1)\pi i t)$ for the other one. On $T^3 = \mathbb{R}^3/\Lambda$, one can then write the Dirac operator as

$$\partial\!\!\!/ = -i\sum_{j=1}^{3}(\tau_j \cdot \underline{\partial})\sigma_j = -i((\tau_1 \cdot \underline{\partial})\sigma_1 + (\tau_2 \cdot \underline{\partial})\sigma_2 + (\tau_3 \cdot \underline{\partial})\sigma_3) \qquad (5.8)$$

$$= \begin{pmatrix} -i\partial_{\tau_3} & -\partial_{\tau_2} - i\partial_{\tau_1} \\ \partial_{\tau_2} - i\partial_{\tau_1} & i\partial_{\tau_3} \end{pmatrix},$$

where $\underline{\partial} = (\partial_1, \partial_2, \partial_3) = (\frac{\partial}{\partial x}, \frac{\partial}{\partial y}, \frac{\partial}{\partial z})$, $\partial_{\tau_j} = \tau_j \cdot \underline{\partial}$, and σ_j are the Pauli matrices. The Dirac operator $\partial\!\!\!/$ for the spin structure $\mathfrak{s}$ has spectrum

$$\mathrm{Spec}(\partial\!\!\!/) = \{\pm 2\pi\|w + \frac{1}{2}\sum_{j=1}^{3}\mathfrak{s}_j w_j\| \mid w \in \Lambda^\vee\}. \qquad (5.9)$$

Consider the spatial section $T_t^3 = \mathbb{R}^3/\Lambda_t$ at time t with the Kasner metric (5.7). Since the Kasner metric (5.7) has $p_1 + p_2 + p_3 = 1$, the torus has volume $\mathrm{Vol}(T_t^3) = t^{p_1 + p_2 + p_3} \mathrm{Vol}(T^3) = t \mathrm{Vol}(T^3)$. The dual lattice $\Lambda_t^\vee$ is spanned by the basis of vectors $\{t^{p_1} e_1, t^{p_2} e_2, t^{p_3} e_3\}$, with e_i the standard orthonormal basis, and the Dirac operator is given by

$$\partial\!\!\!/_t = -i(\sigma_1 t^{-p_1}\frac{\partial}{\partial x} + \sigma_2 t^{-p_2}\frac{\partial}{\partial y} + \sigma_3 t^{-p_3}\frac{\partial}{\partial z}) \qquad (5.10)$$

$$= \begin{pmatrix} -it^{-p_3}\frac{\partial}{\partial z} & -t^{-p_2}\frac{\partial}{\partial y} - it^{-p_1}\frac{\partial}{\partial x} \\ t^{-p_2}\frac{\partial}{\partial y} - it^{-p_1}\frac{\partial}{\partial x} & it^{-p_3}\frac{\partial}{\partial z} \end{pmatrix},$$

with spectrum

$$\mathrm{Spec}(\partial\!\!\!/_t) = \{\pm 2\pi\|(t^{-p_1} k, t^{-p_2} m, t^{-p_3} n) + \frac{1}{2}\sum_{j=1}^{3}\mathfrak{s}_j t^{-p_j} e_j\| \mid (k, m, n) \in \mathbb{Z}^3\}.$$

$$(5.11)$$

The mixmaster tori $T_t^3 = \mathbb{R}^3/\Lambda_t$ described above are spectral triples with $\mathcal{A} = C^\infty(T^3)$, and with $\mathcal{H} = L^2(T_t^3, \mathbb{S}_\mathfrak{s})$, where $\mathbb{S}_\mathfrak{s}$ is the spinor bundle for the spin structure $\mathfrak{s} = (\mathfrak{s}_1, \mathfrak{s}_2, \mathfrak{s}_3)$ as above, with

$$L^2(T_t^3, \mathbb{S}_\mathfrak{s}) = \{\psi : \mathbb{R}^3 \to \mathbb{C}^2 \mid \psi \in L^2, \ \psi(\underline{x} + t^{p_j} v_j) = (-1)^{\mathfrak{s}_j}\psi(\underline{x})\}, \quad (5.12)$$

for $\underline{x} = (x, y, z) \in \mathbb{R}^3$ and $\{\tau_j\}$ the basis of Λ, and with Dirac operator

$$\not{\partial} = -i \sum_{j=1}^{3} t^{-p_j} (\tau_j \cdot \underline{\partial}) \sigma_j = -i(t^{-p_1}(\tau_1 \cdot \underline{\partial})\sigma_1 + t^{-p_2}(\tau_2 \cdot \underline{\partial})\sigma_2 + t^{-p_3}(\tau_3 \cdot \underline{\partial})\sigma_3).$$

$$(5.13)$$

Here, for convenience, we scaled the spatial coordinates $x_i \mapsto t^{p_i} x_i$, so that $\partial_{x_i} \mapsto t^{-p_i} \partial_{x_i}$, and fixed the τ_i, instead of scaling the lattice $\Lambda \mapsto \Lambda_t$ as before. The result is the same.

In a mixmaster dynamics model we have in each Kasner cycle and Kasner epoch a different Kasner metric (5.7), with a different choice of exponents $p_i = p_i(u_n)$, depending on the parameter u_n, which evolves according to the discrete dynamical system T. We denote by $T^3_{u_n}$ the associated family of tori T^3_t for the Kasner metric (5.7) with $p_i = p_i(u_n)$.

5.3.2 *Noncommutative θ-deformations*

We obtain a noncommutative deformation of the mixmaster model with torus sections by deforming all the tori to noncommutative tori.

A noncommutative 3-torus is the universal C^*-algebra $\mathcal{A}_\Theta$ generated by three unitaries U_1, U_2, U_3 with the relations

$$U_j U_k = \exp(2\pi i \Theta_{jk}) U_k U_j,$$

$$(5.14)$$

where $\Theta = (\Theta_{jk})$ is a skew-symmetric matrix

$$\Theta = \begin{pmatrix} 0 & \theta_3 & -\theta_2 \\ -\theta_3 & 0 & \theta_1 \\ \theta_2 & -\theta_1 & 0 \end{pmatrix}.$$

$$(5.15)$$

The algebra of smooth functions is given by

$$\mathcal{A}_\Theta^\infty = \{ X \in \mathcal{A}_\Theta, \ X = \sum_{m,n,k \in \mathbb{Z}} a_{m,n,k} U_1^m U_2^n U_3^k \ | \ a = (a_{m,n,k}) \in \mathcal{S}(\mathbb{Z}^3, \mathbb{C}) \}.$$

$$(5.16)$$

These should be thought of as functions with rapidly decaying Fourier coefficients. It is possible to endow these noncommutative tori with spectral triples, obtained from the spectral triples of the ordinary commutative 3-tori via the θ-deformation method of [Connes, Landi (2001)] that we reviewed briefly in the first chapter. Thus, we consider spectral triples of the form $(\mathcal{A}_\Theta^\infty, L^2(T^3, \mathbb{S}), \not{\partial})$, where $\mathbb{S}$ is the spinor bundle for one of the eight spin structures on the ordinary torus T^3 and the Dirac operator $\not{\partial}$ is of the form (5.8).

In the mixmaster dynamics, we now expect that the same discrete dynamical system that governs how the Kasner metric changes from one Kasner cycle to the next and one Kasner epoch to the next, via the action of the matrices $\gamma_n \in \mathrm{GL}_2(\mathbb{Z})$,

$$\gamma_n = \begin{pmatrix} -[u_n] & 1 \\ 1 & 0 \end{pmatrix} \in \mathrm{GL}_2(\mathbb{Z}), \qquad (5.17)$$

will also tranform the modular parameter of the noncommutative torus, which will then evolve dynamically along the mixmaster trajectory.

Within each Kasner epoch there is a spatial direction that is preferred, in the sense that it is the spatial axis that is driving the expansion of the universe, while the other two directions are oscillating. In the passage to the successive epoch the role of the three space axes is reshuffled with a permutation, as we discussed in the previous sections. The choice of one of the three spatial dimensions determines a choice of an embedding of $\mathrm{GL}_2(\mathbb{Z})$ inside $\mathrm{GL}_3(\mathbb{Z})$. Using this embedding, we can view the transformation (5.17) as acting also on the noncommutative torus moduli, via the action $\underline{\theta}' = M\underline{\theta}$ of $\mathrm{GL}_3(\mathbb{Z})$ on Θ identified with the vector $\underline{\theta} = (\theta_1, \theta_2, \theta_3) \in \mathbb{R}^3$. Thus, for example, if in the n-th Kasner epoch the first coordinate is the dominant direction, the noncommutative torus moduli in the Kasner cycles transform as

$$\underline{\theta} = (\theta_1, \theta_2, \theta_3) \mapsto \underline{\theta}' = (\theta_1, -[u_n]\theta_2 + \theta_3, \theta_2). \qquad (5.18)$$

It is known (see for instance the survey of 3-dimensional noncommutative tori in [Bédos (1999)]) that moduli satisfying $\underline{\theta}' = M\underline{\theta}$ for $M \in \mathrm{GL}_3(\mathbb{Z})$ correspond to isomorphic C^*-algebras $\mathcal{A}_\Theta \simeq \mathcal{A}_{\Theta'}$ and vice versa, while moduli satisfying $\underline{\theta}' = M\underline{\theta}$ for $M \in \mathrm{SL}_3(\mathbb{Z})$ correspond to isomorphic smooth subalgebras $\mathcal{A}_\Theta^\infty \simeq \mathcal{A}_{\Theta'}^\infty$, and vice versa. This means that the 3-dimensional noncommutative tori have an exotic smoothness phenomenon: if M is in $\mathrm{GL}_3(\mathbb{Z})$ but not in $\mathrm{SL}_3(\mathbb{Z})$, then the noncommutative tori are homeomorphic but not diffeomorphic. This exotic smoothness phenomenon happens in the mixmaster dynamics described above, whenever $\det(\gamma_n) = -1$.

5.3.3 *Spectral action and inflation scenario*

Apart from these changes in the smooth structure, the previous observation shows that, along the mixmaster dynamics the noncommutative tous remains the same (in the homeomorphism sense). In particular, a mixmaster dynamics in the early universe that would persist unaltered in the modern universe would imply that noncommutativity also persists in the modern universe. Thus, we need to introduce some other mechanism that can interfere with the mixmaster dynamics and damp it so that the universe can exit the noncommutativity of its oscillating early phase and reach a commutative modern universe. It is natural to try to construct such a model associated to an inflation scenario, especially since we already know that the spectral action functional of gravity produces a scalar field with a slow-roll inflation potential.

In a homogeneous isotropic Friedmann cosmology

$$ds^2 = -dt^2 + a(t)^2(dx^2 + dy^2 + dz^2),$$

the Friedmann equation relates the scale factor to the Hubble parameter

$$\frac{\dot{a}}{a} = H,$$

and the slow-roll models of inflation are based on the relation of the latter to a scalar field ϕ with potential $V(\phi)$ via

$$H^2 = \frac{8\pi G}{3}(\frac{1}{2}\dot{\phi}^2 + V(\phi)).$$

The slow-roll condition then corresponds to the condition that

$$\dot{\phi}^2 << |V(\phi)|,$$

so that the term in $\dot{\phi}$ in the Friedmann equation becomes negligible.

In the case of an anisotropic spacetime line the mixmaster models considered here, one introduces an average scale factor

$$\mathfrak{a}(t) = (a(t)b(t)c(t))^{1/3} \tag{5.19}$$

and directional Hubble parameters

$$H_1 = \frac{\dot{a}}{a}, \quad H_2 = \frac{\dot{b}}{b}, \quad H_3 = \frac{\dot{c}}{c},$$

and an average Hubble parameter

$$H = \frac{1}{3}(H_1 + H_2 + H_3).$$

This satisfies

$$H = \frac{1}{3}(\frac{\dot{a}}{a} + \frac{\dot{b}}{b} + \frac{\dot{c}}{c}) = \frac{1}{3}\frac{(\dot{a}bc + b\dot{a}c + \dot{c}ab)}{(abc)^{2/3}} \cdot \frac{1}{(abc)^{1/3}} = \frac{\dot{\mathfrak{a}}}{\mathfrak{a}}.$$

Thus, we obtain the same picture as in the isotropic case, but for the average Hubble parameter and the average scale factor.

For a Kasner metric (5.7), with $p_1 + p_2 + p_3 = 1$, the average scale factor is given by $\mathfrak{a} = (t^{p_1+p_2+p_3})^{1/3} = t^{1/3}$, with $\dot{\mathfrak{a}}/\mathfrak{a} = (1/3)t^{-1}$, while

$$\frac{\ddot{a}}{a} + \frac{\ddot{b}}{b} + \frac{\ddot{c}}{c} = 0,$$

since, for $a(t) = t^{p_1}$, one has $\ddot{a}/a = p_1(p_1 - 1)t^{-2}$, and similarly for the scale factors $b(t)$ and $c(t)$ so that

$$\frac{\ddot{a}}{a} + \frac{\ddot{b}}{b} + \frac{\ddot{c}}{c} = (p_1(p_1 - 1) + p_2(p_2 - 1) + p_3(p_3 - 1))t^{-2},$$

which vanishes, since $p_1 + p_2 + p_3 = 1$ and $p_1^2 + p_2^2 + p_3^2 = 1$.

Using the Poisson summation method for the computation of the spectral action on tori, as we discussed in the cosmic topology chapter, we find that for a torus T^3 with metric $a(t)^2 dx^2 + b(t)^2 dy^2 + c(t)^2 dz^2$ and Dirac operator

$$\not{D}_t = -i(\sigma_1 \frac{1}{a(t)} \frac{\partial}{\partial x} + \sigma_2 \frac{1}{b(t)} \frac{\partial}{\partial y} + \sigma_3 \frac{1}{c(t)} \frac{\partial}{\partial z}), \qquad (5.20)$$

the spectral action is of the form

$$\mathrm{Tr}(f(\not{D}_t^2/\Lambda^2)) = a(t)b(t)c(t)\frac{\Lambda^3}{4\pi^3} \int_{\mathbb{R}^3} f(u^2 + v^2 + w^2)\, du\, dv\, dw + O(\Lambda^{-k}),$$

$$(5.21)$$

for arbitrary $k > 0$. In particular, for a Kasner metric (5.7), with Dirac operator $\not{D}_t$ as in (5.10), the spectral action on the torus T_t^3 is given by

$$\mathrm{Tr}(f(\not{D}_t^2/\Lambda^2)) = t \cdot \frac{\Lambda^3}{4\pi^3} \int_{\mathbb{R}^3} f(u^2 + v^2 + w^2)\, du\, dv\, dw + O(\Lambda^{-k}). \quad (5.22)$$

As we have seen in the cosmic topology chapter, scalar fluctuations of the Dirac operator $D^2 \mapsto D^2 + \phi^2$ generate a slow-roll potential for $V(\phi)$. In the case of a Dirac operator (5.20) we obtain an expression of the form

$$\frac{\Lambda^4 \beta a(t)b(t)c(t)}{4\pi} \mathcal{V}(\phi^2/\Lambda^2)$$

where $\mathcal{V}(x) = \int_0^\infty u(h(u + x) - h(u))\, du$, with $x = \phi^2/\Lambda^2$, or equivalently

$$V(x) = \frac{\Lambda^4 \beta\, \mathfrak{a}(t)^3}{4\pi} \mathcal{V}(x), \qquad (5.23)$$

with $\mathfrak{a}(t)$ the average scale factor.

In the case of a homogeneous and isotropic expanding Friedmann cosmology with a single scale factor $a(t)$, the energy scale associated to the cosmological timeline behaves like $\Lambda(t) \sim 1/a(t)$. In this case, there is a density function ρ related to the scaling factor and the Hubble parameter by

$$\left(\frac{\dot{a}}{a}\right)^2 = H^2 = \frac{8\pi G}{3}\rho(t), \tag{5.24}$$

In a matter-dominated universe $\rho(t) \sim a(t)^{-3}$, in a radiation-dominated universe $\rho(t) \sim a(t)^{-4}$, while $\rho(t)$ is constant in the inflationary phase. In a slow-roll model of inflation, this constant is the plateau of the slow-roll potential.

In the setting considered here, one assumes that $\Lambda(t) \sim 1/\mathfrak{a}(t)$, based on the average scale factor, and that the parameter β, the radius of the Euclidean compactification, should be proportional to the density function $\beta \sim \rho$. The slow-roll parameters are then obtained as

$$\epsilon = \frac{M_{Pl}^2}{16\pi}\left(\frac{V'}{V}\right)^2, \quad \eta = \frac{M_{Pl}^2}{8\pi}\frac{V''}{V}.$$

The presence of a slow-roll potential disrupts the mixmaster dynamics. For an initial value near the plateau level of the slow-roll potential, the Friedmann equation determines an evolution of the average scale factor $\mathfrak{a}(t) = \mathfrak{a}(t_0)\exp(\gamma(t - t_0))$, with $\gamma = \sqrt{(8\pi G V_\infty)/3}$ and V_∞ the plateau value.

A similar scenario occurs in the context of loop quantum cosmology, where it was shown in [Sloan (2010)] that one finds an oscillatory behavior of mixmaster type as one approaches the singularity, as a simplified system from the equations of motion in the Hamiltonian formulation (see §2.3–2.8 of [Sloan (2010)]). Moreover, on the resulting reduced phase space one can also introduce a scalar field with an inflation slow-roll potential (see §2.7 and Chapter 4 of [Sloan (2010)]). The presence of the scalar field has a damping effect on the mixmaster oscillations in this model (see Appendix A of [Sloan (2010)]).

In the setting introduced here, one can also consider a one-parameter continuous family of transformations

$$\gamma(u) = \begin{pmatrix} -u & 1 \\ 1 & 0 \end{pmatrix} \in \mathrm{GL}_2(\mathbb{R}),$$

which deforms the noncommutative $\mathcal{A}_\Theta$, through a continuous family $\mathcal{A}_{\Theta(u)}$ that undo the θ-deformation and deform the noncommutative to ordinary tori, while at the same time preserving the Hubble parameter $H = \dot{\mathfrak{a}}/\mathfrak{a}$, since the scaling factors transform as

$$\frac{\dot{a}}{a} = \frac{-u}{1+u+u^2}t^{-1} + \frac{\dot{u}\left(u^2 - 1\right)}{(1+u+u^2)^2}\log t$$

$$\frac{\dot{b}}{b} = \frac{1+u}{1+u+u^2} t^{-1} - \frac{\dot{u}\,u\,(u+2)}{(1+u+u^2)^2} \log t$$

$$\frac{\dot{c}}{c} = \frac{u(1+u)}{1+u+u^2} t^{-1} + \frac{\dot{u}\,(2u+1)}{(1+u+u^2)^2} \log t.$$

Through a deformation of this sort one can disrupt the mixmaster dynamics, deform the noncommutative 3-torus $\mathcal{A}_\Theta$ of the early universe to a commutative torus, while maintaining the Friedmann equation unaffected.

5.4   Bianchi IX $SU(2)$-gravitational instantons

This section is based on [Manin, Marcolli (2015)].

We consider here Euclidean Bianchi IX spacetimes with $SU(2)$ symmetry, given by metrics on $\mathbb{R} \times S^3$ of the form

$$g = W_1 W_2 W_3 \, d\mu^2 + \frac{W_2 W_3}{W_1} \sigma_1^2 + \frac{W_1 W_3}{W_2} \sigma_2^2 + \frac{W_1 W_2}{W_3} \sigma_3^2,$$

or more generally

$$g = F\left(d\mu^2 + \frac{\sigma_1^2}{W_1^2} + \frac{\sigma_2^2}{W_2^2} + \frac{\sigma_3^2}{W_3^3}\right) \tag{5.25}$$

with a conformal factor $F \sim W_1 W_2 W_3$. Instead of the three spatial coordinate axes dx, dy, dz, we use here three $SU(2)$-invariant 1-forms $\{\sigma_i\}$ on S^3, satisfying the relations

$$d\sigma_i = \sigma_j \wedge \sigma_k$$

for all cyclic permutations (i, j, k) of $(1, 2, 3)$. More explicitly, one can write the σ_i as

$$\sigma_1 = x_1 \, dx_2 - x_2 \, dx_1 + x_3 \, dx_0 - x_0 \, dx_3 = \frac{1}{2}(d\psi + \cos\theta \, d\phi),$$

$$\sigma_2 = x_2 \, dx_3 - x_3 dx_2 + x_1 \, dx_0 - x_0 \, dx_1 = \frac{1}{2}(\sin\psi \, d\theta - \sin\theta \cos\psi \, d\phi),$$

$$\sigma_3 = x_3 \, dx_1 - x_1 \, dx_3 + x_2 \, dx_0 - x_0 \, dx_2 = \frac{1}{2}(-\cos\psi \, d\theta - \sin\theta \sin\psi \, d\phi),$$

with respect to Euclidean coordinates (x_0, x_1, x_2, x_3) from the standard embedding $S^3 \subset \mathbb{R}^4$, and the Euler angles of the spherical coordinates, $0 \le \theta \le \pi$, $0 \le \phi \le 2\pi$ and $0 \le \psi \le 4\pi$ (for the $SU(2)$ case). Upon identifying S^3 with the group $SU(2)$ of unit quaternions, one sees easily that the metrics

$$\frac{W_2 W_3}{W_1} \sigma_1^2 + \frac{W_1 W_3}{W_2} \sigma_2^2 + \frac{W_1 W_2}{W_3} \sigma_3^2$$

on S^3 are left-invariants under the action of $SU(2)$, but not right-invariant, unlike the standard round metric on S^3.

In particular, we focus on metrics of the form (5.25) on $\mathbb{R} \times S^3$ that are *gravitational instantons*, which means that they satisfy the following conditions:

(1) the metric g is self-dual, that is, the Weyl curvature tensor is self-dual;
(2) the metric g satisfies the Einstein equations (Einstein metrics).

Ordinarily, the vanishing of the anti-self-dual part of the Weyl curvature tensor is a partial differential equations. However, in the particular case of $SU(2)$-symmetric metrics of the form (5.25), they reduce to a system of ordinary differential equations, which in turn is expressible in terms of a Painlevé VI equation for a particular value of the parameters,

$$(\alpha, \beta, \gamma, \delta) = (\frac{1}{8}, -\frac{1}{8}, \frac{1}{8}, \frac{3}{8}),$$

see [Hitchin (1995)], [Okumura (1998)], [Tod (1994)].

5.4.1 *Painlevé VI equation*

The Painlevé transcendents are a famous family of solutions of nonlinear second-order ordinary differential equations in the plane, with the Painlevé property: the only movable singularities are poles. They are not expressible in terms of elementary functions. They are classified in six different types. For a general introduction to Painlevé equations see [Gromak, Laine, Shimomura (2002)].

The Painlevé VI equation is a 4-parameter family, depending on parameters $(\alpha, \beta, \gamma, \delta)$, of the form

$$\frac{d^2 X}{dt^2} = \frac{1}{2} \left(\frac{1}{X} + \frac{1}{X-1} + \frac{1}{X-t} \right) \left(\frac{dX}{dt} \right)^2$$

$$- \left(\frac{1}{t} + \frac{1}{t-1} + \frac{1}{X-t} \right) \frac{dX}{dt} +$$

$$+ \frac{X(X-1)(X-t)}{t^2(t-1)^2} \left(\alpha + \beta \frac{t}{X^2} + \gamma \frac{t-1}{(X-1)^2} + \delta \frac{t(t-1)}{(X-t)^2} \right).$$

It can be reformulated in a way that makes its relation to elliptic curves manifest, by writing it as

$$t(1-t)\left[t(1-t)\frac{d^2}{dt^2}+(1-2t)\frac{d}{dt}-\frac{1}{4}\right]\int_\infty^{(X,Y)}\frac{dx}{\sqrt{x(x-1)(x-t)}}$$

$$=\alpha Y+\beta\frac{tY}{X^2}+\gamma\frac{(t-1)Y}{(X-1)^2}+(\delta-\frac{1}{2})\frac{t(t-1)Y}{(X-t)^2},$$

where now the pair $(X,Y):=(X(t),Y(t))$ defines a (local and/or multi-valued) section $P:=(X(t),Y(t))$ of the generic elliptic curve $E=E(t):$ $Y^2=X(X-1)(X-t)$. The left-hand-side $\mu(P)$ of the equation satisfies $\mu(P+Q)=\mu(P)+\mu(Q)$ with respect to the addition $P+Q$ on the elliptic curve E.

It was shown in [Manin (1998)] that, if one replaces the algebraic description of the elliptic curve E as polynomial equation $Y^2=X(X-1)(X-t)$ with the analytic description as quotient of $\mathbb{C}$ by a lattice, $E_\tau=\mathbb{C}/\Lambda$, with $\Lambda=\mathbb{Z}+\tau\mathbb{Z}$, for some $\tau\in\mathbb{H}$, then the Painlevé equation can be written in the form

$$\frac{d^2z}{d\tau^2}=\frac{1}{(2\pi i)^2}\sum_{j=0}^3\alpha_j\wp_z(z+\frac{T_j}{2},\tau)$$

with $(\alpha_0,\ldots,\alpha_3):=(\alpha,-\beta,\gamma,\frac{1}{2}-\delta)$ and $(T_0,T_1,T_2,T_3):=(0,1,\tau,1+\tau)$. The $\wp$-function

$$\wp(z,\tau):=\frac{1}{z^2}+\sum_{(m,n)\neq(0,0)}\left(\frac{1}{(z-m\tau-n)^2}-\frac{1}{(m\tau+n)^2}\right)$$

satisfies

$$\wp_z(z,\tau)^2=4(\wp(z,\tau)-e_1(\tau))(\wp(z,\tau)-e_2(\tau))(\wp(z,\tau)-e_3(\tau)),$$

for $e_i(\tau)=\wp(\frac{T_i}{2},\tau)$, with $e_1+e_2+e_3=0$.

A multivalued solution $z=z(\tau)$ defines a multi-section of the family, which is a covering of $\mathbb{H}$. If one knows the data of the ramification and monodromy, one can explicitly study the behavior of solutions when restricted to geodesics in $\mathbb{H}$.

5.4.2 *Gravitational instantons and Painlevé*

An explicit parameterization the coefficients W_i for the $SU(2)$ Bianchi IX gravitational instantons, constructed using solutions of Painlevé VI, was given in [Babich, Korotkin (1998)] in terms of theta functions.

The *theta–characteristics* with parameters (p, q) is defined as

$$\vartheta[p, q](z, i\mu) := \sum_{m \in \mathbb{Z}} \exp\left(-\pi(m + p)^2 \mu + 2\pi i(m + p)(z + q)\right).$$

theta-characteristics and theta functions with vanishing characteristics are related by

$$\vartheta[p, q](z, i\mu) = \exp\left(-\pi p^2 \mu + 2\pi i p q\right) \cdot \vartheta[0, 0](z + p i\mu + q, i\mu).$$

With the notation $\vartheta[p, q] := \vartheta[p, q](0, i\mu)$, and

$$\vartheta_2 := \vartheta[1/2, 0], \qquad \vartheta_3 := \vartheta[0, 0], \qquad \vartheta_4 := \vartheta[0, 1/2],$$

the explicit solutions of [Babich, Korotkin (1998)], satisfying the self-dual condition, are of the form (5.25) with

$$W_1 = -\frac{i}{2}\vartheta_3\vartheta_4 \frac{\frac{\partial}{\partial q}\vartheta[p, q + \frac{1}{2}]}{e^{\pi i p}\vartheta[p, q]}, \qquad W_2 = \frac{i}{2}\vartheta_2\vartheta_4 \frac{\frac{\partial}{\partial q}\vartheta[p + \frac{1}{2}, q + \frac{1}{2}]}{e^{\pi i p}\vartheta[p, q]},$$

$$W_3 = -\frac{1}{2}\vartheta_2\vartheta_3 \frac{\frac{\partial}{\partial q}\vartheta[p + \frac{1}{2}, q]}{\vartheta[p, q]}.$$

In the case with non-zero cosmological constant Λ, the conformal factor is given by

$$F = \frac{2}{\pi\Lambda} \frac{W_1 W_2 W_3}{(\frac{\partial}{\partial q}\log\vartheta[p, q])^2}.$$

These metrics also satisfy the Einstein equations if either

(1) $\Lambda < 0$ with $p \in \mathbb{R}$ and $q \in \frac{1}{2} + i\mathbb{R}$
(2) $\Lambda > 0$ with $q \in \mathbb{R}$ and $p \in \frac{1}{2} + i\mathbb{R}$.

In the case with vanishing cosmological constant, one has the simpler form

$$W_1' = \frac{1}{\mu + q_0} + 2\frac{d}{d\mu}\log\vartheta_2, \quad W_2' = \frac{1}{\mu + q_0} + 2\frac{d}{d\mu}\log\vartheta_3,$$

$$W_3' = \frac{1}{\mu + q_0} + 2\frac{d}{d\mu}\log\vartheta_4, \quad F' := C(\mu + q_0)^2\, W_1' W_2' W_3'$$

with $q_0, C \in \mathbb{R}, C > 0$.

It was shown in [Manin, Marcolli (2015)] that asymptotically for $\mu \to \infty$ the gravitational instantons have the following behavior:

(1) $\Lambda < 0$: $W_2 \sim \pm W_3$ and

$$W_1 \sim \pi\langle p\rangle\, e^{\pi i\langle p\rangle - p}, \qquad W_3 \sim -2\pi i\, \langle p + \frac{1}{2}\rangle \cdot e^{\pi i\, \text{sgn}\,\langle p\rangle q} \cdot e^{\pi\mu(|\langle p\rangle| - \frac{1}{2})}$$

(2) $\Lambda > 0$ with $p = \frac{1}{2} + ip_0, p_0 \in \mathbb{R}$: $W_2 \sim -W_3$ and

$$-W_1 \sim \pi p_0 \tan(\pi(q - p_0\mu)) - \frac{1}{2}, \quad W_3 \sim 2\pi p_0 \cdot (\cos(\pi(q - p_0\mu)))^{-1}$$

(3) $\Lambda = 0$: $W_2' \sim W_3'$ and

$$W_1' \sim \frac{\pi}{2}, \quad W_3' \sim \frac{1}{\mu + q_0}$$

One can see this by extracting the dominant terms for the theta-functions expression of [Babich, Korotkin (1998)] recalled above. We refer the reader to [Manin, Marcolli (2015)] for a detailed computation of the asymptotics.

Notice the presence of singularities (poles) on the real time axis. These behave like the Taub-NUT infinity. Also notice that sign changes can be introduced, see [Babich, Korotkin (1998)], to reformulate all the asymptotics in the form $W_2 \sim W_3 \neq W_1$.

We can regard these solutions as gravitational instanton analogs of the Kasner solutions of the mixmaster dynamics, with $i\mu \in \Delta \subset \overline{\mathbb{H}}$ in the vicinity of $i\infty$, but not necessarily on the imaginary axis. The behavior $\mu \to \infty$ of these Bianchi IX cosmologies can then be though of as a possible model of (Wick rotated) time at the singularity in algebro-geometric gluing of spacetimes, as proposed in [Manin, Marcolli (2014)].

5.5   Noncommutativity in the early universe

We return now to discuss the noncommutativity hypothesis: that near the initial singularity spacetime coordinates can acquire noncommutativity as part of quantum gravity effects. As we have already discussed in the previous chapter, this means that there should be noncommutative deformations of spacetime that preserve the metric structure. As discussed in §1.1.6, metric-preserving noncommutative deformations can be obtained by encoding the metric structure in the Dirac operator, using the spectral triple formalism, and then deforming the underlying manifold to a noncommutative space using the method of isospectral deformations of [Connes, Landi (2001)]. For the 3-sphere S^3 with the round metric isospectral deformations can be obtained by deforming all the tori of the Hopf fibration into noncommutative tori. The left-$SU(2)$-invariant Bianchi IX metrics do not always admit such isospectral deformations, but they do admit them in the case that corresponds to the asymptotic behavior at $\mu \to \infty$. This fact can be interpreted as saying that these spacetimes admit noncommutativity near $\mu \to \infty$, as one would expect, but not everywhere on $\mathbb{R} \times S^3$.

More precisely, the Hopf coordinates on S^3 are given by (ξ_1, ξ_2, η), where

$$z_1 := x_1 + ix_2 = e^{i(\psi+\phi)} \cos \frac{\theta}{2} = e^{i\xi_1} \cos \eta,$$

$$z_2 := x_3 + ix_0 = e^{i(\psi-\phi)} \sin \frac{\theta}{2} = e^{i\xi_2} \sin \eta.$$

We identify S^3 with the unit quaternions $SU(2)$ by writing unit quaternions as

$$q := \begin{pmatrix} z_1 & z_2 \\ -\bar{z}_2 & \bar{z}_1 \end{pmatrix} = \begin{pmatrix} e^{i\xi_1} \cos \eta & e^{i\xi_2} \sin \eta \\ -e^{-i\xi_2} \sin \eta & e^{-i\xi_1} \cos \eta \end{pmatrix} \tag{5.26}$$

with $|z_1|^2 + |z_2|^2 = 1$ and (ξ_1, ξ_2, η) Hopf coordinates. The Hopf fibration is of the form

$$S^1 \hookrightarrow S^3 \to S^2,$$

where the subgroup S^1 in $SU(2)$ corresponds to the angle ξ_1 in the Hopf coordinates, see Figure 5.4.

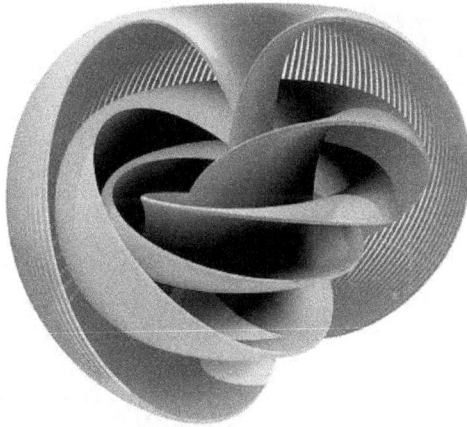

Fig. 5.4 The Hopf fibration of S^3 (image by Niles Johnson, 2012).

The noncommutative deformation of the 3-sphere S^3 obtained by deforming the tori of the Hopf fibration consists of replacing the functions $e^{i\xi_1}$ and $e^{i\xi_2}$ in (5.26) with

$$\begin{pmatrix} U \cos \eta & V \sin \eta \\ -V^* \sin \eta & U^* \cos \eta \end{pmatrix},$$

where U, V are the generators of the noncommutative 2-torus T_θ^2, satisfying

$$VU = e^{2\pi i\theta}UV.$$

This deformation no longer has the group structure of $SU(2)$. The C^*-algebra of the deformed 3-sphere can equivalently be described in terms of generators $\alpha = U\cos\eta$ and $\beta = V\sin\eta$ with the relations $\alpha\beta = e^{2\pi i\theta}\beta\alpha$, $\alpha^*\beta = e^{-2\pi i\theta}\beta\alpha^*$, $\alpha^*\alpha = \alpha\alpha^*$, $\beta^*\beta = \beta\beta^*$, and $\alpha\alpha^* + \beta\beta^* = 1$.

The Riemannian geometry of S^3 is described by the spectral triple $(C^\infty(S^3), L^2(S^3, \mathbb{S}), D\!\!\!\!/_{S^3,g})$, where $D\!\!\!\!/_{S^3,g}$ is the Dirac operator with respect to the metric $SU(2)$-Bianchi IX metric g. According to [Connes, Landi (2001)], a condition that ensures the existence of an isospectral deformation is an action of T^2 by isometries. So in the case of the deformation described above, associated to the Hopf fibration, the question is whether the T^2-action on S^3 that rotates the tori of the Hopf fibration,

$$(t_1, t_2) : (\xi_1, \xi_2) \mapsto (\xi_1 + t_1, \xi_2 + t_2)$$

is isometric. In terms of the Euler angles, this is equivalently expressed as $(u, v) : (\phi, \psi) \mapsto (\phi + u, \psi + v)$, with $t_1 = (u+v)/2$ and $t_2 = (v-u)/2$. The $U(1)$-action $u : \phi \mapsto \phi + u$ leaves the three 1-forms σ_i invariant. This is the action that rotates the circles $S^1 \hookrightarrow S^3$ of the Hopf fibration. The form σ_1 is also invariant under other $U(1)$-action, $v : \psi \mapsto \psi + v$. However, the forms σ_2, σ_3 transform as

$$v^*\sigma_2 = \frac{1}{2}\left(\sin(\psi + \beta)\, d\theta - \cos(\psi + \beta)\sin\theta\, d\phi\right)$$

$$v^*\sigma_3 = \frac{1}{2}\left(-\cos(\psi + \beta)\, d\theta - \sin(\psi + \beta)\sin\theta\, d\phi\right).$$

Thus, for $SU(2)$-Bianchi IX metric g of the form (5.25), the condition $v^*g = g$ is satisfied if and only if $W_2 = W_3$. This class of Bianchi IX metrics include the Taub-NUT and Eguchi-Hanson gravitational instantons, as well as the asymptotic form of the general Bianchi IX gravitational instantons, as discussed above. We refer the reader to [Taub (1951)], [Newman, Tamburino, Unti (1963)], [Eguchi, Hanson (1979)], [Eguchi, Hanson (1979b)]. An explicit Dirac operator on these metrics that admit noncommutative deformations is constructed using the Dirac operators on Berger spheres of [Hitchin (1974)],

$$D_B = -i\begin{pmatrix} \frac{1}{\lambda}X_1 & X_2 + iX_3 \\ X_2 - iX_3 & -\frac{1}{\lambda}X_1 \end{pmatrix} + \frac{\lambda^2 + 2}{2\lambda},$$

with $\{X_1, X_2, X_3\}$ basis of the Lie algebra, for a Berger sphere S^3 with metric $\lambda^2 \sigma_1^2 + \sigma_2^2 + \sigma_3^2$. The Dirac operator on the resulting gravitational instanton is then of the form

$$\mathcal{D} = \frac{1}{W_1^{1/2} W} \left(\gamma^0 \left(\frac{\partial}{\partial \mu} + \frac{1}{2} \left(\frac{\dot{W}}{W} + \frac{1}{2} \frac{\dot{W}_1}{W_1} \right) \right) + W_1 \, D_B |_{\lambda = \frac{W}{W_1}} \right)$$

with $W = W_2 = W_3$.

Chapter 6

The Spectral Action on Bianchi IX Cosmologies

This chapter is based on the results of [Fan, Fathizadeh, Marcolli (2015)] and [Fan, Fathizadeh, Marcolli (2015b)].

As in the previous chapter, we consider here (Euclidean) Bianchi IX mixmaster cosmologies with $SU(2)$ symmetry, which we write as

$$g = W_1 W_2 W_3 \, d\mu^2 + \frac{W_2 W_3}{W_1} \sigma_1^2 + \frac{W_1 W_3}{W_2} \sigma_2^2 + \frac{W_1 W_2}{W_3} \sigma_3^2$$

with time dependent scaling factors $W_i = W_i(\mu)$, or more generally

$$g = F \left(d\mu^2 + \frac{\sigma_1^2}{W_1^2} + \frac{\sigma_2^2}{W_2^2} + \frac{\sigma_3^2}{W_3^3} \right)$$

with also a conformal factor $F \sim W_1 W_2 W_3$. The σ_i are a basis of $SU(2)$–invariant 1-forms satisfying the relations

$$d\sigma_i = \sigma_j \wedge \sigma_k,$$

for all cyclic permutations (i, j, k) of $(1, 2, 3)$.

When we write more explicitly the basis of $SU(2)$–invariant 1-forms, the metric ds^2 is given by

$$W_1 W_2 W_3 \, dt \, dt + \frac{W_1 W_2 \cos(\eta)}{W_3} d\phi \, d\psi + \frac{W_1 W_2 \cos(\eta)}{W_3} d\psi \, d\phi$$
$$+ \left(\frac{W_2 W_3 \sin^2(\eta) \cos^2(\psi)}{W_1} + W_1 \left(\frac{W_3 \sin^2(\eta) \sin^2(\psi)}{W_2} + \frac{W_2 \cos^2(\eta)}{W_3} \right) \right) d\phi \, d\phi$$
$$+ \frac{(W_1^2 - W_2^2) W_3 \sin(\eta) \sin(\psi) \cos(\psi)}{W_1 W_2} d\eta \, d\phi + \frac{(W_1^2 - W_2^2) W_3 \sin(\eta) \sin(\psi) \cos(\psi)}{W_1 W_2} d\phi \, d\eta$$
$$+ \left(\frac{W_2 W_3 \sin^2(\psi)}{W_1} + \frac{W_1 W_3 \cos^2(\psi)}{W_2} \right) d\eta \, d\eta + \frac{W_1 W_2}{W_3} d\psi \, d\psi.$$

As we already discussed in the previous chapter, identifying S^3 with unit quaternions $SU(2)$, the metrics on S^3

$$\frac{W_2 W_3}{W_1} \sigma_1^2 + \frac{W_1 W_3}{W_2} \sigma_2^2 + \frac{W_1 W_2}{W_3} \sigma_3^2$$

are left-invariants under the action of $SU(2)$ but not right-invariant (unlike the round metric on S^3).

In this chapter we discuss the form of the spectral action functional for gravity on these Bianchi IX cosmologies. In particular, we will focus on the case of the Bianchi IX gravitational instantons.

6.1 Pseudodifferential calculus and parametrix method

The approach we follow here for the computation of the spectral action is based on *pseudodifferential calculus*, to obtain the *full* asymptotic expansion of the spectral action.

We work here in the case where the underlying space is an ordinary compact Riemannian spin manifold. Let D be a Dirac operator. The pseudodifferential symbol of D^2 is of the form

$$\sigma(D^2)(x, \xi) = p_2(x, \xi) + p_1(x, \xi) + p_0(x, \xi),$$

where each p_k is homogeneous of order k in ξ.

The Cauchy integral formula gives

$$e^{-tD^2} = \frac{1}{2\pi i} \int_\gamma e^{-t\lambda} (D^2 - \lambda)^{-1} d\lambda.$$

For a manifold M of dimension $m = \dim M$, the Seeley–deWitt coefficients $a_{2n}(D^2)$ of the heat kernel expansion are given by

$$\text{Tr}(e^{-tD^2}) \sim_{t \to 0+} t^{-m/2} \sum_{n=0}^{\infty} a_{2n}(D^2) t^n.$$

The *parametrix method* provides recursive equations for the computation of these coefficients. It is based on the fact that a second order elliptic differential operator like D^2 admits a parametrix R_λ with

$$\sigma(R_\lambda) \sim \sum_{j=0}^{\infty} r_j(x, \xi, \lambda),$$

where $r_j(x, \xi, \lambda)$ is a pseudodifferential symbol order $-2 - j$,

$$r_j(x, t\xi, t^2\lambda) = t^{-2-j} r_j(x, \xi, \lambda).$$

The parametrix R_λ approximates $(D^2 - \lambda)^{-1}$ with $\sigma((D^2 - \lambda)R_\lambda) \sim 1$. Recursive equations are then obtained from this relation in the form

$$\sigma((D^2 - \lambda)R_\lambda) \sim ((p_2(x, \xi) - \lambda) + p_1(x, \xi) + p_0(x, \xi)) \circ \left(\sum_{j=0}^{\infty} r_j(x, \xi, \lambda) \right) \sim 1.$$

The solution for R_λ is constructed recursively with

$$r_0(x, \xi, \lambda) = (p_2(x, \xi) - \lambda)^{-1}$$

$$r_n(x, \xi, \lambda) = -\sum \frac{1}{\alpha!} \partial_\xi^\alpha r_j(x, \xi, \lambda) \, D_x^\alpha p_k(x, \xi) \, r_0(x, \xi, \lambda),$$

where the summation is over all $\alpha \in \mathbb{Z}_{\geq 0}^4, j \in \{0, 1, \ldots, n-1\}, k \in \{0, 1, 2\}$, with $|\alpha| + j + 2 - k = n$.

The Seeley-deWitt coefficients are then obtained via the parametrix method as

$$a_{2n}(x, D^2) = \frac{(2\pi)^{-m}}{2\pi i} \int \int_\gamma e^{-\lambda} \operatorname{tr}\left(r_{2n}(x, \xi, \lambda)\right) \, d\lambda \, d^m\xi.$$

Note that the odd j coefficients vanish since $r_j(x, \xi, \lambda)$ is an odd function of ξ.

As we discussed in the first chapter, in the case of a manifold (or more generally of an almost commutative geometry) the asymptotic expansion of the spectral action can be computed in terms of the Seeley-deWitt coefficients of the heat kernel. Thus, the parametrix method that provides a computation of the Seeley-deWitt coefficients also gives the full expansion of the spectral action.

In order to apply this method to the Bianchi IX gravitational instantons, we first need to compute explicitly the Dirac operator. The general form of the Dirac operator was already mentioned at the end of the previous chapter, but we discuss it in more detail here. In terms of an orthonormal coframe $\{\theta^a\}$, the Dirac operator is given by

$$D = \sum_a \theta^a \nabla_{\theta_a}^S,$$

where the spin connection ∇^S is described in terms of a matrix of 1-forms $\omega = (\omega_b^a)$ with

$$\nabla \theta^a = \sum_b \omega_b^a \otimes \theta^b.$$

The compatibility with the metric and the torsion free conditions that characterize the Levi–Civita connection give

$$\omega_b^a = -\omega_a^b, \qquad d\theta^a = \sum_b \omega_b^a \wedge \theta^b.$$

The Dirac operator is then written as

$$D = \sum_{a,\mu} \gamma^a dx^\mu(\theta_a) \frac{\partial}{\partial x^\mu} + \frac{1}{4} \sum_{a,b,c} \gamma^c \omega_{ac}^b \gamma^a \gamma^b$$

with $\omega_a^b = \sum_c \omega_{ac}^b \theta^c$. The matrices γ^a here correspond to the Clifford action of θ^a on spin bundle and satisfy the relations $(\gamma^a)^2 = -id$ and $\gamma^a \gamma^b + \gamma^b \gamma^a = 0$, for $a \neq b$.

Consider local coordinates $x^\mu = (t, \eta, \phi, \psi)$ with the sphere S^3 parametrized by

$$(\eta, \phi, \psi) \mapsto \left(\cos(\eta/2) e^{i(\phi+\psi)/2}, \sin(\eta/2) e^{i(\phi-\psi)/2} \right)$$

with $0 \leq \eta \leq \pi, 0 \leq \phi < 2\pi, 0 \leq \psi < 4\pi$.

An orthonormal frame is given by

$$\theta^0 = \sqrt{W_1 W_2 W_3} \, dt,$$

$$\theta^1 = \sin(\eta) \cos(\psi) \sqrt{\frac{W_2 W_3}{W_1}} \, d\phi - \sin(\psi) \sqrt{\frac{W_2 W_3}{W_1}} \, d\eta,$$

$$\theta^2 = \sin(\eta) \sin(\psi) \sqrt{\frac{W_1 W_3}{W_2}} \, d\phi + \cos(\psi) \sqrt{\frac{W_1 W_3}{W_2}} \, d\eta,$$

$$\theta^3 = \cos(\eta) \sqrt{\frac{W_1 W_2}{W_3}} \, d\phi + \sqrt{\frac{W_1 W_2}{W_3}} \, d\psi.$$

The corresponding non-vanishing ω_{ac}^b

$$\omega_{11}^0 = -\frac{W_2 \left(W_1 W_3' - W_3 W_1' \right) + W_1 W_3 W_2'}{2(W_1 W_2 W_3)^{3/2}},$$

$$\omega_{22}^0 = -\frac{W_2 \left(W_3 W_1' + W_1 W_3' \right) - W_1 W_3 W_2'}{2(W_1 W_2 W_3)^{3/2}},$$

$$\omega_{33}^0 = -\frac{W_2 \left(W_3 W_1' - W_1 W_3' \right) + W_1 W_3 W_2'}{2(W_1 W_2 W_3)^{3/2}},$$

$$\omega_{23}^1 = -\frac{W_1^2 W_2^2 - W_3^2 \left(W_1^2 + W_2^2 \right)}{2(W_1 W_2 W_3)^{3/2}},$$

$$\omega_{32}^1 = -\frac{W_1^2 \left(W_2^2 - W_3^2 \right) + W_2^2 W_3^2}{2(W_1 W_2 W_3)^{3/2}},$$

$$\omega_{31}^2 = -\frac{W_2^2 W_3^2 - W_1^2 \left(W_2^2 + W_3^2 \right)}{2(W_1 W_2 W_3)^{3/2}}.$$

The pseudo-differential symbol of the Dirac operator is given by

$$\sigma(D)(x,\xi) = \sum_{a,\mu} i\gamma^a e_a^\mu \xi_{\mu+1} + \frac{1}{4\sqrt{W_1 W_2 W_3}} \left(\frac{W_1'}{W_1} + \frac{W_2'}{W_2} + \frac{W_3'}{W_3} \right) \gamma^1$$

$$- \frac{\sqrt{W_1 W_2 W_3}}{4} \left(\frac{1}{W_1^2} + \frac{1}{W_2^2} + \frac{1}{W_3^2} \right) \gamma^2 \gamma^3 \gamma^4$$

$$= - \frac{i\gamma^2 \sqrt{W_1} \left(\csc(\eta) \cos(\psi) \left(\xi_4 \cos(\eta) - \xi_3 \right) + \xi_2 \sin(\psi) \right)}{\sqrt{W_2} \sqrt{W_3}}$$

$$+ \frac{i\gamma^3 \sqrt{W_2} \left(\sin(\psi) \left(\xi_3 \csc(\eta) - \xi_4 \cot(\eta) \right) + \xi_2 \cos(\psi) \right)}{\sqrt{W_1} \sqrt{W_3}}$$

$$+ \frac{i\gamma^1 \xi_1}{\sqrt{W_1}\sqrt{W_2}\sqrt{W_3}} + \frac{i\gamma^4 \xi_4 \sqrt{W_3}}{\sqrt{W_1}\sqrt{W_2}}$$

$$+ \frac{1}{4\sqrt{W_1 W_2 W_3}} \left(\frac{W_1'}{W_1} + \frac{W_2'}{W_2} + \frac{W_3'}{W_3} \right) \gamma^1$$

$$- \frac{\sqrt{W_1 W_2 W_3}}{4} \left(\frac{1}{W_1^2} + \frac{1}{W_2^2} + \frac{1}{W_3^2} \right) \gamma^2 \gamma^3 \gamma^4$$

where the non-vanishing e_a^μ are

$$e_0^0 = \frac{1}{\sqrt{W_1 W_2 W_3}}, \qquad e_1^1 = -\frac{\sqrt{W_1} \sin(\psi)}{\sqrt{W_2 W_3}},$$

$$e_2^1 = \frac{\sqrt{W_2} \cos(\psi)}{\sqrt{W_1 W_3}}, \qquad e_1^2 = \frac{\sqrt{W_1} \csc(\eta) \cos(\psi)}{\sqrt{W_2 W_3}},$$

$$e_2^2 = \frac{\sqrt{W_2} \csc(\eta) \sin(\psi)}{\sqrt{W_1 W_3}}, \qquad e_1^3 = -\frac{\sqrt{W_1} \cot(\eta) \cos(\psi)}{\sqrt{W_2 W_3}},$$

$$e_2^3 = -\frac{\sqrt{W_2} \cot(\eta) \sin(\psi)}{\sqrt{W_1 W_3}}, \qquad e_3^3 = \frac{\sqrt{W_3}}{\sqrt{W_1 W_2}}.$$

From this expression of the symbol, one obtains the homogeneous components $p_k(x,\xi)$ with

$$\sigma(D^2)(x,\xi) = p_2(x,\xi) + p_1(x,\xi) + p_0(x,\xi).$$

For instance, the expression for $p_0(x, \xi)$ is given by

$$
\left(-\frac{W_1'}{8W_1W_2^2} - \frac{W_1'}{8W_1W_3^2} + \frac{3W_1'}{8W_1^3} - \frac{W_2'}{8W_1^2W_2} - \frac{W_3'}{8W_1^2W_3} - \frac{W_2'}{8W_2W_3^2}\right.
$$

$$
\left.+\frac{3W_2'}{8W_2^3} - \frac{W_3'}{8W_2^2W_3} + \frac{3W_3'}{8W_3^3}\right)\gamma^1\gamma^2\gamma^3\gamma^4+
$$

$$
\left(-\frac{W_1''}{4W_1^2W_2W_3} + \frac{W_1'W_2'}{8W_1^2W_2^2W_3} + \frac{W_1'W_3'}{8W_1^2W_2W_3^2} + \frac{5W_1'^2}{16W_1^3W_2W_3} - \frac{W_2''}{4W_1W_2^2W_3}\right.
$$

$$
+\frac{W_2'W_3'}{8W_1W_2^2W_3^2} + \frac{5W_2'^2}{16W_1W_2^3W_3} - \frac{W_3''}{4W_1W_2W_3^2} + \frac{5W_3'^2}{16W_1W_2W_3^3} + \frac{W_2W_3}{16W_1^3}
$$

$$
\left.+\frac{W_3}{8W_1W_2} + \frac{W_1W_3}{16W_2^3} + \frac{W_2}{8W_1W_3} + \frac{W_1}{8W_2W_3} + \frac{W_1W_2}{16W_3^3}\right)I.
$$

There is a similarly manageable expression for $p_2(x, \xi)$ and a longer one for $p_1(x, \xi)$, for which we refer the reader to [Fan, Fathizadeh, Marcolli (2015)].

Applying the parametrix method to this Dirac operator gives

$$
a_{2n}(x, D^2) = \frac{(2\pi)^{-m}}{2\pi i} \int \int_\gamma e^{-\lambda}\, \mathrm{tr}\,(r_{2n}(x, \xi, \lambda))\, d\lambda\, d^m\xi
$$

This is used to obtain explicit expressions for the coefficient a_0, a_2, a_4. The case of a_0 and a_2 give the relatively simple expressions

$$
a_0(D^2) = 4W_1W_2W_3
$$

$$
a_2(D^2) = -\frac{W_1^2}{3} - \frac{W_2^2}{3} - \frac{W_3^2}{3} + \frac{W_1^2W_2^2}{6W_3^2} + \frac{W_1^2W_3^2}{6W_2^2} + \frac{W_2^2W_3^2}{6W_1^2} - \frac{(W_1')^2}{6W_1^2} - \frac{(W_2')^2}{6W_2^2}
$$

$$
-\frac{(W_3')^2}{6W_3^2} - \frac{W_1'W_2'}{3W_1W_2} - \frac{W_1'W_3'}{3W_1W_3} - \frac{W_2'W_3'}{3W_2W_3} + \frac{W_1''}{3W_1} + \frac{W_2''}{3W_2} + \frac{W_3''}{3W_3}.
$$

while one obtains a much longer and more complicated expression for $a_4(D^2)$, for which again we refer the reader to [Fan, Fathizadeh, Marcolli (2015)].

6.2 Wodzicki residues method

One can observe from the computations carried out with this method that all the coefficients in these expressions for a_0, a_2, and a_4 are *rational numbers*. This appears to indicate the existence of an interesting arithmetic structure in the spectral action of the Bianchi IX cosmologies. Indeed, as shown in [Fan, Fathizadeh, Marcolli (2015)] and [Fan, Fathizadeh, Marcolli (2015b)] there is indeed such a structure, with an interesting relation to modular forms. In order to better understand these properties, it is necessary to obtain a method for the computation of the Seeley-deWitt coefficients that is simpler and computationally more efficient than the parametrix method discussed above.

Such a method, based on Wodzicki residues on products with auxiliary tori, was introduced by Fathizadeh. We summarize it here.

The Wodzicki residue is the unique trace functional on the algebra of pseudodifferential operators on smooth sections of vector bundle over smooth manifold. For a classical pseudodifferential operator P_σ of order $d \in \mathbb{Z}$, with local symbol

$$\sigma(x,\xi) \sim \sum_{j=0}^{\infty} \sigma_{d-j}(x,\xi) \qquad (\xi \to \infty),$$

where σ_{d-j} is positively homogeneous of order $d - j$ in ξ, the residue is given by

$$\mathrm{Res}(P_\sigma) = \int_{S^*M} \mathrm{tr}\left(\sigma_{-m}(x,\xi)\right) d^{m-1}\xi \, d^m x,$$

where $S^*M = \{(x,\xi) \in T^*M; \|\xi\|_g = 1\}$ is the cosphere bundle. The *spectral formulation* of the residue is obtained as follows. For a pseudodifferential operator P_σ, with Δ the Laplacian, one has

$$P_\sigma \mapsto \mathrm{Res}_{s=0}\mathrm{Tr}(P_\sigma \Delta^{-s}).$$

This is the same as the above, up to a constant $c_m = 2^{m+1}\pi^m$.

Assuming for simplicity that $\mathrm{Ker}(\Delta) = 0$, the Mellin transform gives

$$\mathrm{Tr}(\Delta^{-s}) = \frac{1}{\Gamma(s)} \int_0^\infty \mathrm{Tr}(e^{-t\Delta}) t^s \frac{dt}{t},$$

and the heat kernel expansion is of the form

$$\mathrm{Tr}\left(e^{-t\Delta}\right) = t^{-m/2} \sum_{n=0}^{N} a_{2n}t^n + O(t^{-m/2+N+1}).$$

For any non-negative integer $n \leq m/2 - 1$, one then has

$$\mathrm{Res}_{s=m/2-n}\mathrm{Tr}(\Delta^{-s}) = \frac{a_{2n}(\Delta)}{\Gamma(m/2 - n)}.$$

In particular

$$\mathrm{Res}_{s=1}\mathrm{Tr}(\Delta^{-s}) = a_{m-2}(\Delta).$$

In terms of Wodzicki residues this means that

$$a_{m-2}(\Delta) = \frac{1}{c_m}\mathrm{Res}(\Delta^{-1}) = \frac{1}{2^{m+1}\pi^m}\mathrm{Res}(\Delta^{-1}).$$

When applied to $\Delta = D^2$, this gives the coefficient $a_2(D^2)$ as a residue

$$a_2(D^2) = \frac{1}{c_4}\mathrm{Res}(D^{-2}) = \frac{1}{32\pi^4}\int_{S^*M} \mathrm{tr}\left(\sigma_{-4}(D^{-2})\right) d^3\xi\, d^4x.$$

In order to obtain the other coefficients, one can introduce an *auxiliary product space* that adjust the counting of dimensions, without affecting the form of the coefficients. This can be achieved by using a flat r-dimensional torus $T^r = (\mathbb{R}/\mathbb{Z})^r$. Then one has

$$\Delta = D^2 \otimes 1 + 1 \otimes \Delta_{T^r},$$

where Δ_{T^r} is the flat Laplacian on $\mathbb{T}^r$, and

$$a_{2+r}(D^2) = \frac{1}{2^5\,\pi^{4+r/2}}\mathrm{Res}(\Delta^{-1}),$$

because the Künneth formula gives

$$a_{2+r}((x,x'),\Delta) = a_{2+r}(x,D^2)a_0(x',\mathbb{T}^r) = 2^{-r}\pi^{-r/2}a_{2+r}(x,D^2),$$

where the volume term is the only non-zero heat coefficient for the flat metric. This leads to residue formulae for all the heat kernel coefficients,

$$a_{2+r}(D^2) = \frac{1}{2^5\pi^{4+r/2}}\int \mathrm{tr}\left(\sigma_{-4-r}(\Delta^{-1})\right) d^{3+r}\xi\, d^4x.$$

6.3 Rationality result

Using this method for the computation of the Seeley-deWitt coefficients, it is shown in [Fan, Fathizadeh, Marcolli (2015)] that a rationality result indeed holds for the expansion of the spectral action. Namely, one writes $\sigma(\Delta^{-1}) \sim \sum_{j=-2}^{-\infty} \sigma_j(x,\xi)$ inductively as

$$\sigma_{-2}(x,\xi) = p_2'(x,\xi)^{-1},$$

$$\sigma_{-2-n}(x,\xi) = -\sum \frac{1}{\alpha!}\partial_\xi^\alpha \sigma_j(x,\xi)\, D_x^\alpha p_k(x,\xi)\,\sigma_{-2}(x,\xi) \qquad (n > 0),$$

with the summation over all multi-indices given by non-negative integers α with $-2 - n < j \leq -2, 0 \leq k \leq 2$, and with $|\alpha| - j - k = n$. One also sets $\zeta_{\mu+1} = \sum_\nu e_\mu^\nu \xi_{\nu+1}$. Then one finds inductively, for $n \geq 2$,

$$\sigma_{-2-n}(x,\xi)|_{S^*(M\times\mathbb{T}^{n-2})} = \sigma_{-2-n}(x,\xi(\zeta))|_{\zeta\in S^{n+1}} = (W_1 W_2 W_3)^{-\frac{3}{2}n}P_n(\zeta),$$

where the $P_n(\zeta)$ are polynomials where the coefficients are functions of the W_i and derivatives. These explicitly give recursive expressions for the Seeley-deWitt coefficients of the form

$$a_{2n}(D^2) = \frac{Q_{2n}\left(W_1, W_2, W_3, W_1', W_2', W_3', \ldots, W_1^{(2n)}, W_2^{(2n)}, W_3^{(2n)}\right)}{(W_1 W_2 W_3)^{3n-1}},$$

where the Q_{2n} are polynomials with *rational coefficients*,

$$Q_{2n} = \frac{1}{2\pi^{n+1}} \int_{\mathbb{S}^{2n+1}} \text{tr}(P_{2n}(\zeta)(\Delta^{-1})) \, d^{2n+1}\zeta.$$

A similar rationality result for the spectral action of a Robertson–Walker metric was conjectured in [Chamseddine, Connes (2012b)] and proved in [Fathizadeh, Ghorbanpour, Khalkhali (2014)].

In the case of the Bianchi IX cosmologies, it is natural to ask whether this rationality result may be a sign of an arithmetic structure that persists in the spectral action. Indeed, as shown in [Fan, Fathizadeh, Marcolli (2015b)], one finds that in the case of the Bianchi IX gravitational instantons, the expansion of the spectral action can be expressed in terms of modular forms. This result depends on the parameterization of Bianchi IX gravitational instantons in terms of solutions to the Painlevé VI equations that we already discussed in the previous chapter.

6.4   Gravitational instantons and the spectral action

We consider again the resulting parameterization of the Bianchi IX gravitational instantons in terms of theta characteristics, as in [Babich, Korotkin (1998)] and [Manin, Marcolli (2015)].

The previously described computations of the full expansion of the spectral action from [Fan, Fathizadeh, Marcolli (2015)] are based on the original triaxial Bianchi IX metric

$$ds^2 = W_1 W_2 W_3 \, dt^2 + \frac{W_2 W_3}{W_1}\sigma_1^2 + \frac{W_3 W_1}{W_2}\sigma_2^2 + \frac{W_1 W_2}{W_3}\sigma_3^2.$$

When we also have a conformal factor F, as in the case of the gravitational instantons,

$$d\tilde{s}^2 = F ds^2 = F \left(W_1 W_2 W_3 \, dt^2 + \frac{W_2 W_3}{W_1}\sigma_1^2 + \frac{W_3 W_1}{W_2}\sigma_2^2 + \frac{W_1 W_2}{W_3}\sigma_3^2 \right)$$

with $F = F(t)$, this modifies the Dirac operator by

$$\tilde{D} = \frac{1}{\sqrt{F}}D + \frac{3F'}{4F^{\frac{3}{2}}W_1 W_2 W_3}\gamma^0$$

where D is the Dirac operator of original triaxial Bianchi IX metric.

The heat kernel expansion

$$\text{Tr}\left(\exp(-t\tilde{D}^2)\right) \sim t^{-2} \sum_{n=0}^{\infty} \tilde{a}_{2n} t^n, \qquad t \to 0^+$$

also gives a similar rationality result for the coefficients of the spectral action, with

$$\tilde{a}_{2n} = \frac{\tilde{Q}_{2n}\left(W_1, W_2, W_3, F, W_1', W_2', W_3', F', \ldots, W_1^{(2n)}, W_2^{(2n)}, W_3^{(2n)}, F^{(2n)}\right)}{F^{2n}(W_1 W_2 W_3)^{3n-1}}$$

where the $\tilde{Q}_{2n}$ are polynomials with rational coefficients.

For instance, this gives as zeroth order coefficient (cosmological term)

$$\tilde{a}_0 = 4F^2 W_1 W_2 W_3$$

and for the second coefficient $\tilde{a}_2$ (Einstein-Hilbert action)

$$-\frac{F}{3}\left(W_1^2 + W_2^2 + W_3^2\right) + \frac{F}{6}\left(\frac{W_1^2 W_2^2 - W_3'^2}{W_3^2} + \frac{W_1^2 W_3^2 - W_2'^2}{W_2^2} + \frac{W_2^2 W_3^2 - W_1'^2}{W_1^2}\right)$$

$$-\frac{F}{3}\left(\frac{W_1' W_2'}{W_1 W_2} + \frac{W_1' W_3'}{W_1 W_3} + \frac{W_2' W_3'}{W_2 W_3}\right) + \frac{F}{3}\left(\frac{W_1''}{W_1} + \frac{W_2''}{W_2} + \frac{W_3''}{W_3}\right) - \frac{F'^2}{2F} + F''$$

An explicit but much longer and more complicated explicit formula for the $\tilde{a}_4$ term (Weyl conformal gravity and Gauss-Bonnet gravity) is given in [Fan, Fathizadeh, Marcolli (2015b)].

Now assuming that the conformally perturbed Bianchi IX is a self-dual Einstein metric (gravitational instanton) and using the explicit parameterization by theta characteristics recalled above, one obtains a two-parameter family, for the case with non-vanishing cosmological constant, of the form

$$W_1[p,q](i\mu) = -\frac{i}{2}\vartheta_3(i\mu)\vartheta_4(i\mu)\frac{\partial_q \vartheta[p, q + \frac{1}{2}](i\mu)}{e^{\pi i p}\vartheta[p,q](i\mu)}$$

$$W_2[p,q](i\mu) = \frac{i}{2}\vartheta_2(i\mu)\vartheta_4(i\mu)\frac{\partial_q \vartheta[p + \frac{1}{2}, q + \frac{1}{2}](i\mu)}{e^{\pi i p}\vartheta[p,q](i\mu)}$$

$$W_3[p,q](i\mu) = -\frac{1}{2}\vartheta_2(i\mu)\vartheta_3(i\mu)\frac{\partial_q \vartheta[p + \frac{1}{2}, q](i\mu)}{\vartheta[p,q](i\mu)}$$

$$F[p,q](i\mu) = \frac{2}{\pi\Lambda}\frac{1}{(\partial_q \ln \vartheta[p,q](i\mu))^2} = \frac{2}{\pi\Lambda}\left(\frac{\vartheta[p,q](i\mu)}{\partial_q \vartheta[p,q](i\mu)}\right)^2$$

and a one-parameter family, for the case with vanishing cosmological constant, of the form

$$W_1[q_0](i\mu) = \frac{1}{\mu + q_0} + 2\frac{d}{d\mu}\log\vartheta_2(i\mu),$$

$$W_2[q_0](i\mu) = \frac{1}{\mu + q_0} + 2\frac{d}{d\mu}\log\vartheta_3(i\mu),$$

$$W_3[q_0](i\mu) = \frac{1}{\mu + q_0} + 2\frac{d}{d\mu}\log\vartheta_4(i\mu),$$

$$F[q_0](i\mu) = C(\mu + q_0)^2,$$

where C is an arbitrary positive constant.

6.5   The spectral action and modular forms

Assuming that the cosmological time variable (here denoted μ) belongs to the right half-plane $\Re(\mu) > 0$, and using behavior of theta functions and derivatives under modular transformations, it is then shown in [Fan, Fathizadeh, Marcolli (2015b)] that, for the two parameter family of solutions, under the action of the two generators

$$T(\tau) = \tau + 1, \qquad S(\tau) = \frac{-1}{\tau}, \qquad \tau \in \mathbb{H}$$

of the modular group $\mathrm{PSL}_2(\mathbb{Z})$, the functions $W_j[p,q]$ and their derivatives of an arbitrary order $n \geq 1$ with respect to the cosmological time variable μ satisfy the identities

$$W_1[p,q](i\mu + 1) = W_1[p, q + p + \tfrac{1}{2}](i\mu),$$

$$W_1^{(n)}[p,q](i\mu + 1) = W_1^{(n)}[p, q + p + \tfrac{1}{2}](i\mu),$$

$$W_2[p,q](i\mu + 1) = W_3[p, q + p + \tfrac{1}{2}](i\mu),$$

$$W_2^{(n)}[p,q](i\mu + 1) = W_3^{(n)}[p, q + p + \tfrac{1}{2}](i\mu),$$

$$W_3[p,q](i\mu + 1) = W_2[p, q + p + \tfrac{1}{2}](i\mu),$$

$$W_3^{(n)}[p,q](i\mu + 1) = W_2^{(n)}[p, q + p + \tfrac{1}{2}](i\mu).$$

Moreover, the $W_j[p,q]$ and their derivatives up to order four satisfy

$$W_1[p,q](\tfrac{i}{\mu}) = \mu^2 W_3[-q, p](i\mu),$$

$$W_1'[p,q](\tfrac{i}{\mu}) = -\mu^4 W_3'[-q, p](i\mu) - 2\mu^3 W_3[-q, p](i\mu),$$

$$W_1''[p,q](\tfrac{i}{\mu}) = \mu^6 W_3''[-q, p](i\mu) + 6\mu^5 W_3'[-q, p](i\mu) + 6\mu^4 W_3[-q, p](i\mu),$$

$$W_1^{(3)}[p,q](\tfrac{i}{\mu}) = -\mu^8 W_3^{(3)}[-q, p](i\mu) - 12\mu^7 W_3''[-q, p](i\mu)$$
$$- 36\mu^6 W_3'[-q, p](i\mu) - 24\mu^5 W_3[-q, p](i\mu),$$

$$W_1^{(4)}[p,q](\tfrac{i}{\mu}) = \mu^{10} W_3^{(4)}[-q, p](i\mu) + 20\mu^9 W_3^{(3)}[-q, p](i\mu)$$
$$+ 120\mu^8 W_3''[-q, p](i\mu) + 240\mu^7 W_3'[-q, p](i\mu)$$
$$+ 120\mu^6 W_3[-q, p](i\mu).$$

$$W_2[p,q](\frac{i}{\mu}) = \mu^2 W_2[-q,p](i\mu),$$

$$W_2'[p,q](\frac{i}{\mu}) = -\mu^4 W_2'[-q,p](i\mu) - 2\mu^3 W_2[-q,p](i\mu),$$

$$W_2''[p,q](\frac{i}{\mu}) = \mu^6 W_2''[-q,p](i\mu) + 6\mu^5 W_2'[-q,p](i\mu) + 6\mu^4 W_2[-q,p](i\mu),$$

$$W_2^{(3)}[p,q](\frac{i}{\mu}) = -\mu^8 W_2^{(3)}[-q,p](i\mu) - 12\mu^7 W_2''[-q,p](i\mu)$$

$$- 36\mu^6 W_2'[-q,p](i\mu) - 24\mu^5 W_2[-q,p](i\mu),$$

$$W_2^{(4)}[p,q](\frac{i}{\mu}) = \mu^{10} W_2^{(4)}[-q,p](i\mu) + 20\mu^9 W_2^{(3)}[-q,p](i\mu)$$

$$+ 120\mu^8 W_2''[-q,p](i\mu) + 240\mu^7 W_2'[-q,p](i\mu)$$

$$+ 120\mu^6 W_2[-q,p](i\mu).$$

$$W_3[p,q](\frac{i}{\mu}) = -\mu^2 W_1[-q,p](i\mu),$$

$$W_3'[p,q](\frac{i}{\mu}) = \mu^4 W_1'[-q,p](i\mu) + 2\mu^3 W_1[-q,p](i\mu),$$

$$W_3''[p,q](\frac{i}{\mu}) = -\mu^6 W_1''[-q,p](i\mu) - 6\mu^5 W_1'[-q,p](i\mu) - 6\mu^4 W_1[-q,p](i\mu),$$

$$W_3^{(3)}[p,q](\frac{i}{\mu}) = \mu^8 W_1^{(3)}[-q,p](i\mu) + 12\mu^7 W_1''[-q,p](i\mu)$$

$$+ 36\mu^6 W_1'[-q,p](i\mu) + 24\mu^5 W_1[-q,p](i\mu),$$

$$W_3^{(4)}[p,q](\frac{i}{\mu}) = -\mu^{10} W_1^{(4)}[-q,p](i\mu) - 20\mu^9 W_1^{(3)}[-q,p](i\mu)$$

$$- 120\mu^8 W_1''[-q,p](i\mu) - 240\mu^7 W_1'[-q,p](i\mu)$$

$$- 120\mu^6 W_1[-q,p](i\mu).$$

and the conformal factor and its derivatives satisfy

$$F[p,q](i\mu + 1) = F[p,q+p+\frac{1}{2}](i\mu),$$

$$F^{(n)}[p,q](i\mu + 1) = F^{(n)}[p,q+p+\frac{1}{2}](i\mu).$$

for derivatives of arbitrary order, while

$$F[p,q](\frac{i}{\mu}) = -\mu^{-2}F[-q,p](i\mu),$$

$$F'[p,q](\frac{i}{\mu}) = F'[-q,p](i\mu) - 2\mu^{-1}F[-q,p](i\mu),$$

$$F''[p,q](\frac{i}{\mu}) = -\mu^2 F''[-q,p](i\mu) + 2\mu F'[-q,p](i\mu) - 2F[-q,p](i\mu),$$

$$F^{(3)}[p,q](\frac{i}{\mu}) = \mu^4 F^{(3)}[-q,p](i\mu),$$

$$F^{(4)}[p,q](\frac{i}{\mu}) = -\mu^6 F^{(4)}[-q,p](i\mu) - 4\mu^5 F^{(3)}[-q,p](i\mu).$$

We refer the reader to [Fan, Fathizadeh, Marcolli (2015b)] for a detailed proof of these identities.

The case of the one-parameter family of solutions is similar, and it is shown in [Fan, Fathizadeh, Marcolli (2015b)] that the transformation under T and S are given by

$$W_1^{(n)}[q_0](i\mu + 1) = W_1^{(n)}[q_0 - i](i\mu),$$

$$W_2^{(n)}[q_0](i\mu + 1) = W_3^{(n)}[q_0 - i](i\mu),$$

$$W_3^{(n)}[q_0](i\mu + 1) = W_2^{(n)}[q_0 - i](i\mu).$$

$$W_1[q_0](\frac{i}{\mu}) = -\mu^2 W_3[\frac{1}{q_0}](i\mu),$$

$$W_1'[q_0](\frac{i}{\mu}) = \mu^4 W_3'[\frac{1}{q_0}](i\mu) + 2\mu^3 W_3[\frac{1}{q_0}](i\mu),$$

$$W_1''[q_0](\frac{i}{\mu}) = -\mu^6 W_3''[\frac{1}{q_0}](i\mu) - 6\mu^5 W_3'[\frac{1}{q_0}](i\mu) - 6\mu^4 W_3[\frac{1}{q_0}](i\mu),$$

$$W_1^{(3)}[q_0](\frac{i}{\mu}) = \mu^8 W_3^{(3)}[\frac{1}{q_0}](i\mu) + 12\mu^7 W_3''[\frac{1}{q_0}](i\mu) + 36\mu^6 W_3'[\frac{1}{q_0}](i\mu)$$
$$+ 24\mu^5 W_3[\frac{1}{q_0}](i\mu),$$

$$W_1^{(4)}[q_0](\frac{i}{\mu}) = -\mu^{10} W_3^{(4)}[\frac{1}{q_0}](i\mu) - 20\mu^9 W_3^{(3)}[\frac{1}{q_0}](i\mu) - 120\mu^8 W_3''[\frac{1}{q_0}](i\mu)$$
$$- 240\mu^7 W_3'[\frac{1}{q_0}](i\mu) - 120\mu^6 W_3[\frac{1}{q_0}](i\mu).$$

$$W_2[q_0](\frac{i}{\mu}) = -\mu^2 W_2[\frac{1}{q_0}](i\mu),$$

$$W_2'[q_0](\frac{i}{\mu}) = \mu^4 W_2'[\frac{1}{q_0}](i\mu) + 2\mu^3 W_2[\frac{1}{q_0}](i\mu),$$

$$W_2''[q_0](\frac{i}{\mu}) = -\mu^6 W_2''[\frac{1}{q_0}](i\mu) - 6\mu^5 W_2'[\frac{1}{q_0}](i\mu) - 6\mu^4 W_2[\frac{1}{q_0}](i\mu),$$

$$W_2^{(3)}[q_0](\frac{i}{\mu}) = \mu^8 W_2^{(3)}[\frac{1}{q_0}](i\mu) + 12\mu^7 W_2''[\frac{1}{q_0}](i\mu) + 36\mu^6 W_2'[\frac{1}{q_0}](i\mu)$$
$$+ 24\mu^5 W_2[\frac{1}{q_0}](i\mu),$$

$$W_2^{(4)}[q_0](\frac{i}{\mu}) = -\mu^{10} W_2^{(4)}[\frac{1}{q_0}](i\mu) - 20\mu^9 W_2^{(3)}[\frac{1}{q_0}](i\mu) - 120\mu^8 W_2''[\frac{1}{q_0}](i\mu)$$
$$- 240\mu^7 W_2'[\frac{1}{q_0}](i\mu) - 120\mu^6 W_2[\frac{1}{q_0}](i\mu).$$

$$W_3[q_0](\frac{i}{\mu}) = -\mu^2 W_1[\frac{1}{q_0}](i\mu),$$

$$W_3'[q_0](\frac{i}{\mu}) = \mu^4 W_1'[\frac{1}{q_0}](i\mu) + 2\mu^3 W_1[\frac{1}{q_0}](i\mu),$$

$$W_3''[q_0](\frac{i}{\mu}) = -\mu^6 W_1''[\frac{1}{q_0}](i\mu) - 6\mu^5 W_1'[\frac{1}{q_0}](i\mu) - 6\mu^4 W_1[\frac{1}{q_0}](i\mu),$$

$$W_3^{(3)}[q_0](\frac{i}{\mu}) = \mu^8 W_1^{(3)}[\frac{1}{q_0}](i\mu) + 12\mu^7 W_1''[\frac{1}{q_0}](i\mu) + 36\mu^6 W_1'[\frac{1}{q_0}](i\mu)$$
$$+ 24\mu^5 W_1[\frac{1}{q_0}](i\mu),$$

$$W_3^{(4)}[q_0](\frac{i}{\mu}) = -\mu^{10} W_1^{(4)}[\frac{1}{q_0}](i\mu) - 20\mu^9 W_1^{(3)}[\frac{1}{q_0}](i\mu) - 120\mu^8 W_1''[\frac{1}{q_0}](i\mu)$$
$$- 240\mu^7 W_1'[\frac{1}{q_0}](i\mu) - 120\mu^6 W_1[\frac{1}{q_0}](i\mu).$$

$$F[q_0](i\mu + 1) = F[q_0 - i](i\mu), \qquad F'[q_0](i\mu + 1) = F'[q_0 - i](i\mu),$$

$$F''[q_0](i\mu + 1) = F''[q_0 - i](i\mu), \qquad F^{(n)}[q_0](i\mu) = 0, \qquad n \geq 3,$$

$$F[q_0](\frac{i}{\mu}) = q_0^2 \mu^{-2} F[\frac{1}{q_0}](i\mu), \qquad F'[q_0](\frac{i}{\mu}) = q_0 \mu^{-1} F'[\frac{1}{q_0}](i\mu).$$

Using these explicit transformation formulae, it is possible to verify by direct computation the modularity of the coefficients $\tilde{a}_0$, $\tilde{a}_2$ and $\tilde{a}_4$, which is given by

$$\tilde{a}_0[p, q](i\mu + 1) = \tilde{a}_0[p, q + p + \frac{1}{2}](i\mu),$$

$$\tilde{a}_2[p,q](i\mu+1) = a_2[p,q+p+\tfrac{1}{2}](i\mu),$$

$$\tilde{a}_4[p,q](i\mu+1) = \tilde{a}_4[p,q+p+\tfrac{1}{2}](i\mu).$$

$$\tilde{a}_0[p,q](\tfrac{i}{\mu}) = -\mu^2 \tilde{a}_0[-q,p](i\mu),$$

$$\tilde{a}_2[p,q](\tfrac{i}{\mu}) = -\mu^2 \tilde{a}_2[-q,p](i\mu),$$

$$\tilde{a}_4[p,q](\tfrac{i}{\mu}) = -\mu^2 \tilde{a}_4[-q,p](i\mu).$$

for the two-parameter family and

$$\tilde{a}_0[q_0](i\mu+1) = \tilde{a}_0[q_0-i](i\mu), \qquad \tilde{a}_0[q_0](\tfrac{i}{\mu}) = -q_0^4 \mu^2 \tilde{a}_0[\tfrac{1}{q_0}](i\mu),$$

$$\tilde{a}_2[q_0](i\mu+1) = \tilde{a}_2[q_0-i](i\mu), \qquad \tilde{a}_2[q_0](\tfrac{i}{\mu}) = q_0^2 \mu^2 \tilde{a}_2[\tfrac{1}{q_0}](i\mu),$$

$$\tilde{a}_4[q_0](i\mu+1) = \tilde{a}_4[q_0-i](i\mu), \qquad \tilde{a}_4[q_0](\tfrac{i}{\mu}) = -\mu^2 \tilde{a}_4[\tfrac{1}{q_0}](i\mu).$$

for the one-parameter family.

A general proof of the modularity properties of the coefficients $\tilde{a}_{2n}$ is obtained in [Fan, Fathizadeh, Marcolli (2015b)] via a different argument, based on a direct comparison of the Dirac operators. Indeed, it is proved that the Dirac operators $\tilde{D}^2[p,q]$, $\tilde{D}^2[p,q+p+\frac{1}{2}]$ and $\tilde{D}^2[-q,p]$ are *isospectral*. Then by expressing the heat kernel $K_t[p,q]$ of $\exp\left(-t\tilde{D}^2[p,q]\right)$ in terms of eigenvalues and eigenspinors establishes the modularity properties

$$K_t[p,q](i\mu_1+1, i\mu_2+1) = K_t[p,q+p+\tfrac{1}{2}](i\mu_1, i\mu_2),$$

$$K_t[p,q](-\tfrac{1}{i\mu_1}, -\tfrac{1}{i\mu_2}) = (i\mu_2)^2 K_t[-q,p](i\mu_1, i\mu_2).$$

This then implies the modularity of the coefficients $\tilde{a}_{2n}$ in the form

$$\tilde{a}_{2n}[p,q](i\mu+1) = \tilde{a}_{2n}[p,q+p+\tfrac{1}{2}](i\mu),$$

$$\tilde{a}_{2n}[p,q](\tfrac{i}{\mu}) = (i\mu)^2 \tilde{a}_{2n}[-q,p](i\mu).$$

Moreover, the coefficients satisfy

$$\tilde{a}_{2n}[p+1,q] = \tilde{a}_{2n}[p,q+1] = \tilde{a}_{2n}[p,q],$$

hence the $\mathrm{PSL}_2(\mathbb{Z})$ action on $(p, q) \in \mathbb{R}/\mathbb{Z}^2$ given by

$$\tilde{S}(p, q) = (-q, p)$$
$$\tilde{T}(p, q) = (p, q + p + \frac{1}{2})$$

has finite orbits $\mathcal{O}_{(p,q)}$ on the rationals.

We can regard the coefficients $\tilde{a}_{2n}[p', q'](i\mu)$, with $(p', q') \in \mathcal{O}_{(p,q)}$ as a *vector-valued modular form* of weight 2 for the modular group $\mathrm{PSL}_2(\mathbb{Z})$.

Summing over orbits

$$\tilde{a}_{2n}\big(i\mu; \mathcal{O}_{(p,q)}\big) = \sum_{(p',q') \in \mathcal{O}_{(p,q)}} \tilde{a}_{2n}[p', q'](i\mu)$$

gives an ordinary modular form of weight 2 for $\mathrm{PSL}_2(\mathbb{Z})$.

It is in general possible to identify explicitly this modular form in terms of known modular forms like Eisenstein series and the modular discriminant, by analyzing the structure of zeros and poles.

For example, for all n, the modular form $\tilde{a}_{2n}\big(i\mu; \mathcal{O}_{(0,\frac{1}{3})}\big)$ is in the one-dimensional space spanned by

$$\frac{G_{14}(i\mu)}{\Delta(i\mu)},$$

where Δ is the modular discriminant (cusp form of weight 12) and G_{14} is the Eisenstein series of weight 14. As another example, for all n, the modular form $\tilde{a}_{2n}\big(i\mu; \mathcal{O}_{(\frac{1}{6}, \frac{5}{6})}\big)$ is in one-dimensional space spanned by

$$\frac{\Delta(i\mu) G_6(i\mu)}{G_4(i\mu)^4}.$$

Again we refer the reader to [Fan, Fathizadeh, Marcolli (2015b)] for more detailed arguments.

Chapter 7

Motives and Periods in Cosmology

This chapter is based on the results of [Fathizadeh, Marcolli (2016)]. We show that the coefficients a_{2n} of the spectral action expansion of the Robertson–Walker spacetimes can be expressed in terms of periods of motives of algebraic varieties.

For a (Riemannian) Robertson-Walker metric $ds^2 = dt^2 + a(t)^2 d\sigma^2$, with $d\sigma^2$ is the round metric on S^3, the spectral action and its asymptotic expansion were studied in [Chamseddine, Connes (2012b)] and in [Fathizadeh, Ghorbanpour, Khalkhali (2014)]. In particular, it was conjectured in [Chamseddine, Connes (2012b)] and proved in [Fathizadeh, Ghorbanpour, Khalkhali (2014)] that the coefficients of the full asymptotic expansions are rational functions, with $\mathbb{Q}$ coefficients, of the scaling factor and its derivatives. This result was generalized in [Fan, Fathizadeh, Marcolli (2015)] to the case of Bianchi IX metrics, as discussed in the previous chapter. As we already observed in the previous discussion of the Bianchi IX gravitational instantons and the occurrence of modular forms, this type of rationality results points to the existence of a hidden *arithmetic structure* in the coefficients of the asymptotic expansion of the spectral action. In this chapter we show that, in the case of the Robertson–Walker metrics, these manifest themselves through the occurrence of *periods* of algebraic varieties as coefficients of the spectral action expansion. This reveals another unexpected connection between Algebraic Geometry and models of modified gravity in cosmology.

The occurrence of periods of algebraic varieties in high-energy physics is not new. In the past decade a considerable amount of attention has been dedicated to the mysterious interplay of Algebraic and Arithmetic Geometry (especially periods and motives of algebraic varieties) and Feynman integrals in Quantum Field Theory. In the case of a Euclidean (massless) scalar field theory, the perturbative expansion of the effective action in Feynman diagrams leads to a formal series of (typically divergent) integrals on momenta flowing through the edges of the Feynman graph, with conservation conditions at vertices. Passing to Feynman parameters, these Feynman integrals can be rewritten as an integration of an algebraic differential form involving the two Symanzik polynomials of the graph (respectively depending on spanning trees and cut-sets) over a simplex, which is a semi-algebraic set, in the complement of a hypersurface (defined by the vanishing of the first Symanzik polynomial) in a projective space, see [Bloch, Esnault, Kreimer (2006)], [Marcolli (2010)]. In this formulation the divergences can be renormalized by algebro-geometric blowups of certain subspaces in the ambient projective space, which have the effect of separating the domain of integration from the hypersurface where the differential form has poles, [Bloch, Esnault, Kreimer (2006)]. The resulting renormalized convergent integrals are then *periods* in the algebro-geometric sense: complex numbers that can be written in the form

$$\int_\sigma \omega$$

with ω an algebraic differential form on an algebraic variety defined over $\mathbb{Q}$ and σ a semi-algebraic set, also defined by algebraic equations over $\mathbb{Q}$. An interesting aspect of period is the fact that the geometry of the underlying algebraic variety constraints what kind of numbers can occur as periods. More precisely, what determines the nature of the period is the *motive* of the algebraic variety. Motives are a universal cohomology theory for algebraic varieties, built out of algebraic cycles, first developed by Grothendieck in the early 1960s. In the context of Quantum Field Theory, one of the interesting question is understanding when the motive associated to a Feynman graph is contained in an especially interesting subcategory of the category of motives, given by the "mixed Tate motives", which are, roughly speaking, motives built out of non-trivial extensions of motives of projective spaces. The resulting periods are then known to be linear combinations of multiple zeta values, with coefficients that are rational numbers and powers of $2\pi i$. The extensive occurrences of multiple zeta values in Feynman integral computation had originally led to the conjecture that all the underlying motives would be mixed Tate. It is now known that this is not the case and the question of understanding the motivic nature of Feynman integrals remains a challenging open problem, see [Marcolli (2010)] for an overview of this topic.

The main result of [Fathizadeh, Marcolli (2016)] can be summarized as follows. Using the same Wodzicki residue method described in the previous chapter to compute the Seeley-deWitt coefficients in the small time aymptotic expansion of the heat kernel, one shows that the coefficients can be computed recursively using a parametrix and pseudodifferential calculus. An explicit computation of the a_2 term, together with the recursive formula can then be used to show that all the coefficients can be explicitly written as periods. Namely, if one neglects the integration in the time variable and considers the scaling factor purely as an affine parameter, one finds a convenient change of coordinates in which the remaining integrations on the cotangent sphere bundle can be expressed as the integration of an algebraic differential form defined over $\mathbb{Q}$ on a locus that is a $\mathbb{Q}$-semi-algebraic set. The differential form is defined on the complement of a quadric hypersurface. For example, in the case of the term a_2 one finds an algebraic differential form defined on the complement of (a cone over) the pencils of quadrics $Q_\alpha = u_1^2 + \alpha^{-2}(u_2^2 + u_3^2 + u_4^2)$, integrated over the semialgebraic set

$$A = \left\{ (u_0, u_1, u_2, u_3, u_4) \in \mathbb{A}^5(\mathbb{R}) : \begin{array}{l} u_1^2 + u_2^2 + u_0 u_3^2 + (1 - u_0)u_4^2 = 1, \\ 0 < u_i < 1, \text{ for } i = 0, 1, 2 \end{array} \right\}$$

The relevant motive is a mixed motive $\mathfrak{m}(\mathbb{A}^5 \setminus CZ_{Q_\alpha}, \Sigma)$ where Σ is a divisor given by a union of coordinate hyperplanes that contains the boundary of the semialgebraic set, and the affine hypersurface in $\mathbb{A}^5$ is a cone over the quadric Z_{Q_α}. These motives can be expressed in terms of motives of quadrics, which are forms of Tate motives, in the sense of [Rost (1998)], [Vishik (2004)], which become Tate motives in a quadratic field extension where the quadric becomes isotropic. We describe the result more in detail in the following sections.

7.1 Robertson–Walker metrics

Topologically, the Robertson–Walker spacetimes are cylinders $S^3 \times \mathbb{R}$. The (Euclidean) Robertson–Walker metric is given by

$$ds^2 = dt^2 + a(t)^2 d\sigma^2$$

with a single scaling factor $a(t)$ and with the round metric $d\sigma^2$ on S^3.

It is convenient to use Hopf coordinates on S^3, given by

$$x = (t, \eta, \phi_1, \phi_2) \mapsto (t, \sin \eta \cos \phi_1, \sin \eta \sin \phi_2, \cos \eta \cos \phi_1, \cos \eta \sin \phi_2),$$

$$0 < \eta < \frac{\pi}{2}, \qquad 0 < \phi_1 < 2\pi, \qquad 0 < \phi_2 < 2\pi.$$

The Robertson-Walker metric in Hopf coordinates takes the form

$$ds^2 = dt^2 + a(t)^2 \left(d\eta^2 + \sin^2(\eta)\, d\phi_1^2 + \cos^2(\eta)\, d\phi_2^2 \right).$$

As in the previous chapter, we write the Dirac operator in an orthonormal coframe $\{\theta^a\}$ as

$$D = \sum_a \theta^a \nabla_{\theta_a}^S$$

with the spin connection ∇^S with matrix of 1-forms $\omega = (\omega_b^a)$ with

$$\nabla \theta^a = \sum_b \omega_b^a \otimes \theta^b.$$

The metric-compatibility and torsion-freeness of the Levi–Civita connection are expressed by

$$\omega_b^a = -\omega_a^b, \qquad d\theta^a = \sum_b \omega_b^a \wedge \theta^b.$$

The Dirac operator is then of the form

$$D = \sum_{a,\mu} \gamma^a dx^\mu(\theta_a) \frac{\partial}{\partial x^\mu} + \frac{1}{4} \sum_{a,b,c} \gamma^c \omega_{ac}^b \gamma^a \gamma^b$$

with $\omega_a^b = \sum_c \omega_{ac}^b \theta^c$. The gamma matrices γ^a give the Clifford action of the θ^a on the spin bundle, with $(\gamma^a)^2 = -1$ and $\gamma^a \gamma^b + \gamma^b \gamma^a = 0$ for $a \neq b$.

The pseudodifferential symbol $\sigma_D(x,\xi)$ of the Dirac operator D is a sum $q_1(x,\xi) + q_0(x,\xi)$, with $\xi = (\xi_1, \xi_2, \xi_3, \xi_4) \in T_x^* M \simeq \mathbb{R}^4$ the cotangent fiber at $x = (t, \eta, \phi_1, \phi_2)$ and with

$$q_1(x,\xi) = \begin{pmatrix} 0 & 0 & \frac{i\sec(\eta)\xi_4}{a(t)} - \xi_1 & \frac{i\xi_2}{a(t)} + \frac{\csc(\eta)\xi_3}{a(t)} \\ 0 & 0 & \frac{i\xi_2}{a(t)} - \frac{\csc(\eta)\xi_3}{a(t)} & -\xi_1 - \frac{i\sec(\eta)\xi_4}{a(t)} \\ -\xi_1 - \frac{i\sec(\eta)\xi_4}{a(t)} & -\frac{i\xi_2}{a(t)} - \frac{\csc(\eta)\xi_3}{a(t)} & 0 & 0 \\ \frac{\csc(\eta)\xi_3}{a(t)} - \frac{i\xi_2}{a(t)} & \frac{i\sec(\eta)\xi_4}{a(t)} - \xi_1 & 0 & 0 \end{pmatrix},$$

$$q_0(\xi) = \begin{pmatrix} 0 & 0 & \frac{3ia'(t)}{2a(t)} & \frac{\cot(\eta)-\tan(\eta)}{2a(t)} \\ 0 & 0 & \frac{\cot(\eta)-\tan(\eta)}{2a(t)} & \frac{3ia'(t)}{2a(t)} \\ \frac{3ia'(t)}{2a(t)} & \frac{\tan(\eta)-\cot(\eta)}{2a(t)} & 0 & 0 \\ \frac{\tan(\eta)-\cot(\eta)}{2a(t)} & \frac{3ia'(t)}{2a(t)} & 0 & 0 \end{pmatrix}.$$

$$(7.1)$$

Correspondingly, the pseudodifferential symbol of the square D^2 of the Dirac operator is given by

$$\sigma_{D^2}(x,\xi) = p_2(x,\xi) + p_1(x,\xi) + p_0(x,\xi),$$

where we have

$$p_2(x,\xi) = q_1(x,\xi)\,q_1(x,\xi) = \left(\sum g^{\mu\nu}\xi_\mu\xi_\nu\right) I_{4\times 4}$$

$$= \left(\xi_1^2 + \frac{\xi_2^2}{a(t)^2} + \frac{\csc^2(\eta)\xi_3^2}{a(t)^2} + \frac{\sec^2(\eta)\xi_4^2}{a(t)^2}\right) I_{4\times 4},$$

$$p_1(x,\xi) = q_0(x,\xi)\,q_1(x,\xi) + q_1(x,\xi)\,q_0(x,\xi) + \sum_{j=1}^{4} -i\frac{\partial q_1}{\partial \xi_j}(x,\xi)\frac{\partial q_1}{\partial x_j}(x,\xi),$$

$$p_0(x,\xi) = q_0(x,\xi)\,q_0(x,\xi) + \sum_{j=1}^{4} -i\frac{\partial q_1}{\partial \xi_j}(x,\xi)\frac{\partial q_0}{\partial x_j}(x,\xi).$$

7.2 The a_2 term period

Using the Wodzicki residue method, with a parametrix $(D^2)^{-1}$ for the Dirac square D^2, we obtain the $a_2(D^2)$ coefficient as

$$a_2 = \frac{1}{2^5\,\pi^4}\,\mathrm{Res}\left((D^2)^{-1}\right),$$

where for a pseudodifferential operator P_σ we have

$$\mathrm{Res}\,(P_\sigma) = \int_M \mathrm{wres}_x P_\sigma$$

for the 1-density given by the unit cotangent sphere bundle integral

$$\mathrm{wres}_x P_\sigma = \left(\int_{|\xi|=1} \mathrm{tr}\,(\sigma_{-m}(x,\xi))\,|\sigma_{\xi,\,m-1}|\right) |dx^0 \wedge dx^1 \wedge \cdots \wedge dx^{m-1}|.$$

The explicit resulting expression for the a_2 coefficient can be obtained by computer calculation, see [Fathizadeh, Marcolli (2016)] for a detailed expression. Although the explicit formula is lengthy, and we do not reproduce it here, its main important properties are the following:

- Each summand with an odd power of the ξ_j variables in the numerator integrates to zero in the integration of the 1-density, hence all such terms can be neglected.

- The numerical coefficients of all summands in the integrand are *rational numbers*.
- If one does not perform the integration in the time variable, and replaces the scaling factor $a(t)$ and its derivatives $a'(t)$, $a''(t)$, etc. with affine variables $\alpha, \alpha_1, \alpha_2, \ldots$, then there is a natural change of coordinates replacing trigonometric functions by polynomials so that the integrand is a rational function in the new coordinates and the α, α_i, with $\mathbb{Q}$-coefficients.

The explicit change of coordinates that expresses the integrand as a rational function is given by

$$u_0 = \sin^2(\eta), \qquad u_1 = \xi_1, \qquad u_2 = \xi_2,$$
$$u_3 = \csc(\eta)\,\xi_3, \qquad u_4 = \sec(\eta)\,\xi_4.$$

One then has the following identities:

$$\xi_1^2 + \frac{\xi_2^2}{a(t)^2} + \frac{\xi_3^2\,\csc^2(\eta)}{a(t)^2} + \frac{\xi_4^2\,\sec^2(\eta)}{a(t)^2} = u_1^2 + \frac{1}{a(t)^2}(u_2^2 + u_3^2 + u_4^2),$$

$$\cot^2(\eta) = \frac{1 - u_0}{u_0}, \qquad \csc^2(\eta) = \frac{1}{u_0}, \qquad \sec^2(\eta) = \frac{1}{1 - u_0}.$$

Together with simple identities of trigonometric functions, these transform the integral expressing the a_2 coefficient (without time integration) into an algebraic form

$$C \cdot \int_{\mathbb{A}_4} \Omega^\alpha_{(\alpha_1,\alpha_2)}$$

with $C \in \mathbb{Q}[(2\pi i)^{-1}]$.

Here $\Omega = \Omega^\alpha_{(\alpha_1,\alpha_2)}$ is an algebraic differential form

$$\Omega = f\,\tilde{\sigma}_3,$$

in affine coordinates $(u_0, u_1, u_2, u_3, u_4) \in \mathbb{A}^5$, $\alpha \in \mathbb{G}_m$, and $(\alpha_1, \alpha_2) \in \mathbb{A}^2$, where the function $f(u_0, u_1, u_2, u_3, u_4, \alpha, \alpha_1, \alpha_2) = f_{(\alpha_1,\alpha_2)}(u_0, u_1, u_2, u_3, u_4, \alpha)$ is a $\mathbb{Q}$-linear combinations of rational functions

$$\frac{P(u_0, u_1, u_2, u_3, u_4, \alpha, \alpha_1, \alpha_2)}{\alpha^{2r} u_0^k (1 - u_0)^m (u_1^2 + \alpha^{-2}(u_2^2 + u_3^2 + u_4^2))^\ell}$$

where the terms

$$P(u_0, u_1, u_2, u_3, u_4, \alpha, \alpha_1, \alpha_2) = P_{(\alpha_1,\alpha_2)}(u_0, u_1, u_2, u_3, u_4, \alpha)$$

are polynomials in $\mathbb{Q}[u_0, u_1, u_2, u_3, u_4, \alpha, \alpha_1, \alpha_2]$, with r, k, m and ℓ non-negative integers. The algebraic differential form $\widetilde{\sigma}_3 = \widetilde{\sigma}_3(u_0, u_1, u_2, u_3, u_4)$ is given by

$$\widetilde{\sigma}_3 = \frac{1}{2}\Big(u_1 \, du_0 \wedge du_2 \wedge du_3 \wedge du_4 - u_2 \, du_0 \wedge du_1 \wedge du_3 \wedge du_4$$

$$+ u_3 \, du_0 \wedge du_1 \wedge du_2 \wedge du_4 - u_4 \, du_0 \wedge du_1 \wedge du_2 \wedge du_3 \Big).$$

The forms $\Omega^\alpha = \Omega^\alpha_{(\alpha_1, \alpha_2)}$, for a fixed value of $\alpha \in \mathbb{A}^1 \smallsetminus \{0\}$ are a two-parameter family of algebraic differential forms, depending on the two parameters $(\alpha_1, \alpha_2) \in \mathbb{A}^2(\mathbb{Q})$.

This algebraic differential form is defined on the complement in $\mathbb{A}^5$ of a union of two affine hyperplanes $H_0 = \{u_0 = 0\}$ and $H_1 = \{u_0 = 1\}$ and the hypersurface $\widehat{CZ}_\alpha$ defined by vanishing of the quadratic form

$$Q_{\alpha,2} = u_1^2 + \alpha^{-2}(u_2^2 + u_3^2 + u_4^2).$$

The domain of integration is a $\mathbb{Q}$-semialgebraic set. These are subsets S of some ambient $\mathbb{R}^n$ of the form

$$S = \{(x_1, \ldots, x_n) \in \mathbb{R}^n : P(x_1, \ldots, x_n) \geq 0\},$$

for some polynomial $P \in \mathbb{Q}[x_1, \ldots, x_n]$, as well as complements, intersections, and unions of such sets. More precisely, the domain of integration is given by the $\mathbb{Q}$-semialgebraic set

$$A_4 = \left\{ (u_0, u_1, u_2, u_3, u_4) \in \mathbb{A}^5(\mathbb{R}) : \begin{array}{l} u_1^2 + u_2^2 + u_0 u_3^2 + (1 - u_0)u_4^2 = 1, \\ 0 < u_i < 1, \text{ for } i = 0, 1, 2 \end{array} \right\}.$$

7.3 The periods of the higher order terms a_{2n}

The higher order terms can be computed as Wodzicki residues

$$a_{2n} = \frac{1}{2^5 \, \pi^{3+n}} \mathrm{Res}(\Delta_{2n}^{-1})$$

using auxiliary flat tori T^{2n-2} and the operators

$$\Delta_{2n} = D^2 \otimes 1 + 1 \otimes \Delta_{T^{2n-2}},$$

where $\Delta_{T^{2n-2}}$ are the flat Laplacian on the tori. We need the term σ_{-2n-2} that is homogeneous of order $-2n - 2$ in the expansion of the pseudodifferential symbol of the parametrix Δ_{2n}^{-1}.

A recursive argument for the structure of the term σ_{-2n-2} is obtained as follows.

Using the change of coordinates

$$u_0 = \sin^2(\eta), \qquad u_3 = \csc(\eta)\,\xi_3, \qquad u_4 = \sec(\eta)\,\xi_4$$

$$u_j = \xi_j, \qquad j = 1, 2, 5, 6, \ldots, 2n+2$$

and neglecting time integration (treating, as before, the scale factor and its derivatives as affine parameters), the term $\operatorname{tr}(\sigma_{-2n-2})$ is given by

$$\sum_{j=1}^{M_n} c_{j,2n}\, u_0^{\beta_{0,1,j}/2}\,(1-u_0)^{\beta_{0,2,j}/2}\, \frac{u_1^{\beta_{1,j}}\, u_2^{\beta_{2,j}} \,\cdots\, u_{2n+2}^{\beta_{2n+2,j}}}{Q_{\alpha,2n}^{\rho_{j,2n}}}\, \alpha^{k_{0,j}}\, \alpha_1^{k_{1,j}} \,\cdots\, \alpha_{2n}^{k_{2n,j}},$$

where

$$\alpha = a(t), \qquad \alpha_1 = a'(t), \qquad \alpha_2 = a''(t), \qquad \cdots \qquad \alpha_{2n} = a^{2n}(t),$$

$$Q_{\alpha,2n} = u_1^2 + \frac{1}{\alpha^2}(u_2^2 + u_3^2 + u_4^2) + u_5^2 + \cdots + u_{2n+2}^2, \qquad (7.2)$$

$$c_{j,2n} \in \mathbb{Q}, \quad \beta_{0,1,j}, \beta_{0,2,j}, k_{0,j} \in \mathbb{Z}, \quad \beta_{1,j}, \ldots, \beta_{2n+2,j}, \rho_{j,2n}, k_{1,j}, \ldots, k_{2n,j} \in \mathbb{Z}_{\geq 0}.$$

One can then write the associated period integral as

$$C \cdot \int_{A_{2n}} \Omega^\alpha_{\alpha_1,\ldots,\alpha_{2n}},$$

with an algebraic differential form

$$\Omega^\alpha_{\alpha_1,\ldots,\alpha_{2n}}(u_0, u_1, \ldots, u_{2n+2})$$

whose domain of definition is the complement

$$\mathbb{A}^{2n+3} \setminus (\widehat{CZ}_{\alpha,2n} \cup H_0 \cup H_1)$$

with hyperplanes $H_0 = \{u_0 = 0\}$ and $H_1 = \{u_0 = 1\}$ and $\widehat{CZ}_{\alpha,2n}$ the hypersurface defined by the vanishing of the quadric

$$Q_{\alpha,2n} = u_1^2 + \frac{1}{\alpha^2}(u_2^2 + u_3^2 + u_4^2) + u_5^2 + \cdots + u_{2n+2}^2.$$

The domain of integration is the $\mathbb{Q}$-semialgebraic set A_{2n+2} given by

$$\left\{ (u_0, \ldots, u_{2n+2}) \in \mathbb{A}^{2n+3}(\mathbb{R}) : \begin{array}{l} u_1^2 + u_2^2 + u_0 u_3^2 + (1-u_0)u_4^2 + \sum_{i=5}^{2n+2} u_i^2 = 1 \\ 0 < u_i < 1, \quad i = 0, 1, 2, 5, 6, \ldots, 2n+2 \end{array} \right\}.$$

7.4   The mixed motives of Robertson–Walker gravity

In this section we review the results of [Fathizadeh, Marcolli (2016)] on the motive underlying the period integrals described in the previous section, associated to the terms a_{2n} in the asymptotic expansion of the spectral action of Robertson–Walker metrics.

We will not give here an overview of the theory of motives of algebraic varieties. It is a very broad subject of high technical complexity. The interested reader can find a comprehensive introduction in the book [André (2004)]. The theory of motives is subdivided into two main cases: *pure motives*, which provide a universal cohomoloy theory for smooth projective varieties, and *mixed motives*, a more general theory that includes variety that are not necessarily smooth and projective. The difference reflects the different formal properties of the cohomology of spaces that are smooth and compact and those that are not. In the smooth and compact case, one expects a good cohomology theory to satisfy properties like Poincaré duality, while in the non-compact case one expects long exact sequences for inclusions, open coverings, etc. The theory of mixed motives is technically more complicated than the theory of pure motives, see [André (2004)]. In the case we are considering, since the period integrals live on the complement of a hypersurface inside an affine space, one necessarily has to deal with mixed motives. Explicit computations in the category of mixed motives are difficult to obtain, but it is often possible to obtain useful information about a motive by computing a simpler associated invariant, the class in the Grothendieck ring of varieties. This is a universal Euler characteristic of varieties, just as motives are a universal cohomology theory. Euler characteristics are weaker invariants than cohomologies, but they are also generally more computable, and they contain useful information about the cohomology.

For our purposes here, we will only review some general results about the Voevodsky triangulated category of mixed motives of [Voevodsky (2000)] and the Grothendieck ring of varieties, whose interpretation as universal Euler characteristic of algebraic varieties can be found in [Bittner (2004)]. The structure of triangulated category captures the formal categorical properties of cohomology theories for varieties that are not necessarily smooth and compact (smooth and projective). The distinguished triangles of the triangulated structures correspond to natural long exact sequences in cohomology (Mayer–Vietoris, Gysin, etc.).

7.4.1 *Triangulated category of motives*

A triangulated category of mixed motives of algebraic varieties was constructed in [Voevodsky (2000)]. We will write $\mathfrak{m}(X)$ for the Voevodsky motive of an algebraic variety X. While we do not recall here the precise definition of the motive $\mathfrak{m}(X)$, and we simply refer the reader to [Voevodsky (2000)] and [André (2004)] for more information, it is useful to think of $\mathfrak{m}(X)$ heuristically as a more geometric form of the cohomology of X. The main formal property we need to recall about the triangulated category of motives is the existence of certain distinguished triangles.

- The Mayer–Vietoris distinguished triangles are given by objects and morphisms

$$\mathfrak{m}(U \cap V) \to \mathfrak{m}(U) \oplus \mathfrak{m}(V) \to \mathfrak{m}(X) \to \mathfrak{m}(U \cap V)[1],$$

  where $X = U \cap V$ is an open covering of a smooth variety X, the morphisms are induced by inclusions, and $\mathfrak{m}[1] = T\mathfrak{m}$ is the image of a motive $\mathfrak{m}$ under the shift functor T of the triangulated category. This type of distinguished triangle corresponds to the Mayer–Vietoris long exact sequence in cohomology, relating cohomologies of the open sets U, V, and those of their union $U \cup V = X$ and of their intersection $U \cap V$.

- The Gysin distinguished triangle is given by

$$\mathfrak{m}(X \smallsetminus Z) \to \mathfrak{m}(X) \to \mathfrak{m}(Z)(c)[2c] \to \mathfrak{m}(X \smallsetminus Z)[1],$$

  where $Z \hookrightarrow X$ is a closed subvariety of codimension c, with $\mathfrak{m}[n] = T^n \mathfrak{m}$ the shift and $\mathfrak{m}(n) = \mathfrak{m} \otimes \mathbb{Q}(n)$ the twist by the Tate motive, where $\mathbb{Q}(1)[2]$ is the mixed motive that corresponds to the cohomology $H^2(\mathbb{P}^1)$ and $\mathbb{Q}(n) = \mathbb{Q}(1)^{\otimes n}$.

- Homotopy invariance: the morphism $\mathfrak{m}(X \times \mathbb{A}^1) \to \mathfrak{m}(X)$ is an isomorphism.

- The triangulated category of *mixed Tate motives* is the smallest triangulated subcategory of the Voevodsky category generated by the Tate motives $\mathbb{Q}(n)$. Since it is a triangulated subcategory, if in a distinguished triangle

$$\mathfrak{m}(X) \to \mathfrak{m}(Y) \to \mathfrak{m}(Z) \to \mathfrak{m}(X)[1]$$

  in the Voevodsky category two of the three terms belong to the mixed Tate subcategory, then the third one also belongs to it. This is often a very convenient way to show that a motive is mixed Tate by using the formal properties of the category of mixed motives.

7.4.2 *Grothendieck classes of RW spacetimes*

In addition to the objects $\mathrm{m}(X)$ in the triangulated category of motives, we also consider the class $[X]$ of a variety X in the Grothendieck ring of varieties. Let $\mathbb{K}$ be a number field, a finite algebraic extension of $\mathbb{Q}$. The Grothendieck ring $K_0(\mathcal{V}_{\mathbb{K}})$ of varieties over $\mathbb{K}$ is generated by isomorphism classes $[X]$ of varieties X defined over the field $\mathbb{K}$, satisfying the inclusion-exclusion relation $[X] = [Y] + [X \smallsetminus Y]$, for closed embeddings $Y \hookrightarrow X$, and the product given by $[X_1] \cdot [X_2] = [X_1 \times X_2]$. One often refers to the Grothendieck class $[X]$ of X as the *virtual motive* of the variety X. The Tate subring of $K_0(\mathcal{V}_{\mathbb{K}})$ is the ring $\mathbb{Z}[\mathbb{L}]$ generated by the Lefschetz motive $\mathbb{L} = [\mathbb{A}^1]$, the virtual motive of the affine line. It is often easier to check that the virtual motive is Tate, by explicitly computing the class in the Grothendieck ring. Since this is implies by the motive being mixed Tate (and conjecturally implies it), this is often a more explicit way to approach the question of whether a given motive is mixed Tate.

The following identities follow directly from the inclusion-exclusion relation for Grothendieck classes. Let $Z \subset \mathbb{P}^{N-1}$ be a projective variety and let $\hat{Z} \subset \mathbb{A}^N$ be its affine cone. We also denote by CZ the projective cone of Z in $\mathbb{P}^N$. Then we have the following expressions for the Grothendieck classes of complements.

- $[\mathbb{A}^N \smallsetminus \hat{Z}] = (\mathbb{L} - 1)[\mathbb{P}^{N-1} \smallsetminus Z]$
- $[\mathbb{A}^{N+1} \smallsetminus \widehat{CZ}] = (\mathbb{L} - 1)[\mathbb{P}^N \smallsetminus CZ]$
- $[CZ] = \mathbb{L}[Z] + 1$
- $[\mathbb{A}^{N+1} \smallsetminus \widehat{CZ}] = \mathbb{L}^{N+1} - \mathbb{L}(\mathbb{L} - 1)[Z] - \mathbb{L}$
- $[\mathbb{A}^{N+1} \smallsetminus (\widehat{CZ} \cup H \cup H')] = \mathbb{L}^{N+1} - 2\mathbb{L}^N - (\mathbb{L} - 2)(\mathbb{L} - 1)[Z] - (\mathbb{L} - 2)$,

with H and H' two affine hyperplanes with $H \cap H' = \emptyset$, and with intersections $\widehat{CZ} \cap H$ and $\widehat{CZ} \cap H'$ given by sections $\hat{Z}$ of the cone.

Let $Z_{\alpha,2n}$ be the quadric hypersurface in $\mathbb{P}^{2n+1}$ determined by the quadratic form $Q_{\alpha,2n}$ of (7.2). The quadratic form $Q_{\alpha,2n}$, hence the variety $Z_{\alpha,2n}$ is defined over $\mathbb{Q}$. However, passing to a quadratic field extension allows for a change of variable that puts the quadric in a standard form, for which we can compute the Grothendieck class explicitly. Consider the number field $\mathbb{K} = \mathbb{Q}(\sqrt{-1})$, an imaginary quadratic extension of $\mathbb{Q}$, and assume that the affine scaling parameter is a given $\alpha \in \mathbb{Q}^*$. Then, starting with the first case

$$Q_{\alpha,2} = u_1^2 + \alpha^{-2}(u_2^2 + u_3^2 + u_4^2)$$

we can introduce the change of variables

$$X = u_1 + \frac{i}{\alpha} u_2, \quad Y = u_1 - \frac{i}{\alpha} u_2, \quad Z = \frac{i}{\alpha}(u_3 + iu_4), \quad W = \frac{i}{\alpha}(u_3 - iu_4).$$

This change of variables is not defined over $\mathbb{Q}$ but it is defined over $\mathbb{K}$. With the new variables, the quadric Z_α can be identified with the Segre quadric

$$\{XY - ZW = 0\} \simeq \mathbb{P}^1 \times \mathbb{P}^1. \tag{7.3}$$

Thus, the Grothendieck class $[Z_\alpha] = [\mathbb{P}^1 \times \mathbb{P}^1] = (1+\mathbb{L})^2$ in the Grothendieck ring $K_0(\mathcal{V}_\mathbb{K})$. Similarly, for the a_{2n}-term case, the quadric

$$Q_{\alpha,2n} = u_1^2 + \frac{1}{\alpha^2}(u_2^2 + u_3^2 + u_4^2) + u_5^2 + u_6^2 + \cdots + u_{2n+1}^2 + u_{2n+2}^2$$

can be written inductively, using the change of coordinates

$$X = u_{2n+1} + iu_{2n+2}, \quad Y = u_{2n+1} - iu_{2n+2},$$

in the form

$$Q_{\alpha,2n} = Q_{\alpha,2n-2}(u_1, \ldots, u_{2n}) + XY.$$

The general identities listed above for classes of complements of affine and projective cones, together with the inductive description of the quadric hypersurfaces over the field extension $\mathbb{K}$ give the following expression for the Grothendieck classes in $K_0(\mathcal{V}_\mathbb{K})$ (see [Fathizadeh, Marcolli (2016)]):

$$[\mathbb{P}^{2n+1} \smallsetminus Z_{\alpha,2n}] = \mathbb{L}^{2n+1} - \mathbb{L}^n$$

$$[\mathbb{A}^{2n+3} \smallsetminus \widehat{CZ}_{\alpha,2n}] = \mathbb{L}^{2n+3} - \mathbb{L}^{2n+2} - \mathbb{L}^{n+2} + \mathbb{L}^{n+1}$$

$$[\mathbb{A}^{2n+3} \smallsetminus (\widehat{CZ}_{\alpha,2n} \cup H_0 \cup H_1)] = \mathbb{L}^{2n+3} - 3\mathbb{L}^{2n+2} + 2\mathbb{L}^{2n+1} - \mathbb{L}^{n+2} + 3\mathbb{L}^{n+1} - 2\mathbb{L}^n.$$

In particular, this shows that the Grotendieck class of the complement $\mathbb{A}^{2n+3} \smallsetminus (\widehat{CZ}_{\alpha,2n} \cup H_0 \cup H_1)$ is in the Tate part of the Grothendieck ring, by providing an explicit expression as a polynomial in the Lefschetz motive $\mathbb{L}$.

7.4.3   *Mixed Tate motives of RW spacetimes*

In order to obtain information directly about the motive underlying the period integrals described above, we need to consider a relative motive

$$\mathfrak{m}(\mathbb{A}^{2n+3} \smallsetminus (\widehat{CZ}_{\alpha,2n} \cup H_0 \cup H_1), \Sigma), \tag{7.4}$$

where Σ is a divisor that contains boundary of domain of integration A_{2n}. This relative motive corresponds to a relative cohomology, which is needed since the domain of integration is not a closed cycle, namely the semi-algebraic set A_{2n} has a non-empty boundary. Thus, the period integral should be seen as a paring of relative cohomology classes, between the differential form and the domain of integration. In order to understand the nature of the motive (7.4) we can start with the motives $\mathfrak{m}(Z_{\alpha,2n})$ of the quadrics and then arrive at (7.4) through repeated application of distinguishes triangles of Mayer–Vietoris and Gysin type, where each time one can control the nature of two of the three terms. If at each stage two of the terms are mixed-Tate, then by the general argument recalled above we can conclude that the motive (7.4) is also mixed Tate, without relying on the explicit computation of the Grothendieck class.

We again consider the field extension $\mathbb{K} = \mathbb{Q}(\sqrt{-1})$, so that we can use the change of coordinates described above, which puts the quadratic form $Q_{\alpha,2n}$ in the standard form

$$Q_{\alpha,2n}|_{\mathbb{Q}(\sqrt{-1})} = (n+1) \cdot Q_{\mathbb{H}},$$

where $Q_{\mathbb{H}}$ is the hyperbolic quadratic form $Q_{\mathbb{H}} := \langle 1, -1 \rangle$ in two dimensions, which defines the quadric (7.3), and $d \cdot Q_{\mathbb{H}}$ is the hyperbolic form of dimension $2d$, given by d blocks equal to $Q_{\mathbb{H}}$.

The motives of quadrics were computed by [Rost (1998)], [Vishik (2004)]. The motive of the quadric of the hyperbolic form $Q = d \cdot Q_{\mathbb{H}}$ of dimension $2d$ is a Tate motive given by

$$\mathfrak{m}(Z_{d\mathbb{H}}) = \mathbb{Q}(d-1)[2d-2] \oplus \mathbb{Q}(d-1)[2d-2] \oplus \bigoplus_{i=0,\ldots,d-2,d,\ldots,2d-2} \mathbb{Q}(i)[2i]$$

Similarly, for $Q = d \cdot Q_{\mathbb{H}} \perp \langle 1 \rangle$ in dimension $2d+1$

$$\mathfrak{m}(Z_{d\mathbb{H} \perp \langle 1 \rangle}) = \bigoplus_{i=0,\ldots,2d-1} \mathbb{Q}(i)[2i].$$

If there is a quadratic field extension $\mathbb{K}$ where the quadratic form Q becomes hyperbolic, the motive over the original field can be written as

$$\mathfrak{m}(Z_Q) = \begin{cases} \mathfrak{m}_1 \oplus \mathfrak{m}_1(1)[2] & m = 2 \mod 4 \\ \mathfrak{m}_1 \oplus \mathcal{R}_{Q,\mathbb{K}} \oplus \mathfrak{m}_1(1)[2] & m = 0 \mod 4 \end{cases}$$

where the terms of this decomposition involve *forms of Tate motives*, namely motives that become Tate motives after passing to the quadratic field extension. This result gives the explicit form of the motives $\mathfrak{m}(Z_{\alpha,2n})$,

$$\mathfrak{m}(Z_{\alpha,2n}|_{\mathbb{K}}) = \mathbb{Q}(n)[2n] \oplus \mathbb{Q}(n)[2n] \oplus \bigoplus_{i=0,\ldots,n-1,n+1,\ldots 2n} \mathbb{Q}(i)[2i].$$

For simplicity, we show in the case $n = 1$ how to go from the motive of $Z_\alpha = Z_{\alpha,2}$ to the motive $\mathfrak{m}(\mathbb{A}^5 \smallsetminus (\widehat{CZ}_{\alpha,2} \cup H_0 \cup H_1), \Sigma)$. We refer the reader to [Fathizadeh, Marcolli (2016)] for the complete argument for arbitrary n.

One first observes that the motive $\mathfrak{m}(\mathbb{P}^3 \smallsetminus Z_\alpha)$ is mixed Tate because it fits into a Gysin distinguished triangle

$$\mathfrak{m}(\mathbb{P}^3 \smallsetminus Z_\alpha) \to \mathfrak{m}(\mathbb{P}^3) \to \mathfrak{m}(Z_\alpha)(1)[2] \to \mathfrak{m}(\mathbb{P}^3 \smallsetminus Z_\alpha)[1]$$

for the closed codimension one embedding $Z_\alpha \hookrightarrow \mathbb{P}^3$. In this triangle the two terms $\mathfrak{m}(\mathbb{P}^3)$ and $\mathfrak{m}(Z_\alpha)$ are Tate, hence the third term must also be mixed Tate.

Consider next the projective cone CZ_α in $\mathbb{P}^4$. Homotopy invariance for the $\mathbb{A}^1$-fibration $\mathbb{P}^4 \smallsetminus CZ_\alpha \to \mathbb{P}^3 \smallsetminus Z_\alpha$ gives

$$\mathfrak{m}(\mathbb{P}^4 \smallsetminus CZ_\alpha) \simeq \mathfrak{m}(\mathbb{P}^3 \smallsetminus Z_\alpha)(-1)$$

hence the motive $\mathfrak{m}(\mathbb{P}^4 \smallsetminus CZ_\alpha)$ is also mixed Tate.

The motive $\mathfrak{m}(\mathbb{A}^5 \smallsetminus \widehat{CZ}_\alpha)$ is also mixed Tate for the following reason. Consider the $\mathbb{P}^1$-bundle $\mathcal{P}$ compactification of the $\mathbb{G}_m$-bundle

$$\mathcal{T} = \mathbb{A}^5 \smallsetminus \widehat{CZ}_\alpha \to X = \mathbb{P}^4 \smallsetminus CZ_\alpha$$

and the Gysin distinguished triangle

$$\mathfrak{m}(\mathcal{T}) \to \mathfrak{m}(\mathcal{P}) \to \mathfrak{m}_c(\mathcal{P} \smallsetminus \mathcal{T})^*(1)[2] \to \mathfrak{m}(\mathcal{T})[1].$$

The motive with compact support $\mathfrak{m}_c(\mathcal{P} \smallsetminus \mathcal{T})$ is mixed Tate since $\mathcal{P} \smallsetminus \mathcal{T}$ can be identified with two copies of X, hence the motive $\mathfrak{m}(\mathcal{T})$ must also be mixed Tate, since the other two terms in the triangle are mixed Tate.

Finally, we want to check that the union $\widehat{CZ}_\alpha \cup H_0 \cup H_1$ is also mixed Tate. The motives $\mathfrak{m}(\mathbb{A}^5 \smallsetminus (H_0 \cup H_1))$ and $\mathfrak{m}(\mathbb{A}^5 \smallsetminus \widehat{CZ}_\alpha)$ and the motive of the intersection $\mathfrak{m}(\widehat{CZ}_\alpha \cap (H_0 \cup H_1))$ are all mixed Tate, hence in the Mayer-Vietoris distinguished triangle

$$\mathfrak{m}(U \cap V) \to \mathfrak{m}(U) \oplus \mathfrak{m}(V) \to \mathfrak{m}(U \cup V) \to \mathfrak{m}(U \cap V)[1]$$

with $U = \mathbb{A}^5 \smallsetminus \widehat{CZ}_\alpha$ and $V = \mathbb{A}^5 \smallsetminus (H_0 \cup H_1)$ we have the following:

- $\mathfrak{m}(\mathbb{A}^5 \smallsetminus \widehat{CZ}_\alpha)$ is mixed Tate by the previous steps in the argument.
- $\mathfrak{m}(\mathbb{A}^5 \smallsetminus (H_0 \cup H_1))$ also mixed Tate since $\mathfrak{m}(H_0 \cup H_1)$ is mixed Tate.
- $\mathfrak{m}(\widehat{CZ}_\alpha \cap (H_0 \cup H_1))$ mixed Tate because the intersection $\widehat{CZ}_\alpha \cap (H_0 \cup H_1)$ gives two sections of the cone and we know that $\mathfrak{m}(\hat{Z}_\alpha)$ is a Tate motive.

• Then we obtain that $\mathfrak{m}(\mathbb{A}^5 \smallsetminus (\widehat{CZ_\alpha} \cap (H_0 \cup H_1)))$ is also mixed Tate because the other two terms in the distinguished triangle are mixed Tate.

Finally, we are interested in the relative motive. The divisor Σ in $\mathbb{A}^5$ is a union of coordinate hyperplanes and their translates, hence it is mixed Tate. Then the relative motive is also mixed Tate because it fits into a distinguished triangle (corresponding to the exact sequence for relative cohomology) where the other two terms are $\mathfrak{m}(\mathbb{A}^5 \smallsetminus (\widehat{CZ_\alpha} \cap (H_0 \cup H_1)))$ and $\mathfrak{m}(\Sigma)$, which are both mixed Tate.

Thus the conclusion we can draw from the results of [Fathizadeh, Marcolli (2016)] is that the terms in the asymptotic expansion of the spectral action on the Robertson–Walker spacetime is determined by a sequence of periods of mixed Tate motives. Unlike the quantum field theory case, here all the integrals are convergent, hence no renormalization is needed, and all the motives are mixed Tate throughout the entire asymptotic expansion. The crucial difference with respect to the quantum field theory setting lies in the fact that here, each integrand involves a single quadric hypersurface and a union with hyperplanes that are placed in a good position with respect to the cones. In the quantum field theory setting, one can also write the form in terms of quadratic forms, but what one obtains is a union of quadric hypersurfaces, one for each internal edge of the Feynman graphs, which intersect non-transversely. This complicates the structure of the motives and one can not apply a similar argument to check when the resulting motive is mixed Tate.

A similar result holds for the Bianchi IX gravitational instantons where the coefficients of the heat kernel expansion of the Dirac-Laplacian have a similar motivic structure, [Fan, Fathizadeh, Marcolli (2017)]. More generally, it may be interesting to identify which classes of (Euclidean) spacetimes admit an arithmetic structure based on motives and periods in the coefficients of the spectral action, or equivalently in the heat kernel expansion.

Chapter 8

Fractal and Multifractal Structures in Cosmology

In this chapter we analyze cosmological models exhibiting fractal and multifractal structures and we approach their mathematical modeling using Noncommutative Geometry methods. We first consider the fractal Packed Swiss Cheese Cosmology model, and we study it from the point of view of the spectral action as an action functional for gravity. The results for this part of the chapter are based on [Ball, Marcolli (2016)]. We then consider multifractal models of eternal inflation based on the "Eternal Symmetree" of [Harlow, Shenker, Stanford, Susskind (2012)], [Susskind (2012)], which we again revisit from the point of view of operator algebras and Noncommutative Geometry. This second part of the chapter is based on [Marcolli, Tedeschi (2014)].

8.1 Packed Swiss Cheese Cosmology and the spectral action

We present here the results of [Ball, Marcolli (2016)].

The origin of the Packed Swiss Cheese model of cosmology goes back to a construction of [Rees, Sciama (1968)], of a cosmological model that would be isotropic but non-homogeneous.

The standard paradigm of cosmology is the hypothesis that the universe is on large scales both homogeneous and isotropic. The typical example of a homogeneous and isotropic cosmology is given by the usual Friedmann universe, which is topologically a cylinder $\mathbb{R} \times S^3$, with a (Lorentzian or Riemannian) metric of the form

$$\pm dt^2 + a(t)^2 \left(\sigma_1^2 + \sigma_2^2 + \sigma_3^2 \right)$$

with the round metric on S^3 with σ_i the $SU(2)$–invariant 1-forms satisfying the relations

$$d\sigma_i = \sigma_j \wedge \sigma_k$$

for all cyclic permutations (i, j, k) of $(1, 2, 3)$. Homogeneity refers to the property that the universe should look the same at different locations while isotropy is the property that it should look the same in different directions.

We have already seen in the last two chapters that, if one drops the isotropy conditions, while maintaining the homogeneity condition, then one obtains some very interesting classes of cosmological models like the Bianchi IX and mixmaster cosmologies, given by cylinders $\mathbb{R} \times S^3$ with metrics

$$F(t)\,(\pm dt^2 + \frac{\sigma_1^2}{W_1^2(t)} + \frac{\sigma_2^2}{W_2^2(t)} + \frac{\sigma_3^2}{W_3^3(t)})$$

with different non-isotropic scaling factors $W_i(t)$ and with a conformal factor $F(t) \sim W_1(t)W_2(t)W_3(t)$.

On the other hand, if one retains the isotropy condition but drops homogeneity, it is also possible to obtain some very interesting classes of cosmological models. The original construction of [Rees, Sciama (1968)] was based on cutting off one (or a finite number) of 4-balls from a homogeneous and isotropic Friedmann–Robertson–Walker spacetime and replacing them with different density smaller regions, with the inside and outside regions patched across the boundary with vanishing Weyl curvature tensor, so that isotropy is preserved. In a variant of this model, discussed in §8 of [Mureika, Dyer (2004)], instead of compressing the matter inside each spherical region, at each stage of the construction process the matter is expanded to lie along the spherical shell. The Packed Swiss Cheese Cosmology is obtained when this construction is iterated, by removing more and more balls, so that one obtains an *Apollonian sphere packing* of 3-dimensional spheres and a model of gravity interacting with matter, supported on the resulting fractal residual set. This type of fractal cosmology was proposed as a possible explanation for a fractal distribution of matter in galaxies, clusters, and superclusters, see [Sylos Labini, Montuori, Pietroneo (1998)] and [Mureika, Dyer (2004)].

8.1.1　*Apollonian sphere packings*

The best known and best understood Apollonian sphere packing is certainly the case of the Apollonian circle packings, see Figure 8.1. These are configurations of mutually tangent circles in the plane, iterated on smaller

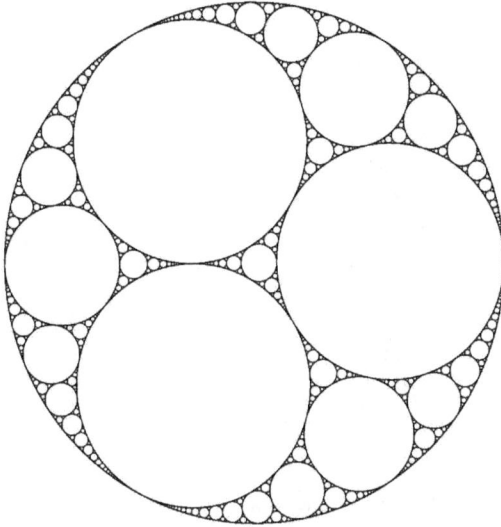

Fig. 8.1   Apollonian packing of circles.

scales, filling a full volume region in the unit $2D$ ball. The residual set has 2-dimensional volume zero and is a fractal of Hausdorff dimension 1.30568...

There are many very interesting mathematical results on Apollonian circle packings, including geometric, arithmetic, and analytic properties, see for example [Graham, Lagarias, Mallows, Wilks, Yan (2003)] [Kontorovich, Oh (2011)].

For the higher dimensional Apollonian packings, we refer the reader to [Graham, Lagarias, Mallows, Wilks, Yan (2006)]. We review here the main properties that we will need in our setting.

For more general higher dimensional packings of S^{D-1} spheres inside a D-dimensional space, a general procedure for the construction of an Apollonian packing starts with an initial *Descartes configuration*. This is an arrangement of $D + 2$ mutually tangent $(D - 1)$-dimensional spheres inside a D-dimensional space. For example, one can obtain such a Descartes configuration by taking $D + 1$ equal size mutually tangent spheres S^{D-1} centered at the vertices of the standard D-simplex and one more smaller sphere in the center of the simplex that is tangent to all the other ones. In particular, in the case $D = 4$ which we will be considering, the 4-simplex is illustrated in the figure.

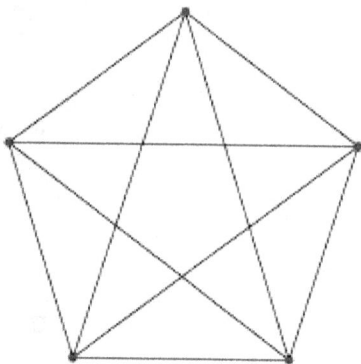

The radii a_k of the $D + 2$ spheres in a Descartes configuration satisfy a relation, known as the *quadratic Soddy–Gosset relation*,

$$\left(\sum_{k=1}^{D+2} \frac{1}{a_k} \right)^2 = D \sum_{k=1}^{D+2} \left(\frac{1}{a_k} \right)^2 .$$

In order to account for possibly degenerate configurations in which at least one of the spheres becomes a hyperplane, it is convenient to use the *curvature-center coordinates*:

$$w = \left(\frac{\|x\|^2 - a^2}{a}, \frac{1}{a}, \frac{1}{a} x_1, \ldots, \frac{1}{a} x_D \right),$$

where the first coordinate is the curvature after inversion in the unit sphere, and the x_i are the coordinates of the centers of the spheres.

It is then possible to obtain a *configuration space* $\mathcal{M}_D$, the parameterizing space of all Descartes configurations in D dimensions. It is given by all the possible solutions $\mathcal{W}$ to the equation

$$\mathcal{W}^t Q_D \mathcal{W} = \begin{pmatrix} 0 & -4 & 0 \\ -4 & 0 & 0 \\ 0 & 0 & 2 I_D \end{pmatrix},$$

with a left and a right action of the Lorentz group $O(D + 1, 1)$.

This configuration space classifies all the possible "initial conditions" for the construction of Apollonian sphere packings. Given one of these configurations, the Apollonian packing is then constructed by an iteration process that reproduces nested copies of the same configuration under the action of sequences of reflections about one of the spheres. This is formally described by introducing a group of transformations, the *dual Apollonian group* $\mathcal{G}_D^\perp$, which is generated by reflections given by the inversions with respect to the j-th sphere in the configuration,

$$S_j^\perp = I_{D+2} + 2\,1_{D+2}e_j^t - 4\,e_je_j^t$$

where e_j is the j-th unit coordinate vector.

When the dimension is $D \neq 3$ it is known that the only relations in $\mathcal{G}_D^\perp$ are those describing the fact that each $S_j^\perp$ is a reflection, namely $(S_j^\perp)^2 = 1$. Moreover, $\mathcal{G}_D^\perp$ is a discrete subgroup of $\mathrm{GL}(D+2, \mathbb{R})$.

An Apollonian packing $\mathcal{P}_D$ is then an orbit of $\mathcal{G}_D^\perp$ on the configuration space $\mathcal{M}_D$. Equivalently, one can describe $\mathcal{P}_D$ by an iterative construction where at the n-th step one adds spheres obtained from the initial Descartes configuration by applying to it all the possible transformations

$$S_{j_1}^\perp S_{j_2}^\perp \cdots S_{j_n}^\perp, \quad j_k \neq j_{k+1}, \forall k.$$

There are N_n spheres in the n-th level of the construction, with

$$N_n = (D+2)(D+1)^{n-1}.$$

The first few steps in the iterative construction of a sphere packing of S^2 spheres are illustrated in the figure. We will be especially interested in the case of packings of S^3 spheres, which are less directly visualizable.

8.1.2 *Length spectrum and zeta functions*

We associate to an Apollonian packing of S^{D-1} spheres a *length spectrum*. It is the sequence of radii of the spheres in the packing $\mathcal{P}_D$,

$$\mathcal{L} = \mathcal{L}(\mathcal{P}_D) = \{a_{n,k} : n \in \mathbb{N}, 1 \leq k \leq (D+2)(D+1)^{n-1}\},$$

where the $a_{n,k}$ are the radii of the spheres $S_{a_{n,k}}^{D-1}$ in the n-th level of the iterative construction of the packing $\mathcal{P}_D$.

The *Melzak's packing constant* $\sigma_D(\mathcal{P}_D)$ of the Apollonian packing $\mathcal{P}_D$ is the exponent of convergence of the series

$$\zeta_\mathcal{L}(s) = \sum_{n=1}^{\infty} \sum_{k=1}^{(D+2)(D+1)^{n-1}} a_{n,k}^s,$$

which is the generating function for the length spectrum.

The *residual set* $\mathcal{R}(\mathcal{P}_D) = B^D \smallsetminus \cup_{n,k} B_{a_{n,k}}^D$ of the Apollonian packing, with $\partial B_{a_{n,k}}^D = S_{a_{n,k}}^{D-1} \in \mathcal{P}_D$, has $\mathrm{Vol}_D(\mathcal{R}(\mathcal{P}_D)) = 0$ in the D-dimensional

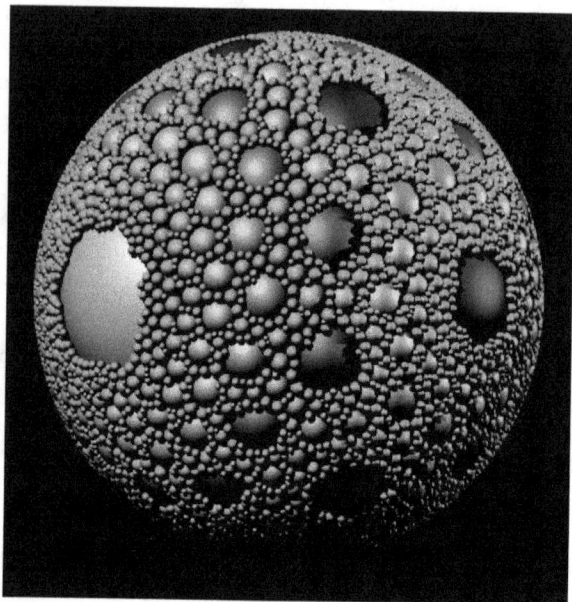

Fig. 8.2 Apollonian sphere packing (image by Paul Bourke, 2006)

volume of the ambient space. The convergence $\sum_{\mathcal{L}} a_{n,k}^D < \infty$ implies that $\sigma_D(\mathcal{P}_D) \leq D$.

While it is in general not possible to compute the exact value of the Hausdorff dimension of the residual set, it is possible to estimate it in terms of the packing constant,

$$\dim_H(\mathcal{R}(\mathcal{P}_D)) \leq \sigma_D(\mathcal{P}_D).$$

For the simpler case of the Apollonian circles, these are known to be the same.

Given an Apollonian packing $\mathcal{P}_D$, consider all the spheres in the packing that have a given bound on the curvature. This determines a *sphere counting function*

$$\mathcal{N}_\alpha(\mathcal{P}_D) = \#\{S_{a_{n,k}}^{D-1} \in \mathcal{P}_D : a_{n,k} \geq \alpha\}$$

where the curvatures are bounded by $c_{n,k} = a_{n,k}^{-1} \leq \alpha^{-1}$.

In the case of the Apollonian circles it is known by [Kontorovich, Oh (2011)] that there is a power law

$$\mathcal{N}_\alpha(\mathcal{P}_2) \sim_{\alpha \to 0} \alpha^{-\dim_H(\mathcal{R}(\mathcal{P}_2))}.$$

For the higher dimensional cases, [Boyd (1973)] showed that the packing constant satisfies

$$\limsup_{\alpha \to 0} -\frac{\log \mathcal{N}_\alpha(\mathcal{P}_D)}{\log \alpha} = \sigma_D(\mathcal{P}_D)$$

and if the limit exists $\mathcal{N}_\alpha(\mathcal{P}_D) \sim_{\alpha \to 0} \alpha^{-(\sigma_D(\mathcal{P}_D)+o(1))}$.

In general, the zeta functions $\zeta_{\mathcal{L}_D}(s)$ does not necessarily admit an analytic continuation to a meromorphic function on the whole complex plane $\mathbb{C}$.

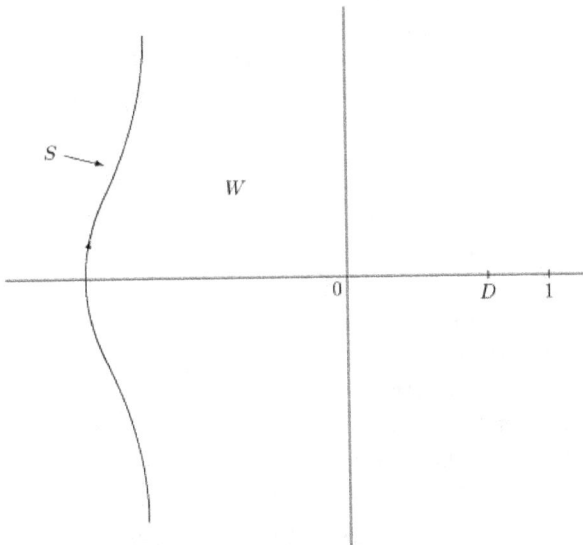

Fig. 8.3   Analytic continuation region with boundary screen curve.

However, it is shown in [Lapidus, van Frankenhuijsen (2013)] that there exists a *screen* curve $\mathcal{S}$, that is, a curve $S(t)+it$ with $S : \mathbb{R} \to (-\infty, \sigma_D(\mathcal{P}_D)]$ and a *window* region $\mathcal{W}$, as shown in Figure 8.3, given by the region to the right of the screen curve $\mathcal{S}$, where the zeta function admits an analytic continuation.

As shown in [Lapidus, van Frankenhuijsen (2013)], for general zeta functions $\zeta_{\mathcal{L}}(s)$ of length spectra, the distribution of the non-real poles can be very complicated. The best possible case corresponds to fractals with an exact self-similarity, where the contraction ratios are all integer powers of a fixed scale $0 < r < 1$. This is called the *lattice property*. In this special case, the non-real poles are periodically spaced on finitely many vertical lines.

For the purpose of the type of cosmological model we want to construct, we restrict our attention to a class of "especially good" Apollonian sphere packings, where the good properties we need are identified as follows.

We say that an Apollonian packing $\mathcal{P}_D$ of $(D-1)$-spheres is *analytic* if

(1) $\zeta_{\mathcal{L}}(s)$ has an analytic continuation to a meromorphic function on a region $\mathcal{W}$ containing the non-negative real axis $\mathbb{R}_+$;
(2) $\zeta_{\mathcal{L}}(s)$ has only one pole on $\mathbb{R}_+$ at $s = \sigma_D(\mathcal{P}_D)$;
(3) the pole at $s = \sigma_D(\mathcal{P}_D)$ is simple;
(4) There is a family $\mathcal{L}_n$, $n \in \mathbb{N}$, of self-similar fractal strings with the lattice property, and with increasingly large periods, such that the complex poles of $\zeta_{\mathcal{L}}(s)$ are approximated by the complex poles of $\zeta_{\mathcal{L}_n}(s)$.

It is also convenient to assume the existence of the limit

$$\lim_{\alpha \to 0} -\frac{\log \mathcal{N}_\alpha(\mathcal{P}_D)}{\log \alpha} = \sigma_D(\mathcal{P}_D).$$

The period of a self-similar fractal string $\mathcal{L}_n$ with the lattice property is the length $\pi_n := \frac{2\pi}{-\log r_n}$ with the property that all the poles of $\zeta_{\mathcal{L}_n}(s)$ off the real line lie on finitely may vertical lines $\Re(s) = \sigma_j$ with periodic spacing by $\frac{2\pi}{-\log r_n}$. The last listed condition above allows for some control on the distribution of poles of the zeta function $\zeta_{\mathcal{L}}(s)$ in the complex plane. As we will see below, this condition becomes useful in describing the log oscillatory terms in the asymptotic expansion of the spectral action associated to the behavior of the heat kernel on fractals that we discussed in a simple case in Chapter 1.

While it is not presently known how large is the class of Apollonian packing satisfying these conditions, it is expected that especially symmetric initial Descartes configurations will yield these properties.

We now focus more specifically on $D = 4$ cases with these conditions. One can obtain a rough estimate of the packing constant by considering averages of the curvature at each level of the iterative construction of an Apollonian packing $\mathcal{P} = \mathcal{P}_4$ of 3-spheres $S^3_{a_{n,k}}$. At level n, the average curvature is

$$\frac{\gamma_n}{N_n} = \frac{1}{6 \cdot 5^{n-1}} \sum_{k=1}^{6 \cdot 5^{n-1}} \frac{1}{a_{n,k}}.$$

One can then estimate $\sigma_4(\mathcal{P}_4)$ with an averaged version $\sum_n N_n(\frac{\gamma_n}{N_n})^{-s}$, which gives

$$\sigma_{4,av}(\mathcal{P}) = \lim_{n\to\infty} \frac{\log(6 \cdot 5^{n-1})}{\log\left(\frac{\gamma_n}{6 \cdot 5^{n-1}}\right)}.$$

The generating function of the γ_n is known by [Mallows (2009)],

$$G_{D=4} = \sum_{n=1}^{\infty} \gamma_n\, x^n = \frac{(1-x)(1-4x)u}{1 - \frac{22}{3}x - 5x^2},$$

where u is the sum of the curvatures of the initial Descartes configuration. For $u = 1$, this gives

$$\gamma_n = \frac{(11 + \sqrt{166})^n(-64 + 9\sqrt{166}) + (11 - \sqrt{166})^n(64 + 9\sqrt{166})}{3^n \cdot 10 \cdot \sqrt{166}}$$

and one obtains a value

$$\sigma_{4,av}(\mathcal{P}) = 3.85193\ldots$$

In the Apollonian circle case it is known that this method gives a strictly larger value, so one expects that for this higher dimensional cases also the estimate will not be sharp, and one will only have $\sigma_4(\mathcal{P}) < \sigma_{4,av}(\mathcal{P})$. Thus, the constraints on the packing constant are

$$3 < \dim_H(\mathcal{R}(\mathcal{P})) \leq \sigma_4(\mathcal{P}) < \sigma_{4,av}(\mathcal{P}) = 3.85193\ldots$$

8.1.3 *Models of fractal spacetimes*

In the homogeneous and isotropic case, we can consider a model of (compactified, Euclidean) spacetime that is of the form $S_\beta^1 \times S_a^3$ or $S^1 \times Y_a$, where $Y = S^3/\Gamma$ is a spherical space form, with a is the radius, as we did in the previous chapter on cosmic topology. Here we modify this space by replacing the single sphere S_a^3 with an Apollonian packing $\mathcal{P}$ of 3-spheres $S_{a_{n,k}}^3$, so that we have a space of the form $S_\beta^1 \times \mathcal{P}$.

Similarly, we can consider fractal arrangements based on some nontrivial cosmic topology Y, instead of a 3-sphere. For example, one of the most promising candidates for a non-trivial cosmic topology is the Poincaré homology sphere, the dodecahedral space $Y = S^3/\mathcal{I}_{120}$, which has fundamental domains given by spherical dodecahedra, as in the figure, where the spherical dodecahedron has $Vol(Y) = Vol(S_a^3/\mathcal{I}_{120}) = \frac{\pi^2}{60}a^3$, see Figure 8.4.

Fig. 8.4   Poincaré homology sphere (drawn by Jenn3D)

In this case, one can consider a fractal packing of dodecahedra, in the form of the Sierpinski dodecahedra of the figure, where all the dodecahedra are then simultaneously folded up in the corresponding Poincaré spaces, Figure 8.5.

This case with non-trivial cosmic topology is in fact simpler geometrically than the case of the Apollonian packing of spheres, because the Sierpinski dodecahedron is a fractal with exact self-similarity, unlike the Apollonian packings, and is obtained by a single uniform scaling at each step of the iterative construction, giving 20^n new dodecahedra, each scaled by a factor of $(2 + \phi)^{-n}$. For instance, the Hausdorff dimension can be computed exactly, and it is given by

$$\dim_H(\mathcal{P}_{\mathcal{I}_{120}}) = \frac{\log(20)}{\log(2 + \phi)} = 2.32958...$$

When closing up all dodecahedra in the fractal, by identifying opposite faces with the action $\mathcal{I}_{120}$ one gets a fractal arrangement $\mathcal{P}_Y$ of Poincaré spheres $Y_{a(2+\phi)^{-n}}$. The zeta function of the length spectrum has analytic continuation to all of $\mathbb{C}$,

$$\zeta_{\mathcal{L}}(s) = \sum_n 20^n (2 + \phi)^{-ns} = \frac{1}{1 - 20(2 + \phi)^{-s}}$$

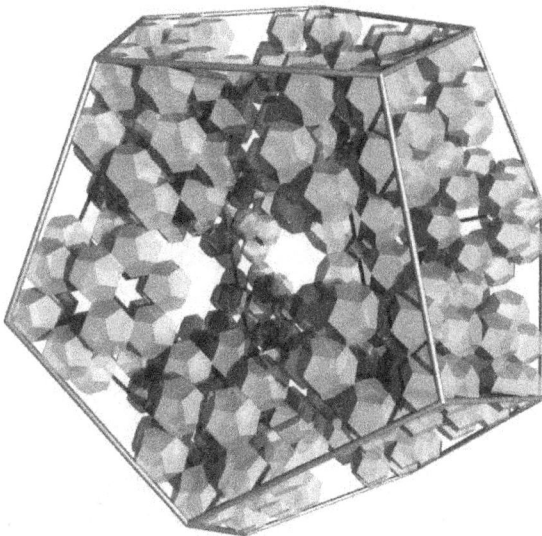

Fig. 8.5 Fractal arrangement of dodecahedra (image by Paul Bourke, 2006).

with exponent of convergence $\sigma = \dim_H(\mathcal{P}_{\mathcal{I}_{120}}) = \frac{\log(20)}{\log(2+\phi)}$ and with poles along a single vertical line

$$\sigma + \frac{2\pi i m}{\log(2+\phi)}, \quad m \in \mathbb{Z}.$$

8.1.4 *The spectral action functional on fractal cosmologies*

In both these cases (Apollonian packings of 3-spheres and Sierpiski fractals of Dodecahedral spaces), we compute the spectral action and we show how this, viewed as an action functional of gravity, detects the presence of fractality, by comparing its behavior to the case of a single sphere or dodecahedral space.

The first step is assigning a spectral triple to a fractal geometry of the type described above. This can be done following a general procedure developed in [Christensen, Ivan, Lapidus (2008)] and [Christensen, Ivan, Schrohe (2012)], for the case of Sierpinski gaskets.

The cases of $\mathcal{P}$ and $\mathcal{P}_Y$ are similar in this respect. In the first case, we are considering a D-dimensional sphere packing (in particular $D = 4$),

$$\mathcal{P}_D = \{S^{D-1}_{a_{n,k}} : n \in \mathbb{N}, 1 \leq k \leq (D+2)(D+1)^{n-1}\}$$

we consider a spectral triple which is a sum of spectral triples associated to all the spheres in the packing,

$$(\mathcal{A}_{\mathcal{P}_D}, \mathcal{H}_{\mathcal{P}_D}, \mathcal{D}_{\mathcal{P}_D}) = \oplus_{n,k}(\mathcal{A}_{\mathcal{P}_D}, \mathcal{H}_{S^{D-1}_{a_{n,k}}}, \mathcal{D}_{S^{D-1}_{a_{n,k}}}).$$

In the second case, with the dodecahedral space $Y_a = S^3/\mathcal{I}_{120}$, we are considering a spectral triple that is a sum of the triples associated to each dodecahedral space in the packing,

$$(\mathcal{A}_{\mathcal{P}_Y}, \mathcal{H}_{\mathcal{P}_Y}, \mathcal{D}_{\mathcal{P}_Y}) = (\mathcal{A}_{\mathcal{P}_Y}, \oplus_n \mathcal{H}_{Y_{a_n}}, \oplus_n \mathcal{D}_{Y_{a_n}})$$

with $a_n = a(2 + \phi)^{-n}$.

The zeta function of the Dirac operator, for the Apollonian packing of 3-spheres, is closely related to the zeta function of the length spectrum

$$\zeta_{\mathcal{L}}(s) := \sum_{n \in \mathbb{N}} \sum_{k=1}^{6 \cdot 5^{n-1}} a_{n,k}^s$$

with $\mathcal{L} = \mathcal{L}_4 = \{a_{n,k} \,|\, n \in \mathbb{N}, k \in \{1, \ldots, 6 \cdot 5^{n-1}\}\}$. Namely, the zeta function of Dirac operator of the spectral triple is given by

$$\text{Tr}(|\mathcal{D}_{\mathcal{P}}|^{-s}) = \sum_{n=1}^{\infty} \sum_{k=1}^{6 \cdot 5^{n-1}} \text{Tr}(|D_{S^3_{a_{n,k}}}|^{-s})$$

where each term is of the form

$$\text{Tr}(|D_{S^3_{a_{n,k}}}|^{-s}) = a_{n,k}^s (2\zeta(s-2, \tfrac{3}{2}) - \tfrac{1}{2}\zeta(s, \tfrac{3}{2}))$$

hence the total zeta function gives

$$\text{Tr}(|\mathcal{D}_{\mathcal{P}}|^{-s}) = \left(2\zeta(s-2, \tfrac{3}{2}) - \tfrac{1}{2}\zeta(s, \tfrac{3}{2})\right) \sum_{n,k} a_{n,k}^s$$

$$= \left(2\zeta(s-2, \tfrac{3}{2}) - \tfrac{1}{2}\zeta(s, \tfrac{3}{2})\right) \zeta_{\mathcal{L}}(s),$$

where the first factor $2\zeta(s-2, \tfrac{3}{2}) - \tfrac{1}{2}\zeta(s, \tfrac{3}{2})$ is simply the zeta function of the Dirac operator on a single S^3 of radius one. The positive dimension spectrum, that is, the part of the dimension spectrum that lies on the non-negative real line is $\Sigma^+_{ST_{PSC}} = \{1, 3, \sigma_4(\mathcal{P})\}$, where $\sigma = \sigma_4(\mathcal{P})$ is the packing constant. In addition to these points, there are all the off-real poles of the zeta function that contribute to the spectral action.

Assuming the good conditions of an "analytic packing" specified above for the zeta function $\zeta_{\mathcal{L}}(s)$, the asymptotic of the spectral action is given by

$$\mathrm{Tr}(f(\mathcal{D}_{\mathcal{P}}/\Lambda)) \sim \Lambda^3 \, \zeta_{\mathcal{L}}(3) \, f_3 - \Lambda \frac{1}{4} \, \zeta_{\mathcal{L}}(1) \, f_1$$

$$+\Lambda^\sigma \left(\zeta(\sigma - 2, \frac{3}{2}) - \frac{1}{4}\zeta(\sigma, \frac{3}{2}) \right) \mathcal{R}_\sigma \, f_\sigma + \mathcal{S}_\Lambda^{osc},$$

where the residue is given by $\mathcal{R}_\sigma = \mathrm{Res}_{s=\sigma}\zeta_{\mathcal{L}}(s)$, and the f_β are momenta $f_\beta = \int_0^\infty v^{\beta-1} f(v) dv$. The additional term $\mathcal{S}_\Lambda^{osc}$ consists of a series of *oscillatory terms* coming from the contributions of the poles of the zeta function that are off the real line. Under the assumption that the zeta function $\zeta_{\mathcal{L}}(s)$ is well approximated by a family of zeta functions $\zeta_{\mathcal{L}_n}(s)$ with the lattice property, the oscillatory terms can also be approximated by the type of oscillatory terms that we discussed in §1.2.8 of Chapter 1.

In the case of the Sierpinski dodecahedral spaces $\mathcal{P}_Y$ that have exact self-similarity, the computation is simpler and direct and does not require an approximation. In this case the poles of $\zeta_{\mathcal{L}}(s)$ that are off the real line are given by the points $s = \sigma + \frac{2\pi i m}{\log \ell}$, $m \in \mathbb{Z}$. The heat kernel expansion on the fractal is exactly of the form discussed in §1.2.8 of Chapter 1, and the spectral action has additional log-oscillatory terms in the expansion. We refer the reader to [Ball, Marcolli (2016)] for a more detailed discussion and more explicit expressions for these terms.

Given the zeta function and the spectral action expansion for the fractal spaces $\mathcal{P}$ and $\mathcal{P}_Y$, we consider the effect of taking the product with a compactified time direction S_β^1.

In the case of $S_\beta^1 \times S_a^3$, as we saw in the cosmic topology chapter, we have

$$D_{S_\beta^1 \times S_a^3} = \begin{pmatrix} 0 & D_{S_a^3} \otimes 1 + i \otimes D_{S_\beta^1} \\ D_{S_a^3} \otimes 1 - i \otimes D_{S_\beta^1} & 0 \end{pmatrix}$$

with the spectral action of the form

$$\mathrm{Tr}(h(D_{S_\beta^1 \times S_a^3}^2/\Lambda)) \sim 2\beta\Lambda\mathrm{Tr}(\kappa(D_{S_a^3}^2/\Lambda)),$$

for a test function $h(x)$ and a test function

$$\kappa(x^2) = \int_{\mathbb{R}} h(x^2 + y^2)dy.$$

In the case of $S^1_\beta \times \mathcal{P}$ we have

$$\mathcal{S}_{S^1_\beta \times \mathcal{P}}(\Lambda) \sim 2\beta \left(\Lambda^4 \, \zeta_{\mathcal{L}}(3) \, \mathfrak{h}_3 - \Lambda^2 \frac{1}{4} \zeta_{\mathcal{L}}(1) \, \mathfrak{h}_1 \right)$$

$$+ 2\beta \, \Lambda^{\sigma+1} \left(\zeta(\sigma - 2, \frac{3}{2}) - \frac{1}{4} \zeta(\sigma, \frac{3}{2}) \right) \mathcal{R}_\sigma \, \mathfrak{h}_\sigma$$

with momenta

$$\mathfrak{h}_3 := \pi \int_0^\infty h(\rho^2) \rho^3 d\rho, \quad \mathfrak{h}_1 := 2\pi \int_0^\infty h(\rho^2) \rho \, d\rho$$

$$\mathfrak{h}_\sigma = 2 \int_0^\infty h(\rho^2) \rho^\sigma d\rho.$$

The interpretation of this modified expression is the following.

The term $2\Lambda^4 \beta a^3 \mathfrak{h}_3 - \frac{1}{2}\Lambda^2 \beta a \mathfrak{h}_1$ that corresponds to the cosmological and Einstein–Hilbert terms on a space of the form $S^1_\beta \times S^3_a$, are replaced on $S^1_\beta \times \mathcal{P}$ by a term of the form

$$2\Lambda^4 \beta \zeta_{\mathcal{L}}(3) \mathfrak{h}_3 - \frac{1}{2}\Lambda^2 \beta \zeta_{\mathcal{L}}(1) \mathfrak{h}_1,$$

which we can view as a zeta regularization of the divergent series of the spectral actions of the individual 3-spheres of the packing. The additional terms in this gravity action functional are the ones that detect the presence of fractality: they are corrections to gravity coming from fractality, in the form

$$2\beta \, \Lambda^{\sigma+1} \left(\zeta(\sigma - 2, \frac{3}{2}) - \frac{1}{4} \zeta(\sigma, \frac{3}{2}) \right) \mathcal{R}_\sigma \mathfrak{h}_\sigma.$$

In the case of the fractal dodecahedral space $\mathcal{P}_Y$, we have

$$\zeta_{\mathcal{L}(\mathcal{P}_Y)}(s) = \sum_{n \geq 0} 20^n (2 + \phi)^{-ns}$$

$$\zeta_{\mathcal{D}_{\mathcal{P}_Y}}(s) = \frac{a^s}{120} \left(2\zeta(s - 2, \frac{3}{2}) - \frac{1}{2}\zeta(s, \frac{3}{2}) \right) \zeta_{\mathcal{L}(\mathcal{P}_Y)}(s)$$

and the spectral action gives

$$\mathrm{Tr}(f(\mathcal{D}_{\mathcal{P}_Y}/\Lambda)) \sim (\Lambda a)^3 \frac{\zeta_{\mathcal{L}(\mathcal{P}_Y)}(3)}{120} f_3 - \Lambda a \frac{\zeta_{\mathcal{L}(\mathcal{P}_Y)}(1)}{120} f_1$$

$$+ (\Lambda a)^\sigma \frac{\zeta(\sigma - 2, \frac{3}{2}) - \frac{1}{4}\zeta(\sigma, \frac{3}{2})}{120 \log(2 + \phi)} f_\sigma + \mathcal{S}^{osc}_{Y,\Lambda}$$

$\sigma = \dim_H(\mathcal{P}_Y) = \frac{\log(20)}{\log(2+\phi)} = 2.3296...$ with oscillatory terms as in §1.2.8 of Chapter 1.

On the product geometry $S^1_\beta \times \mathcal{P}_Y$ the spectral action becomes

$$\mathcal{S}_{S^1_\beta \times \mathcal{P}_Y}(\Lambda) \sim 2\beta \left(\Lambda^4 \frac{a^3 \zeta_{\mathcal{L}(\mathcal{P}_Y)}(3)}{120} \mathfrak{h}_3 - \Lambda^2 \frac{a \zeta_{\mathcal{L}(\mathcal{P}_Y)}(1)}{120} \mathfrak{h}_1 \right)$$

$$+ 2\beta \, \Lambda^{\sigma+1} \frac{a^\sigma (\zeta(\sigma - 2, \frac{3}{2}) - \frac{1}{4}\zeta(\sigma, \frac{3}{2}))}{120 \log(2 + \phi)} \mathfrak{h}_\sigma + \mathcal{S}^{osc}_{S^1_\beta \times Y, \Lambda}.$$

8.1.5   *Slow-roll inflation potentials with fractality*

A way in which the modification caused by fractality to the gravity action functional given by the spectral action can determine observable consequences is through the same mechanism that we described in the chapter on cosmic topology, by which the spectral action determines a slow-roll inflation potential. The shape of this potential, which in turn affects measurable quantities like the slow-roll parameters, is affected by the change to the spectral action caused by fractality.

More precisely, we find that there is an additional term in the slow-roll potential, which is of the form

$$\mathcal{U}_\sigma(x) = \int_0^\infty u^{(\sigma-1)/2}(h(u+x) - h(u))du$$

and depends on σ (hence on the fractal dimension). At leading term, the size of this correction depends on

$$(\zeta(\sigma - 2, \tfrac{3}{2}) - \tfrac{1}{4}\zeta(\sigma, \tfrac{3}{2}))\mathcal{R}_\sigma,$$

with a series of additional corrections coming from the oscillatory terms in the spectral action.

Thus, in a model of gravity based on the spectral action as the action functional, the possible presence of fractality in the structure of spacetime is detectable, and can be read off the slow-roll potential V, hence the slow-roll coefficients, which depend on V, V', V''.

8.2   A *p*-adic model of eternal inflation

The content of this section is based on [Marcolli, Tedeschi (2014)]. It is based on an elaboration in operator algebraic terms, using a multifractal analysis approach, of the discrete model of eternal inflation recently constructed in [Harlow, Shenker, Stanford, Susskind (2012)], [Susskind (2012)]. In this model, a multiverse landscape arises through a stochastic process, which assigns certain likelihoods to transitions between different types of vacua. The latter are labeled by letters $\{0, 1\}$ of more generally belonging to a finite alphabet $\mathfrak{A} = \{0, \ldots, p-1\}$. Each vacuum represents a collection of microstates with assigned entropies $S = (S_a)_{a \in \mathfrak{A}}$. The causal future of a node in the tree is the oriented subtree that branches off from that node to the boundary at infinity. Time is discretized: the proper time lapse along an edge connecting adjacent nodes is the inverse of the Hubble constant, which can vary with the label attached to the edge. The possibility of terminal vacua is introduced by pruning the tree.

8.2.1 *Bruhat–Tits tree and Bethe tree*

Let $\mathbb{K}$ be a finite extension of the field of p-adic numbers $\mathbb{Q}_p$ and let $\mathcal{O} = \mathcal{O}_{\mathbb{K}} \subset \mathbb{K}$ be its ring of integers, with $\mathfrak{m} \subset \mathcal{O}$ the maximal ideal. The residue field $k = \mathcal{O}/\mathfrak{m}$ is a finite field of cardinality $q = \#\mathcal{O}/\mathfrak{m}$ given by some power $q = p^r$. Let $\mathcal{T}_{\mathbb{K}}^0$ denote the set of equivalence classes of free rank 2 $\mathcal{O}$-modules, with respect to the relation

$$M_1 \sim M_2 \Leftrightarrow \exists \lambda \in \mathbb{K}^*, \quad M_1 = \lambda M_2.$$

The group $\mathrm{PGL}_2(\mathbb{K})$ acts on $\mathcal{T}_{\mathbb{K}}^0$ by $gM = \{gm \mid m \in M\}$. Given $M_2 \subset M_1$, one has $M_1/M_2 \simeq \mathcal{O}/\mathfrak{m}^l \oplus \mathcal{O}/\mathfrak{m}^k$, for some $l, k \in \mathbb{N}$. Since the action of $\mathbb{K}^*$ preserves the inclusion $M_2 \subset M_1$, one can assign a well defined metric on the set $\mathcal{T}_{\mathbb{K}}^0$ by setting $d(M_1, M_2) = |l - k|$. The Bruhat–Tits tree $\mathcal{T} = \mathcal{T}_{\mathbb{K}}$ of $\mathrm{PGL}_2(\mathbb{K})$ is the infinite graph with set of vertices $\mathcal{T}_{\mathbb{K}}^0$, and an edge connecting two vertices M_1, M_2 whenever $d(M_1, M_2) = 1$. It is an infinite tree where all vertices have the same valence $q + 1$, with $q = \#\mathcal{O}/\mathfrak{m}$. By construction, the group $\mathrm{PGL}_2(\mathbb{K})$ acts on $\mathcal{T}_{\mathbb{K}}$ by isometries. A ray is a half-infinite path without backtracking and an infinite geodesic is an infinite path without backtracking. We denote by $\partial \mathcal{T}$ the boundary at infinity of the tree, which is the set of equivalence classes of rays, where two rays are equivalent if they have an infinite number of vertices in common. Any choice of two distinct points on the boundary determines a unique infinite geodesic in $\mathcal{T}$ that connects them. The boundary at infinity $\partial \mathcal{T}_{\mathbb{K}}$ of the Bruhat–Tits tree is naturally identified with the projective space $\mathbb{P}^1(\mathbb{K})$. We refer the reader to [Gerritzen, van der Put (1980)], [Manin (1976)], [Mumford (1972)] for a detailed exposition of the p-adic geometry of Bruhat–Tits trees and their quotients by discrete subgroups of $\mathrm{PGL}_2(\mathbb{K})$. The choice of a coordinate function z on $\mathbb{P}^1(\mathbb{K}) = \partial \mathcal{T}$ corresponds to fixing the choice of points $\{0, 1, \infty\}$ in $\mathbb{P}^1(\mathbb{K})$. This in turn determines a unique choice of a vertex v_0 of the tree $\mathcal{T}$, as the unique origin of three non-overlapping rays with endpoints $\{0, 1, \infty\}$. Let v_0 be the base vertex in the tree $\mathcal{T}$ obtained in this way. We choose an orientation of the tree $\mathcal{T}$ with all the edges pointing outwards from v_0, so that at each vertex $v \neq v_0$ we have one incoming and q outgoing edges. In particular, having fixed a projective coordinate on $\mathbb{P}^1(\mathbb{K})$ and a corresponding base vertex in $\mathcal{T}$, the Bethe tree is the subtree $\mathcal{T}'$ of the Bruhat–Tits tree $\mathcal{T}$, with root v_0

and with boundary $\partial \mathcal{T}' = \mathcal{O}_{\mathbb{K}}$. The use of Bruhat–Tits trees as physics models, in the context of the AdS/CFT holographic correspondence, has been investigated recently by several authors, see for instance [Heydeman, Marcolli, Saberi, Stoica (2016)], from which Figure 8.6 is taken.

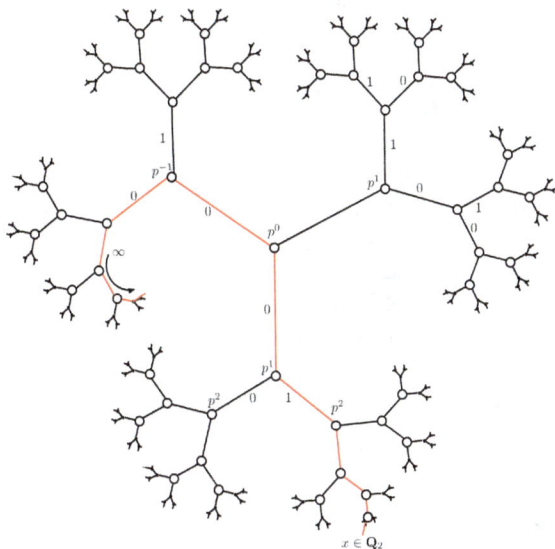

Fig. 8.6　The Bruhat–Tits tree of $\mathbb{Q}_2$.

8.2.2　*Multifractals, symbolic dynamics, and stochastics*

The boundary at infinity of the Bethe tree $\mathcal{T}'$ is topologically a Cantor set. After fixing a root vertex v_0 in the tree, it can be identified with the ω-language in an alphabet $\mathfrak{A}$ with the same cardinality $\#\mathfrak{A} = q = \#\mathcal{O}/\mathfrak{m}$ as the residue field, namely with the set of infinite sequences in the alphabet

$$\mathcal{W}_{\mathfrak{A}}^{\omega} = \{a_0 a_1 \dots a_n \dots : a_i \in \mathfrak{A}\},$$

with the topology generated by the cylinder set $\mathcal{W}_{\mathfrak{A}}^{\omega}(w)$ of all sequences that start with a given (finite) word w. Indeed, one can see this easily by looking at the case where $\mathbb{K} = \mathbb{Q}_p$ (the case of an extension is analogous). The p-adic integers in $\mathbb{Z}_p$ can be written as infinite series $x = \sum_{k=0}^{\infty} x_k p^k$, in powers of p, with coefficients in $\{0, \dots, p-1\}$. This corresponds to labeling the outgoing edges at each vertex of $\mathcal{T}'$ with a set of labels $\{e_i\}_{i=0,\dots,p-1}$.

Thus, one can identify rays starting at v_0 with arbitrary infinite words in the alphabet $\mathfrak{A} = \{e_i\}_{i=0,\dots,p-1}$.

The Cantor set $\mathcal{W}_{\mathfrak{A}}^{\omega}$ is endowed with a natural dynamical system, the one-sided (non-invertible) shift $\sigma : \mathcal{W}_{\mathfrak{A}}^{\omega} \to \mathcal{W}_{\mathfrak{A}}^{\omega}$, which shifts the sequence one step to the left and drops the first letter, $\sigma : a_0 a_1 \cdots a_n \cdots \mapsto a_1 a_2 \cdots a_{n+1} \cdots$. On the Cantor set $\mathcal{W}_{\mathfrak{A}}^{\omega}$, we then consider $\mathbb{R}_+$-valued potentials W_{β} satisfying the *Keane condition*:

$$\sum_{a \in \mathfrak{A}} W_{\beta}(ax) = 1, \quad \forall x \in \mathcal{W}_{\mathfrak{A}}^{\omega}. \tag{8.1}$$

This condition ensures that the function $f(x) \equiv 1$ is fixed point of the Ruelle transfer operator

$$\mathcal{R}_{\sigma, W, \beta} f(x) = \sum_{\sigma(y) = x} W_{\beta}(y) f(y) = \sum_{a \in \mathfrak{A}} W_{\beta}(ax) f(ax),$$

for the dynamical system given by the shift map σ and with weight given by the potential W_{β}.

A Keane potential W_{β} determines a stochastic process and a multifractal measure on $\mathcal{W}_{\mathfrak{A}}^{\omega}$ obtained by choosing a base point $x_0 \in \mathcal{W}_{\mathfrak{A}}^{\omega}$, and setting

$$\mu_{W, \beta, x_0}(\mathcal{W}_{\mathfrak{A}}^{\omega}(w)) = W_{\beta}(a_1 x_0) W_{\beta}(a_2 a_1 x_0) \cdots W_{\beta}(a_m \cdots a_2 a_1 x_0) \tag{8.2}$$

for a cylinder set

$$\mathcal{W}_{\mathfrak{A}}^{\omega}(w) = \{x = a_0 a_1 \cdots a_n \cdots \in \mathcal{W}_{\mathfrak{A}}^{\omega} \mid a_0 \cdots a_m = w\},$$

specified by a finite word $w = a_0 \cdots a_m \in \mathcal{W}_{\mathfrak{A}, m}$. The Keane condition ensures that (8.2) satisfies the additivity property of a measure, see [Dutkay, Jorgensen (2006)], [Marcolli, Paolucci (2011)]. This construction of measures via Keane potentials includes the cases of Bernoulli and Markov processes, considered in [Manin, Marcolli (2011)] in relation to coding theory:

- *Bernoulli*: for $x = a_0 a_1 a_2 \cdots a_n \cdots \in \mathcal{W}_{\mathfrak{A}}^{\omega}$ let $W_{\beta}(x) = e^{-\beta \lambda_{a_0}}$, with $\{\lambda_a\}_{a \in \mathfrak{A}}$ satisfying

$$\sum_{a \in \mathfrak{A}} e^{-\beta \lambda_a} = 1.$$

Then for $w = a_0 \cdots a_m \in \mathcal{W}_{\mathfrak{A}, m}$, the multifractal measure is given by

$$\mu_{W, \beta, x}(\mathcal{W}_{\mathfrak{A}}^{\omega}(w)) = \prod_{j=0}^{m} e^{-\beta \lambda_{a_j}}.$$

- *Markov:* for $x = a_0 a_1 a_2 \cdots a_n \cdots \in \mathcal{W}_{\mathfrak{A}}^\omega$ let $W_\beta(x) = e^{-\beta \lambda_{a_0 a_1}}$ with the stochastic matrix conditions

$$\sum_{a \in \mathfrak{A}} e^{-\beta \lambda_{ab}} = 1, \quad \forall b \in \mathfrak{A}.$$

Then for $w = a_0 \cdots a_m \in \mathcal{W}_{\mathfrak{A},m}$, the multifractal measure is given by

$$\mu_{W,\beta,x}(\mathcal{W}_{\mathfrak{A}}^\omega(w)) = e^{-\beta \lambda_{x_0 a_0}} \prod_{j=1}^{m} e^{-\beta \lambda_{a_{j-1} a_j}}.$$

In the Eternal Symmetree model of [Harlow, Shenker, Stanford, Susskind (2012)], the letters of the alphabet $\mathfrak{A}$ are "color" labels for different types of vacua, with associated entropies $\{S_a\}_{a \in \mathfrak{A}}$, and a collection of probabilities γ_{ab} at each node, for transitions between an incoming color a and an outgoing color b. In the case without pruning (without terminal vacua), it was shown in [Marcolli, Tedeschi (2014)] that the stochastic process considered in the Eternal Symmetree model is a variant of the second example above. More precisely, in the case without terminal vacua, the stochastic process of [Harlow, Shenker, Stanford, Susskind (2012)] is constructed in the following way. The probability $P_a(k)$ of obtaining a vacuum of type $a \in \mathfrak{A}$ after k steps from the root vertex in the Bethe tree is given by $P_a = e^{S_a/2} \Phi_a$, with S_a the entropy and the Φ_a satisfying the process

$$\Phi(k+1) = \mathcal{S} \, \Phi(k), \tag{8.3}$$

where $\mathcal{S}$ is a positive stochastic matrix with Perron–Frobenius eigenvalue $\lambda_{\mathcal{S}} = 1$ and positive Perron–Frobenius eigenvector $v_{\mathcal{S}}$. The probabilities γ_{ab} are related to $\mathcal{S}$ by

$$\delta_{ab} - \sum_c \gamma_{ca} \delta_{ab} + \gamma_{ab} = (ZSZ^{-1})_{ab},$$

where Z is the diagonal matrix with entries $e^{S_a/2}$. To fit this construction in the general setting described above, for a symmetric positive stochastic matrix $\mathcal{S} = (\mathcal{S}_{ab})$ with Perron–Frobenius eigenvalue $\lambda_{\mathcal{S}} = 1$ and positive (left) Perron–Frobenius eigenvector $v_{\mathcal{S}}$, normalized by $\sum_{a \in \mathfrak{A}} v_{\mathcal{S},a} = 1$, set

$$W(x) = v_{\mathcal{S},a_0} \, \mathcal{S}_{a_0,a_1} \, v_{\mathcal{S},a_1}^{-1}. \tag{8.4}$$

This satisfies the Keane condition. The entropies are inputed by $v_{\mathcal{S},a} = e^{-S_a/2}$. The resulting multifractal measure is then

$$\mu_{W,x}(\mathcal{W}_{\mathfrak{A}}^\omega(w)) = e^{(S_{x_1} - S_{w_m})/2} \mathcal{S}_{w_m w_{m-1}} \cdots \mathcal{S}_{w_2 w_1} \mathcal{S}_{w_1,x_1}.$$

The stochastic process (8.3) is then obtained by setting $\Phi_{x,a}(m+1) = e^{S_a/2} \sum_w \mu_{W,x}(\mathcal{W}_{\mathfrak{A}}^\omega(w))$. The dependence on the choice of a basepoint x

can be eliminated by averaging with respect to the normalized Hausdorff measure on $\mathcal{W}_{\mathfrak{A}}^\omega$,

$$\Phi_a(m) = \int_{\mathcal{W}_{\mathfrak{A}}^\omega} \Phi_{x,a}(m)\, d\mu_H(x).$$

This can be seen as a special case of the Markov processes mentioned above, by setting

$$W_\beta(x) = e^{-\beta\lambda_{x_1 x_2}} = e^{-\beta\lambda_{x_1}}\, \mathcal{S}_{x_1 x_2}\, e^{\beta\lambda_{x_2}},$$

with $\mathcal{S} = (\mathcal{S}_{ab})_{a,b\in\mathfrak{A}}$ symmetric positive stochastic as above, and with λ_a such that $e^{-\beta\lambda_a} := (v_S)_a = e^{-S_a/2}$, with v_S the normalized (left) Perron-Frobenius eigenvector.

Pruning can be accounted for in this approach in a special case where the pruning of the tree is performed by consistently cutting off certain sequences of vertices, replacing the fully shift dynamical system $(\mathcal{W}_{\mathfrak{A}}^\omega, \sigma)$ with a subshift of finite type $(\mathcal{W}_{\mathfrak{A},A}^\omega, \sigma)$, determined by a matrix $A = (A_{ab})_{a,b\in\mathfrak{A}}$ with entries $A_{a,b} \in \{0,1\}$, where the zero entries determine which edges are pruned and the entries equal to one determine which edges remain. The Cantor set $\mathcal{W}_{\mathfrak{A},A}^\omega$ is given by

$$\mathcal{W}_{\mathfrak{A},A}^\omega = \{a_0 a_1 \cdots a_n \cdots \in \mathcal{W}_{\mathfrak{A}}^\omega \mid A_{a_k a_{k+1}} = 1,\ \forall k \geq 0\}.$$

It is clearly invariant under the action of the shift map σ. The Keane condition for a potential on $\mathcal{W}_{\mathfrak{A},A}^\omega$ becomes, correspondingly, of the form

$$\sum_{a\in\mathfrak{A}} A_{a x_1}\, W_\beta(\sigma_a(x)) = 1, \qquad (8.5)$$

which is equivalent to the condition $\sum_{y:\sigma(y)=x} W_\beta(y) = 1$, written in terms of the shift map $\sigma : \mathcal{W}_{\mathfrak{A},A}^\omega \to \mathcal{W}_{\mathfrak{A},A}^\omega$. As in the unpruned case, let $\mathcal{S} = (\mathcal{S}_{ab})$ be a positive stochastic matrix, and set $\tilde{\mathcal{S}} = A\mathcal{S}$. This is a non-negative matrix with Perron–Frobenius $\lambda_{\tilde{\mathcal{S}}} < 1$ and Perron–Frobenius eigenvector $v_{\tilde{\mathcal{S}}} = (v_{\tilde{\mathcal{S}},a})_{a\in\mathfrak{A}}$, with $v_{\tilde{\mathcal{S}},a} > 0$. Let β and $\{\lambda_a\}_{a\in\mathfrak{A}}$ be chosen so that $e^{-\beta\lambda_a} = v_{\tilde{\mathcal{S}},a}$. Then the potential

$$W_\beta(x) = e^{-\beta\lambda_{x_1 x_2}} := \frac{1}{\lambda_{\tilde{\mathcal{S}}}}\, e^{-\beta\lambda_a}$$

satisfies the Keane condition (8.5),

$$\sum_{a\in\mathfrak{A}} A_{ab}\, e^{-\beta\lambda_{ab}} = 1, \quad \forall b \in \mathfrak{A}.$$

The stochastic process in the pruned tree is given by

$$\Phi_a(m+1) = \lambda_{\tilde{\mathcal{S}}}^m e^{S_a/2} \sum_{w\in\mathcal{W}_{\mathfrak{A},A,m}} \nu_{W,\beta}(\mathcal{W}_{\mathfrak{A},A}^\omega(aw)),$$

satisfying $\Phi_a(m+1) = \sum_b S_{ab}\Phi_b(m)$ for entropies S_a satisfying $e^{-S_a/2}/\mathcal{N} = e^{-\beta\lambda_a}$, with normalization factor $\mathcal{N} = \sum_a e^{-S_a/2}$. Unlike [Harlow, Shenker, Stanford, Susskind (2012)], here we are maintaining the normalization of the measure, by maintaining the Keane condition on the potential. This results in dividing by increasingly large powers of the Perron–Frobenius eigenvalue $\lambda_{\tilde{S}}$, along the stochastic process. This is equivalent to the observation of [Harlow, Shenker, Stanford, Susskind (2012)], [Susskind (2012)] that, in the presence of terminal vacua, the eternal inflation is concentrated on a fractal set of increasingly small volume with respect to the original stochastic process of the unpruned tree, which scales by a power of $\lambda_{\tilde{S}}$.

Let β and $\{\lambda_a\}_{a\in\mathfrak{A}}$ be chosen so that $e^{-\beta\lambda_a} = v_{\tilde{S},a}$ are the components of the normalized (right) Perron–Frobenius eigenvector of $\tilde{S} = AS$, with eigenvalue $\lambda_{\tilde{S}}$. Let

$$W_\beta(x) = e^{-\beta\lambda_{x_1 x_2}} = \frac{1}{\lambda_{\tilde{S}}} e^{-\beta\lambda_a} S_{ab} e^{\beta\lambda_b}. \tag{8.6}$$

Then W_β satisfies the Keane condition. Let $T_{ab} = A_{ab}e^{-\beta\lambda_{ab}}$ with Perron–Frobenius eigenvector v_T with eigenvalue $\lambda_T = 1$. The components of v_T satisfy $v_{T,a} = q^{-1}$ for all $a \in \mathfrak{A}$, with $q = \#\mathfrak{A}$. The measure $\nu_{W,\beta}$ associated to the potential (8.6) is given by

$$\nu_{W,\beta}(\Lambda_{\mathfrak{A},A}(w)) = \frac{e^{\beta(\lambda_{w_k}-\lambda_{w_1})}}{q\,\lambda_{\tilde{S}}^{k-1}} S_{w_1 w_2}\cdots S_{w_{k-1}w_k}. \tag{8.7}$$

For entropies S_a satisfying $e^{-S_a/2}/\mathcal{N} = e^{-\beta\lambda_a}$, with the normalization factor $\mathcal{N} = \sum_a e^{-S_a/2}$, consider the process

$$\Phi_a(m+1) = \lambda_{\tilde{S}}^m e^{S_a/2} \sum_{w\in\mathcal{W}_{\mathfrak{A},A,m}} \nu_{W,\beta}(\Lambda_{\mathfrak{A},A}(aw)).$$

This determines a stochastic process on the pruned tree satisfying $\Phi_a(m+1) = \sum_b S_{ab}\Phi_b(m)$. For this random process with $e^{\beta\lambda_a} = e^{S_a/2}$, the propagator $C_{a,b}(x,y)$ is given by

$$C_{a,b}(x,y) = \frac{e^{(S_a+S_b)/2}}{\mathcal{N}q\lambda_{\tilde{S}}^{k-1}} \sum_{c\in\mathfrak{A}} S_{c,b}^{k-2} S_{c,a}^{k-2}.$$

Again, here we have a power of $\lambda_{\tilde{S}}$ in the denominator (instead of the numerator as in [Harlow, Shenker, Stanford, Susskind (2012)]) because we are measuring volumes with respect to a measure that remains normalized on a fractal that scales in size as a power of $\lambda_{\tilde{S}}$, with respect to the original measure on the unpruned tree.

It is also shown in [Marcolli, Tedeschi (2014)] that the construction can be generalized to quotients of the Bruhat–Tits tree by the action of p-adic Schottky groups (Mumford curves). The resulting quotient graphs have a central finite graph with infinite trees attached to the vertices, so asymptotically they again look like trees but they contain some trapped region which is a generalization of a BTZ black hole. We refer the reader to [Manin, Marcolli (2001)] and [Heydeman, Marcolli, Saberi, Stoica (2016)] for a more detailed discussion of these p-adic black holes. A more general theory of the AdS/CFT holographic correspondence over p-adic Bruhat–Tits trees and their quotient was more recently developed in [Heydeman, Marcolli, Saberi, Stoica (2016)], in relation to tensor networks and entanglement entropy.

Chapter 9

Noncommutative Quantum Cosmology

In this chapter we discuss various aspects of the relation between the spectral action model of gravity and the Hartle–Hawking approach to quantum cosmology. We first review some basic ideas of the Hartle–Hawking formulation of Euclidean quantum gravity and quantum cosmology, and how these ideas can be transposed in a model of gravity based on the spectral action functional. We then describe a construction of categories and algebras of geometries with dynamical evolutions based on gravity action functionals and other physical and topological invariants, based on [Marcolli, Zainy al-Yasry (2008)]. We then use a discretization of this construction to enrich the formalism of spin networks and spin foams in quantum gravity with topological data that render the background topology dynamical, through monodromy data, in addition to the holonomy data describing the metric structure. We again consider categories and algebras of these topological spin networks and foams and their dynamical evolution, based on [Denicola, Marcolli, Zainy al-Yasry (2010)]. We present also a discretization of almost commutative geometries, via gauge networks, consisting of systems of finite spectral triples on a graph, and we show that they recover the usual spectral geometry and Dirac operator in a suitable continuum limit, based on [Marcolli, van Suijlekom (2014)]. Finally we present an approach to path integrals of noncommutative geometries, based on finite spectral triples, developed in [Barrett (2015)], [Barrett, Glaser (2015)].

9.1 Hartle-Hawking quantum cosmology

Some of the main ideas in the Hartle–Hawking approach to quantum cosmology [Hartle, Hawking (1983)] can be summarized as follows:

- The quantum state of the universe is described by a path integral over 4-dimensional geometries.
- This path integral is treated via perturbative calculations around a mini-superspace background.
- The wave function of the universe satisfies a Wheeler-DeWitt equation (as a zero-energy Schröedinger equation).
- The oscillatory part of the wave function gives an ensemble of classical spacetime evolutions.
- The most probable evolution of the universe arises via quantum tunneling from gravitational instantons.
- Bianchi IX anisotropies are damped away by Planck inflation.

We have listed here some quick slogans pointing to some aspects of Hartle–Hawking quantum cosmology that will be relevant in our setting. We refer the reader to the introductory survey of the volume [Fang, Ruffini (1987)] for a more careful overview, and to the papers collected in the same volume for a detailed account of the development of the field. We discuss here how some of the main aspects of Hartle–Hawking quantum cosmology fit with the approach to gravity and cosmology via Noncommutative Geometry and the spectral action functional that we present in this book.

9.1.1 *Path integral and wave function of the universe*

The Hartle–Hawking path integral is defined over all Riemannian 4-dimensional geometries,

$$\Psi[h] = \int e^{-S(g)} \, \mathcal{D}[g],$$

where the classical Euclidean action for gravity (including cosmological term) is

$$S(g) = -\frac{1}{16} \left(\int_W (R - 2\Lambda) \sqrt{g} \, d^4x + \int_M 2K \sqrt{h} \, d^3x \right)$$

where W is a 4-manifold with boundary $\partial W = M$, g and h are compatible Riemannian metrics on W and on M, while K is the trace of the II-fundamental form, which describes the extrinsic geometry of ∂W. The sum is over all 4-dimensional geometries g with fixed boundary value h.

In our setting, we consider the spectral action as an action functional for gravity, and we also consider the possibility of enlarging the 4-dimensional smooth spacetime geometry with an almost-commutative geometry whose noncommutative extra dimensions are a finite space describing the presence of a matter sector.

In [Chamseddine, Connes (2011)] a formulation of the spectral action with boundary is obtained. The approach is based on identifying suitable boundary conditions for the Dirac operator on a manifold with boundary and extend the formulation to more general (possibly noncommutative) spaces, just as the original notion of spectral triple reflects the analytic properties of the Dirac operator on a smooth compact Riemannian spin manifold without boundary. The requirement that the hermitian property $\langle \psi, D\psi \rangle = \langle D\psi, \psi \rangle$ of the Dirac operator is preserved determines a choice of boundary conditions

$$\Pi_- \psi|_{\partial W} = 0,$$

where $\Pi_- = (1 - \chi)/2$, with χ the $\mathbb{Z}/2\mathbb{Z}$-grading related to the grading γ_5 by $\gamma_5 = \chi \gamma^\perp$, with $\gamma^\perp = \gamma^\mu n_\mu$, where $g_{\mu\nu} = h_{ab} e^a_\mu e^b_\nu + n_\mu n_\nu$. Setting $\gamma^a = \gamma^\mu e^a_\mu$ gives

$$\chi = -\frac{\sqrt{h}}{3!} \epsilon^{abc} \gamma_a \gamma_b \gamma_c,$$

with $\chi^2 = 1$, $\chi \gamma_a = \gamma_a \chi$, $\chi \gamma^\perp = -\gamma^\perp \chi$, and $\chi \gamma_5 = -\gamma_5 \chi$. For a Dirac operator on a 4-dimensional manifold W with boundary $\partial W = M$, the Seeley-DeWitt coefficients can be computed explicitly, [Branson, Gilkey (1992)]. The local expressions involve terms that depend on the extrinsic curvature K_{ab}. It is shown in [Chamseddine, Connes (2011)] that this leads to an asymptotic expansion for the spectral action on a 4-dimensional manifold with boundary, where the a_2 term involves the correct combination of the Einstein–Hilbert action and the trace of the II-fundamental form that is needed for the Hartle–Hawking formulation,

$$a_2(D^2, \chi) = -\frac{1}{24\pi^2} \left(\int_W \frac{1}{2} R \sqrt{g} \, d^4x + \int_M K \sqrt{h} \, d^3x \right).$$

Thus, the spectral action, defined with the appropriate boundary conditions, is compatible with the Hartle–Hawking form of quantum cosmology.

In [Hartle, Hawking (1983)], see also [Hawking (1984b)], one also considers more generally the case of gravity coupled to matter, with a Euclidean action functional of the form

$$S(g, \phi) = -\frac{M_{Pl}^2}{16\pi} \left(\int_M 2K \sqrt{h} \, d^3x + \int_W \left(R - 2\Lambda + \frac{16\pi}{M_{Pl}^2} \mathcal{L}(g, \phi) \right) \sqrt{g} \, d^4x \right),$$

with $M = \partial W$, where $\mathcal{L}(g, \phi)$ is the Lagrangian for the matter field ϕ coupled to gravity. The Hartle–Hawking wave function, in this case, is given by

$$\Psi[h, \xi] = \int e^{-S(g,\phi)} \, \mathcal{D}[g] \, \mathcal{D}[\phi],$$

where the sum is over all 4-geometries g with boundary h and over the matter field configurations ϕ with assigned boundary conditions $\phi|_{\partial W} = \xi$.

As we have discussed in the previous chapters of this book, the presence of a matter sector can be accounted for in the spectral action model by considering the same action functional for gravity (the spectral action) but modifying the underlying geometry to an almost-commutative geometry, where a finite noncommutative space determines the matter sector and the Lagrangian for matter coupled to gravity. In the case of scalar field ϕ conformally coupled to gravity, this can be obtained by scalar fluctuations of the Dirac operator, as we have discussed in relation to slow-roll inflation models.

9.1.2   *Hamiltonian constraint, Wheeler-DeWitt equation*

The Hartle–Hawking wave function satisfies a functional differential equation, the Wheeler–DeWitt equation, which is obtained by writing the metric in a neighborhood of the 3-manifold M in the form

$$ds^2 = (N^2 - N_i N^i)d\tau^2 + 2N_i dx^i d\tau + h_{ij} dx^i dx^j,$$

where τ is a coordinate that is constant on M. The constraint

$$\int \frac{\delta S(g(N, N_i, \tau, h), \phi)}{\delta N} \, e^{-S(g,\phi)} \, \mathcal{D}[g] \, \mathcal{D}[\phi] = 0,$$

expressing the fact that the wave function is independent of time, gives the Wheeler–DeWitt equation, which corresponds to the Hamiltonian constraint of General Relativity

$$\frac{\delta S(g(N, N_i, \tau, h), \phi)}{\delta N} = -\sqrt{h} \left(K^{ij} K_{ij} - K^2 + R(h) - 2\Lambda + \frac{8\pi}{M_{Pl}^2} \right) = 0.$$

The Wheeler–DeWitt equation is usually written in the form

$$\left(G_{ijkl} \frac{\delta^2}{\delta h_{ij} \delta h_{kl}} + \sqrt{h} \left(R(h) - 2\Lambda + \frac{8\pi}{M_{Pl}^2} T^\perp (\frac{\delta}{\delta \psi}, \psi) \right) \right) \Psi[h, \psi] = 0,$$

where G_{ijkl} is the metric

$$G_{ijkl} = \frac{1}{2} \sqrt{h} \left(h_{ik} h_{jl} + h_{il} h_{jk} - h_{ij} h_{kl} \right),$$

on the superspace, that is, the space of all 3-dimensional geometries h, and where $T^\perp$ is the energy-momentum tensor of the matter sector projected in the direction normal to M. The metric G_{ijkl} on the superspace has signature $(-, +, +, +, +, +)$, which gives $\sqrt{h}$ the meaning of a time coordinate on superspace. The first term of the equation is just reformulating the extrinsic curvature term of the Hamiltonian constraint through

$$\frac{\delta}{\delta h_{ij}} S(g(N, N_i, \tau, h), \phi) = \frac{M_{Pl}^2}{16\pi} \sqrt{h}(K_{ij} - h_{ij}K).$$

In our setting, with a model of gravity based on the spectral action, the wave function of the universe would be expressed in terms of a Hartle–Hawking path integral

$$\Psi[D_\partial, \Lambda] = \int e^{-\mathcal{S}(D/\Lambda)} \mathcal{D}[D],$$

where $\mathcal{S}(D/\Lambda)$ is the spectral action functional and the sum is over all commutative (or almost-commutative in the case coupled to matter) geometries, expressed as a sum over Dirac operators. Here we assume to work with a versions of spectral triples with boundary, as in [Chamseddine, Connes (2011)], and we assume the boundary geometry to be fixed, so that the resulting wave function is a function of the boundary spectral triple (through its Dirac operator D_∂).

It is natural to expect that, in the case of the spectral action, an analog of the Wheeler–DeWitt equation for the Hartle–Hawking wave function, which corresponds to the Hamiltonian constraint of General Relativity, would be given by a compatibility condition relating the a_3 coefficient of the expansion of the spectral action on the 4-dimensional geometry with boundary and the a_3 and a_1 coefficients of the spectral action expansion for the boundary spectral geometry. To see this more precisely, we consider first the case without matter sector, where we are simply considering the spectral action over a commutative 4-dimensional geometry W with 3-dimensional boundary $M = \partial W$. The case with matter sector will be obtained by replacing both the 4-dimensional geometry and its 3-dimensional boundary with compatible almost-commutative geometries.

More precisely, the computation of [Chamseddine, Connes (2011)] of the spectral action on a 4-dimensional manifold with boundary gives for the spectral action $\mathcal{S}(D/\Lambda)$ the expansion

$$2(f_4 \Lambda^4 a_0(D^2, \chi) + f_2 \Lambda^2 a_2(D^2, \chi) + f_1 \Lambda a_3(D^2, \chi) + f_0 a_4(D^2, \chi)),$$

up to terms of order $O(\Lambda^{-2})$, where

$$a_0(D^2,\chi) = \int_W \sqrt{g}\, d^4x,$$

$$a_2(D^2,\chi) = -\frac{1}{24\pi^2}\left(\int_W \frac{1}{2}R\,\sqrt{g}\,d^4x + \int_M K\,\sqrt{h}\,d^3x\right)$$

$$a_3(D^2,\chi) = \frac{1}{32(4\pi)^{3/2}}\int_M (K^2 - 2K_{ab}K^{ab})\,\sqrt{h}\,d^3x$$

$$\begin{aligned}
a_4(D^2,\chi) = \frac{1}{360}\frac{1}{16\pi^2}\Bigg(&\int_W (5R^2 - 8R_{\mu\nu}^2 - 7R_{\mu\nu\rho\sigma}^2 - 12R_{;\mu}^\mu)\sqrt{g}\,d^4x\\
&+4\int_M \frac{2}{21}(17K^3 + 39KK_{ab}K^{ab} - 116K_a^b K_b^c K_c^a)\sqrt{h}\,d^3x\\
&+4\int_M (5RK + 4KR_{nan}^a + 4K_{ab}R_{acb}^c + 18R_{anbn}K^{ab})\sqrt{h}\,d^3x\Bigg)
\end{aligned}$$

while the a_1 coefficient vanishes and does not appear in the expansion above. On the other hand, for a 3-dimensional geometry M with Dirac operator D_∂ the spectral action would be of the form

$$\mathcal{S}(D_\partial/\Lambda) \sim 2(f_3\Lambda^3 a_1(D_\partial) + f_1\Lambda a_3(D_\partial))$$

up to terms that are small for large Λ. Here the a_1-term contributes a volume term while the a_3 term contributes a curvature term for h. Imposing a compatibility condition between the $a_3(D^2,\chi)$ term and the $a_1(D_\partial)$ and $a_3(D_\partial)$ terms would then produce an equation that closely resembles the constraint used for the Wheeler–DeWitt equation.

Note that the expansion of the spectral action, as we have discussed already, includes higher derivative terms, like the Weyl curvature and the Gauss–Bonnet gravity term. The inclusion of modified gravity models with higher derivative terms is compatible with the Hartle–Hawking approach to quantum cosmology and was discussed in [Hawking, Luttrell (1984)].

9.1.3   *Minisuperspace models*

In the Hartle–Hawking setting, solutions of the Wheeler–DeWitt equation with given boundary conditions can be found in a special setting where one restricts the degrees of freedom of superspace, by restricting to a certain class of *minisuperspace* models, see [D'Eath (1996)], [Fang, Ruffini (1987)].

Minisuperspaces include the (Riemannian) Robertson–Walker

$$ds^2 = dt^2 + a(t)^2 d\sigma^2,$$

with $d\sigma^2$ the standard round metric on S^3, and also the Bianchi IX models with $SU(2)$-symmetry

$$ds^2 = F(t)dt^2 + \frac{W_2(t)W_3(t)}{W_1(t)}\sigma_1^2 + \frac{W_3(t)W_1(t)}{W_2(t)}\sigma_2^2 + \frac{W_1(t)W_2(t)}{W_3(t)}\sigma_3^2,$$

with conformal factor $F(t) \sim W_1(t)W_2(t)W_3(t)$, that we discussed at length in Chapters 5 and 6. These minisuperspace models have homogeneous (though not necessarily isotropic) 3-dimensional spatial sections: this high degree of symmetry makes it possible to encode the metric structure as a finite dimensional problem. In particular, the gravitational instanton solutions, as we have discussed in Chapter 5 and 6, are expressed as solutions of a system of ordinary differential equations, the Painlevé VI equations. In these minisuperspace models, the Hartle–Hawking path integral becomes an integral over these restricted degrees of freedom. The wave function is then $\Psi[a, \psi]$ or $\Psi[W_i, F, \psi]$, for the cases of minisuperspace mentioned above, where a and W_i, F are the scaling and conformal factors restricted to the boundary 3-geometry M, and $\psi = \phi|_M$ is a coupled matter field as before. The Wheeler–DeWitt equation becomes a differential equation for this wave function. In the isotropic case, with ϕ a conformally invariant scalar field that is constant on spatial sections, the equation takes the form

$$\frac{1}{2}\left(\frac{1}{a^p}\frac{\partial}{\partial a}a^p\frac{\partial}{\partial a} - a^2 + \lambda a^4 - \frac{\partial^2}{\partial \psi^2} + \psi^2\right)\Psi[a, \psi] = 0,$$

where the exponent p is an arbitrary power that accounts for the ambiguity in the ordering of operators in the quantization of the Hamiltonian constraint. The equation uncouples, with solutions of the form $\Psi[a, \psi] = C(a)f(\psi)$, with $f(\psi)$ behaving like a harmonic oscillator, and $C(a)$ behaving like a wave function $\Psi[a]$ in the case without matter, see [Hawking (1984b)] for a more detailed account. The case with a spatially constant massive scalar field is discussed in [Hawking (1984)]. The boundary conditions of the Wheeler–deWitt equation are obtained from the semiclassical approximation of the path integral, which becomes a combination of exponential terms with the value of the action at classical gravitational instanton solutions with assigned values of the parameters on the 3-dimensional boundary M.

The spectral action of the Riemannian Friedmann–Robertson–Walker metric was computed in [Chamseddine, Connes (2012b)] and [Fathizadeh,

Ghorbanpour, Khalkhali (2014)], while the spectral action of the Bianchi IX minisuperspace models was computed in [Fan, Fathizadeh, Marcolli (2015)] and [Fan, Fathizadeh, Marcolli (2015b)], as discussed in Chapter 6 above.

Following the point of view proposed above, we can think of the boundary conditions for the wave functions $\Psi[a, \psi]$ or $\Psi[W_i, F, \psi]$ in these cases coming from the semiclassical approximation of the path integral, with a combination of exponentials of the values of the spectral action at classical gravitational instanton solutions. The appropriate Wheeler–DeWitt equation should again arise from a compatibility condition for the spectral action on the 4-dimensional manifold with boundary and the spectral action on the boundary 3-dimensional geometry. As in the minisuperspace model it will become an equation in the scale factors on the 3-dimensional slice, which occur in the coefficients of the expansion of the spectral action. The details of this argument will be treated elsewhere.

9.1.4 *Exotic smoothness*

In the usual formulation, the Hartle–Hawking path integral is a sum over four-dimensional geometries. In 4-dimensions, however, the interesting and very delicate phenomenon of exotic smoothness occurs, namely there are smooth 4-manifolds that are homeomorphic but not diffeormorphic. There are also topological 4-manifolds that do not admit any smooth structures, but it is reasonable to assume that those would not contribute to the path integral. It is more subtle to examine how homeomorphic but non-diffeomorphic manifolds should be accounted for, for the purposes of the path integrals of Euclidean quantum gravity. We refer the reader to [Scorpan (2005)] for a detailed introduction to the geometry and topology of 4-manifolds and the exotic smoothness phenomena.

The contribution of exotic smoothness to Euclidean quantum gravity was analyzed in an explicit example in [Duston (2011)]. A specific family of simply-connected, compact, smooth 4-manifolds is chosen, which were constructed and studied in [Braungardt, Kotschick (2005)], [Salvetti (1989)]. The manifolds Y_r are algebraic surfaces obtained as iterated branched coverings of $\mathbb{P}^2(\mathbb{C})$, of degrees d_i, branched along smooth curves of degrees $n_i = d_i m_i$, for a set of positive integers $\{d_i, m_i\}_{i=1,\ldots,r}$. The canonical class and the signature were computed in [Braungardt, Kotschick (2005)] and are given by

$$K_{Y_r} = d_1 \cdots d_r \left(\sum_{j=1}^{r} (d_j - 1)m_j - 3 \right)^2 = d_1 \cdots d_r n^2$$

$$\sigma(Y_r) = -\frac{1}{3} d_1 \cdots d_r \left(\sum_{j=1}^{r} (d_j^2 - 1)m_j^2 - 3 \right).$$

In the same paper they also proved that, for every $k \in \mathbb{N}$ there are manifolds Y_r that have at least k distinct smooth structures supporting Einstein metrics.

As Einstein manifolds, these can be considered gravitational instantons around which one can expand the Hartle–Hawking path integral

$$Z = \int e^{-S(g)} \, \mathcal{D}[g]$$

using a semiclassical approximation, so that it is replaced by a finite sum

$$Z = \sum_i e^{-S(g_i)}.$$

In [Duston (2011)] an explicit expectation value is computed, the expectation value of the volume

$$\langle V \rangle = \frac{\sum_i V_i e^{-S(g_i)}}{\sum_i e^{-S(g_i)}}.$$

This is done in both the case where the action $S(g)$ is the Einstein–Hilbert action, and in the case where it is the spectral action. The result is shown to depend on the total degree of the branched coverings, in a way that weights the quantum effects more in the Einstein-Hilbert rather than in the spectral action case. The correlation functions are also computed, by a similar technique, and the computation is applied to specific manifolds Y_r in the given family. The expectation value of volume is dominated by the larger volume, while in the one-loop corrections, it is shown that the result is strongly dependent on the conformal scale of the metric. This shows in an explicit case that exotic structures should be accounted for in the path integral. We refer the reader to [Duston (2011)] for a more detailed account and explicit results.

9.2   Categories and algebras of geometries

This section is based on [Marcolli, Zainy al-Yasry (2008)]. It presents a general construction of categories and 2-categories of 3 and 4-dimensional geometries and associated algebras with dynamical evolutions. As we see in the following sections, this approach suggests discretizations of (commutative and almost-commutative) spacetime geometries that can be used as a possible approach to a quantization based on noncommutative geometry.

The Hilden–Montesinos theorem, in its weak form, states that every compact oriented (smooth) 3-manifold M is a branched cover of S^3 branched along a link $L \subset S^3$. In its strong form, it states that every compact oriented (smooth) 3-manifold M is a 3-fold branched cover of S^3 branched along a knot $K \subset S^3$. It is known that there are *universal knots* (Hilden–Lozano–Montesinos), namely knots $K \subset S^3$ such that every M is a branched cover of S^3 branched along the same universal knot K.

The description of 3-manifolds as branched coverings of the 3-sphere is very non-unique. For example, M is the Poincaré homology sphere (the dodecahedral space we have encountered several times already in previous chapters), them M can be obtained as a 5-fold cover branched along the trefoil knot, or as a 3-fold cover branched along the $(2,5)$ torus knot, or also as a 2-fold cover branched along $(3,5)$ torus knot.

This ambiguity in the description of a 3-manifold as a branched covering suggests using these branched coverings as a form of *correspondences*. We define a set of *geometric correspondences* $\mathcal{C}(K, K')$ between two *embedded graphs* to be given by data of the form

$$K \subset L \subset S^3 \xleftarrow{\pi_K} M \xrightarrow{\pi_{K'}} S^3 \supset L' \supset K'$$

where K, K', L, L' are embedded *graphs* in S^3. As a particular case, this includes the possibility that L, L' and K, K' are links and knots. The set $\mathcal{C}(K, K)$ contains the trivial unbranched identity map $id : S^3 \to S^3$ (which is the only unbranched covering since S^3 is simply connected). If K is a universal knot, then $\mathcal{C}(K, K)$ contains all smooth compact 3-manifolds.

We define the set of *virtual correspondences* to be

$$\mathrm{Hom}_R(K, K') = R[\mathcal{C}(K, K')],$$

where R is a commutative ring. Elements of this set are formal finite linear combinations

$$\phi = \sum_M a_M \, M$$

with coefficients $a_M \in R$.

Branched coverings can be described in terms of representations

$$\sigma : \pi_1(S^3 \setminus L) \to S_n,$$

where S_n is the symmetric group of permutations of n elements. The data $K \subset L \subset S^3 \leftarrow M \to S^3 \supset L' \supset K'$ of geometric correspondences in $\mathcal{C}(K, K')$ determine a subset of

$$\bigcup_{n,m,L,L'} \mathrm{Hom}(\pi_1(S^3 \backslash L), S_n) \times \mathrm{Hom}(\pi_1(S^3 \backslash L'), S_m)$$

with the condition that (σ_1, σ_2) define isomorphic 3-manifolds. This ensures that the categories we will be discussing are small categories.

The composition of correspondences

$$\circ : \mathcal{C}(K, K') \times \mathcal{C}(K', K'') \to \mathcal{C}(K, K'')$$

is given by the fibered product. Given two correspondences

$$K \subset L \subset S^3 \xleftarrow{\pi_K} M \xrightarrow{\pi_1} S^3 \supset L_1' \supset K'$$

with generic fibers $\#\pi_K^{-1}(x) = n$, $\#\pi_1^{-1}(x) = m$, and

$$K' \subset L_2' \subset S^3 \xleftarrow{\pi_2} \tilde{M} \xrightarrow{\tilde{\pi}_{K''}} S^3 \supset L'' \supset K''$$

with generic fibers $\#\pi_2^{-1}(x) = \tilde{n}$, $\#\pi_{K''}^{-1}(x) = \tilde{m}$, the composition is given by $\hat{M} = M \circ \tilde{M} := M \times_{K'} \tilde{M}$

$$= \{(x, y) \in M \times \tilde{M} \,|\, \pi_1(x) = \pi_2(y)\}$$

which gives a branched cover

$$L \cup \pi_K \pi_1^{-1}(L_2) \subset S^3 \xleftarrow{\hat{\pi}_K} \hat{M} \xrightarrow{\hat{\pi}_{K''}} S^3 \supset L'' \cup \pi_{K''} \pi_2^{-1}(L_1)$$

with generic fibers $\#\hat{\pi}_K^{-1}(x) = n\tilde{n}$, $\#\hat{\pi}_{K''}^{-1}(x) = m\tilde{m}$.

For example, if M_n denotes the n-fold cyclic branched cover of S^3 branched along the unknot O, then $M_m \circ M_n = M_{mn}$ in $\mathcal{C}(O, O)$.

For technical reasons related to this composition operation, it is preferable to extend the construction by allowing not only smooth compact 3-manifolds, but also *orbifolds*. However, we will not discuss this further here for simplicity. The composition defined in this way is associative, with identity element $\mathbb{U}_K \in \mathcal{C}(K, K)$ given by the trivial unbranched $id : S^3 \to S^3$.

Thus, we obtain a *semigroupoid* (which is the same thing as a small category) $\mathcal{G}$, with units $\gamma \in \mathcal{U}(\mathcal{G})$ satisfying $\gamma\alpha = \alpha$ and $\beta\gamma = \beta$ when defined, and source and target maps $s(\alpha)$, $t(\alpha)$ in $\mathcal{U}(\mathcal{G})$

$$\mathcal{G}_\gamma = \{\alpha \in \mathcal{G} | s(\alpha) = \gamma\}$$

with $s(\alpha\beta) = s(\alpha)$ and $t(\alpha\beta) = t(\beta)$. Composition is defined when $s(\beta) = t(\alpha)$. Morphisms are not invertible, hence the semigroupoid is not a groupoid.

There is a *convolution algebra* associated to a semigroupoid. Let R be a commutative ring, and let $R[\mathcal{G}]$ be the ring of functions $f : \mathcal{G} \to R$ with finite support with the product

$$(f_1 \star f_2)(\alpha) = \sum_{\alpha_1, \alpha_2 \in \mathcal{G}: \alpha_1 \alpha_2 = \alpha} f_1(\alpha_1) f_2(\alpha_2)$$

where $f = \sum_{\alpha \in \mathcal{G}} a_\alpha \delta_\alpha$, a finite sum with coefficients $a_\alpha \in R$.

We consider an additive category $\mathcal{C}(\mathcal{G})$ obtained as follows. For $\gamma, \gamma' \in \mathcal{U}(\mathcal{G})$ let

$$\mathrm{Hom}_R(\gamma, \gamma') \ni \phi = \sum_\alpha a_\alpha \delta_\alpha,$$

for $a_\alpha \in R$, $\alpha \in \mathcal{G}$, $s(\alpha) = \gamma$, $t(\alpha) = \gamma'$. The additive envelope $Mat(\mathcal{C}(\mathcal{G}))$ has objects that are formal direct sums $\oplus_i \gamma_i$ and morphisms that are matrices of morphisms $\phi_{ij} \in \mathrm{Hom}_R(\gamma_i, \gamma_j)$.

In our case, we obtain an additive category $\mathcal{K}$ of embedded graphs and correspondences, with $Obj(\mathcal{K})$ being the set of embedded graphs $K \subset S^3$ (considered up to ambient isotopy) and the morphisms $\mathrm{Hom}_\mathcal{K}(K, K')$ given by finite sums with $\mathbb{Z}$-coefficients

$$\phi = \sum_M a_M M$$

where here M stands for the datum of branched covers

$$K \subset L \subset S^3 \leftarrow M \to S^3 \supset L' \supset K'.$$

If M is multi-connected, say with connected components M_1 and M_2, we impose the identification

$$M = M_1 \amalg M_2 \Leftrightarrow M_1 + M_2.$$

The category $Mat(\mathcal{K})$ has objects $\oplus_i K_i$ and morphisms $\Phi = (\phi_{ij})$ with $\phi_{ij} \in \mathrm{Hom}_\mathcal{K}(K_i, K_j')$.

The convolution algebra of 3-manifolds is obtained by considering the semigroupoid with arrows

$$\alpha = \left(K \subset L \subset S^3 \xleftarrow{\pi_K} M \xrightarrow{\pi_{K'}} S^3 \supset L' \supset K' \right)$$

which we write with the shorthand notation $\alpha = (M, K, K')$, with

$$s(\alpha) = (\mathbb{U}_K, K, K) \quad t(\alpha) = (\mathbb{U}_{K'}, K', K').$$

The set $\mathcal{G}_K$ consists of 3-manifolds that are branched covers along K. We then consider the ring $\mathbb{C}[\mathcal{G}]$ os functions $f : \mathcal{G} \to \mathbb{C}$ with finite support with product

$$(f_1 \star f_2)(M) = \sum_{M_1 \circ M_2 = M} f_1(M_1) f_2(M_2).$$

We can also define an involution by taking $\alpha = (M, K, K') \mapsto \alpha^\vee = (M, K', K)$

$$\alpha = \left(K \subset L \subset S^3 \xleftarrow{\pi_K} M \xrightarrow{\pi_{K'}} S^3 \supset L' \supset K' \right)$$

$$\alpha^\vee = \left(K' \subset L' \subset S^3 \xleftarrow{\pi_{K'}} M \xrightarrow{\pi_K} S^3 \supset L \supset K \right)$$

and set $f^\vee(\alpha) = \overline{f(\alpha^\vee)}$.

A time evolution on this algebra can be assigned by taking $\sigma : \mathbb{R} \to \mathrm{Aut}(\mathbb{C}[\mathcal{G}])$ given by

$$\sigma_t(f)(M) := \left(\frac{n}{m} \right)^{it} f(M)$$

for

$$K \subset L \subset S^3 \xleftarrow{\pi_K} M \xrightarrow{\pi_{K'}} S^3 \supset L' \supset K'$$

where the generic fibers of the two branched covering maps have $\#\pi_K^{-1}(x) = n$, $\#\pi_{K'}^{-1}(x) = m$. This time evolution measures the asymmetry between the orders of the branched coverings in the correspondence.

One can construct representation of this algebra with time evolution. Consider the vector space $\mathcal{H}_K$ of finitely supported $\xi : \mathcal{G}_K \to \mathbb{C}$ with

$$(\rho(f)\xi)(M) = \sum_{M_1 \in \mathcal{G}, M_2 \in \mathcal{G}_K : M_1 \circ M_2 = M} f(M_1) \xi(M_2)$$

and with Hamiltonian satisfying $\rho(\sigma_t(f)) = e^{itH} \rho(f) e^{-itH}$ given by

$$(H\,\xi)(M) = \log(n)\ \xi(M)$$

where $\pi_K : M \to S^3 \supset L \supset K$ is a branched covering map of order n. There is a clear problem with this setting, coming from the presence of infinite multiplicities (for example, due to the existence of universal knots).

9.2.1 *Cobordisms: equivalences and 2-categories*

In order to obtain a more tractable algebra, with a better behaved time evolution and Hamiltonian, one can consider passing to an equivalence relation that reduces the number of objects considered. A natural such choice is the cobordism relation. The 4-dimensional manifolds we consider here will be PL or smooth. We will not consider topological 4-manifolds due to the problem of possible non-existence of smooth structures, since we want to eventually think of these 4-dimensional geometries as being spacetime geometries.

Given correspondences M_1 and M_2 in $\mathcal{C}(K, K')$

$$K \subset L_1 \subset S^3 \xleftarrow{\pi_{K,1}} M_1 \xrightarrow{\pi_{K',1}} S^3 \supset L_1' \supset K'$$

$$K \subset L_2 \subset S^3 \xleftarrow{\pi_{K,2}} M_2 \xrightarrow{\pi_{K',2}} S^3 \supset L_2' \supset K'.$$

A *branched-cover-cobordism* is a 4-dimensional manifold W with boundary $\partial W = M_1 \cup -M_2$ and with branched covering maps

$$S \subset S^3 \times [0,1] \xleftarrow{q} W \xrightarrow{q'} S^3 \times [0,1] \supset S'$$

branched along surfaces with boundary S, S' in $S^3 \times [0,1]$, with

$$M_1 = q^{-1}(S^3 \times \{0\}) = q'^{-1}(S^3 \times \{0\})$$

$$M_2 = q^{-1}(S^3 \times \{1\}) = q'^{-1}(S^3 \times \{1\})$$

$$q|_{M_1} = \pi_{K,1}, \quad q'|_{M_1} = \pi_{K',1}$$

$$q|_{M_2} = \pi_{K,2}, \quad q'|_{M_2} = \pi_{K',2}$$

The boundary is $\partial S = L_1 \cup -L_2$ and $\partial S' = L_1' \cup -L_2'$

$$L_1 = S \cap (S^3 \times \{0\}) \quad L_2 = S \cap (S^3 \times \{1\})$$

$$L_1' = S' \cap (S^3 \times \{0\}) \quad L_2' = S' \cap (S^3 \times \{1\})$$

This defines an equivalence relation according to which two correspondences $M_1 \sim M_2$ are equivalent if and only if there is a branched-cover-cobordism W with $\partial W = M_1 \cup -M_2$. It is easy to check it is indeed an equivalence relation. The *gluing* of cobordisms is given by

$$W = W_1 \cup_M W_2$$

with $\partial W_1 = M_1 \cup -M$, $\partial W_2 = M \cup -M_2$ and with

$$S_1 \subset S^3 \times [0,1] \xleftarrow{q_1} W_1 \xrightarrow{q_1'} S^3 \times [0,1] \supset S_1'$$

$$S_2 \subset S^3 \times [0,1] \xleftarrow{q_2} W_2 \xrightarrow{q_2'} S^3 \times [0,1] \supset S_2'$$

which gives a branched cover cobordism

$$W_1 \cup_M W_2$$

$$q_1 \# q_2 \qquad\qquad q_1' \# q_2'$$

$$S_1 \cup_L S_2 \subset S^3 \times [0,1] \qquad\qquad S^3 \times [0,1] \supset S_1' \cup_{L'} S_2'$$

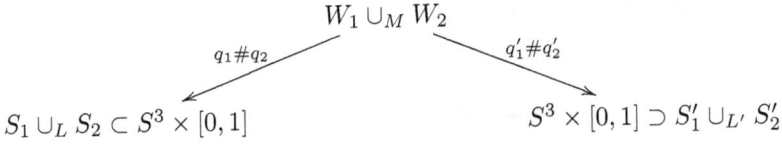

The compatibility of this equivalence relation with the composition of correspondences is also satisfied, since for

$$M_1 \sim M_2 \in \mathcal{C}(K, K'), \quad M_1' \sim M_2' \in \mathcal{C}(K', K'')$$

we have

$$M_1 \circ M_1' \sim M_2 \circ M_2' \in \mathcal{C}(K, K'')$$

with the fibered product of cobordisms

$$W_1 \circ W_2 := \{(x, y) \in W_1 \times W_2 | q_1'(x) = q_2(y)\}$$

given by a branched-cover-cobordism

$$\partial(W_1 \circ W_2) = \partial W_1 \circ \partial W_2 = (M_1 \circ M_1') \cup -(M_2 \circ M_2')$$

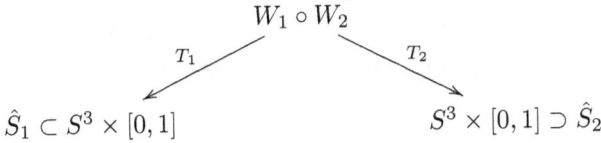

$$W_1 \circ W_2$$

$$T_1 \qquad\qquad T_2$$

$$\hat{S}_1 \subset S^3 \times [0,1] \qquad\qquad S^3 \times [0,1] \supset \hat{S}_2$$

where we have

$$\hat{S}_1 = S_1 \cup q_1(q_1'^{-1}(S_2)) \quad \hat{S}_2 = S_2' \cup q_2'(q_2^{-1}(S_1'))$$

$$\partial \hat{S}_1 = I_1 \cup -I_3 \quad \partial \hat{S}_2 = I_2 \cup -I_4$$

$$T_1^{-1}(S^3 \times \{0\}) = M_1 \circ M_1' = T_2^{-1}(S^3 \times \{0\})$$

$$T_1^{-1}(S^3 \times \{1\}) = M_2 \circ M_2' = T_2^{-1}(S^3 \times \{1\}).$$

We obtain in this way an additive category $\mathcal{K}_\sim$, where the set of objects $Obj(\mathcal{K}_\sim)$ is given by embedded graphs $K \subset S^3$ (up to ambient isotopy) and the set of morphisms: $Mor(\mathcal{K}_\sim)$ is given by cobordism classes $\mathrm{Hom}_\sim(K, K') = \mathbb{Z}[\mathcal{C}_\sim(K, K')] = \mathbb{Z}[\mathcal{C}(K, K')/\sim]$

$$\phi = \sum_{[M]} a_{[M]} [M],$$

and with composition $[M] = [M_1] \circ [M_2]$

$$\mathrm{Hom}_\sim(K, K') \times \mathrm{Hom}_\sim(K', K'') \to \mathrm{Hom}_\sim(K, K'').$$

Let $\bar{\mathcal{G}}$ be the corresponding semigroupoid $\alpha = ([M], K, K')$ with cobordism equivalence classes $[M]$, and $\mathbb{C}[\bar{\mathcal{G}}]$ its convolution algebra of finitely supported functions $f : \bar{\mathcal{G}} \to \mathbb{C}$ with product

$$(f_1 * f_2)[M] = \sum_{[M_1]\circ[M_2]=[M]} f_1[M_1] f_2[M_2].$$

The time evolution is then given by

$$\sigma_t(f)[M] := \left(\frac{n}{m}\right)^{it} f[M].$$

Note that the covering order is well defined on the classes. The Hamiltonian is again of the form

$$(H\,\xi)[M] = \log(n)\ \xi[M].$$

Now it is possible to show that the Hamiltonian has finite spectral multiplicities.

The classifying spaces for branched coverings have been studied in [Brand (1980)]. The set of k-fold branched coverings of M up to branched-cover-cobordism is classified by the homotopy classes of maps

$$B_k(M) = [M, B_k].$$

In particular, $B_k(S^3) = \pi_3(B_k)$. The rational homotopy type satisfies

$$B_k \otimes \mathbb{Q} \simeq \bigvee_\alpha K(\mathbb{Q}, 4)$$

where α are partitions of k. There is a fibration

$$K(\pi, j-1) \to \bigvee^{t-1} \Sigma K(\pi, j-1) \to \bigvee^{t} K(\pi, j)$$

$$S^3 \otimes \mathbb{Q} \to \bigvee^{p(k)-1} S^4 \otimes \mathbb{Q} \to B_k \otimes \mathbb{Q}$$

that can be used to compute the groups $\pi_n(B_k) \otimes \mathbb{Q}$. One obtains $\pi_n(B_k) \otimes \mathbb{Q} = \mathbb{Q}^D$, where

$$D = \begin{cases} p(k) & n = 4 \\ Q(\frac{n-1}{3}, p(k)-1) & n = 1, 4, 10 \mod 12, n \neq 1, 4 \\ Q(\frac{n-1}{3}, p(k)-1) + Q(\frac{n-1}{6}, p(k)-1) & n \equiv 7 \mod 12 \\ 0 & \text{otherwise} \end{cases}$$

where $p(k)$ is the number of partitions and $Q(a,b) = \frac{1}{a} \sum_{d|a} \mu(d) b^{a/d}$, where $\mu(d)$ is the Möbius function.

In particular, $\pi_3(B_k)$ is *finite* for all k, and this implies that the Hamiltonian has finite multiplicities $N_n(K)$, bounded by

$$1 \leq N_n(K) \leq \#\pi_3(B_n).$$

We refer the reader to [Marcolli, Zainy al-Yasry (2008)] for more details.

The partition function of the time evolution

$$Z_K(\beta) = \mathrm{Tr}(e^{-\beta H_K}) = \sum_n \exp(-\beta \log(n)) N_n(K)$$

$$\zeta(\beta) \leq Z_K(\beta) \leq \sum_n \#\pi_3(B_n) n^{-\beta}$$

is controlled by the generating function for the $\#\pi_3(B_n)$. There is evidence from the results of [Brand, Brumfiel (1980)] in favor of some strong constraints on the growth of the numbers $\#\pi_3(B_n)$ (hence of the $\#\pi_3(B_n(4))$), based on the periodicities along certain arithmetic progressions of the localizations at primes.

If finite summability holds then the low temperature extremal KMS$_\beta$ states are Gibbs sums

$$\varphi_{\beta,K}(f) = Z_K(\beta)^{-1} \mathrm{Tr}(\rho_K(f) e^{-\beta H_K}).$$

Their weak limits as $\beta \to \infty$ (zero temperature limit)

$$\lim_{\beta \to \infty} \varphi_{\beta,K}(f) = f(\mathbb{U}_K)$$

recover the embedded graph K.

For the purpose of obtaining a Hamiltonian with finite spectral multiplicities, we have used cobordisms as an equivalence relation. However, from the point of view of noncommutative geometry, it is more natural to treat equivalence relations not by passing to the quotient, but by retaining the identifications explicitly as arrows in a groupoid. Since here we are considering an equivalence relations on the morphisms of a category, the correct structure to consider in the role of the non-commutative quotient is a 2-category.

A 2-category $\mathcal{C}$ consists of compatible data of

- Objects $Obj(\mathcal{C}) \ni X$,
- 1-morphisms $\mathcal{C}(X,Y) \ni \varphi$, with associative composition $\mathcal{C}(X,Y) \times \mathcal{C}(Y,Z) \to \mathcal{C}(X,Z)$, with unit, as usual with morphisms in a category;

- 2-morphisms $\mathcal{C}^{(2)}(\varphi, \psi) \ni \Phi$, which are morphisms between 1-morphisms, with two types of composition, with compatible associativity conditions and identity
    - Vertical composition $\bullet : \mathcal{C}^{(2)}(\varphi, \psi) \times \mathcal{C}^{(2)}(\varphi, \psi) \to \mathcal{C}^{(2)}(\varphi, \psi)$
    - Horizontal composition $\circ : \mathcal{C}^{(2)}(\varphi, \psi) \times \mathcal{C}^{(2)}(\psi, \eta) \to \mathcal{C}^{(2)}(\varphi, \eta)$
- the interchange law of horizontal and vertical composition (for composable 2-morphisms)

$$(\alpha \circ \beta) \bullet (\gamma \circ \delta) = (\alpha \bullet \gamma) \circ (\beta \bullet \delta).$$

Here we consider a *2-category* $\mathcal{G}^2$ constructed as follows:

- Objects: embedded graphs $K \subset S^3$;
- 1-morphisms: branched coverings

$$K \subset L \subset S^3 \xleftarrow{\pi_K} M \xrightarrow{\pi_{K'}} S^3 \supset L' \supset K';$$

- 2-morphisms: branch-cover-cobordisms

$$S \subset S^3 \times [0,1] \xleftarrow{q} W \xrightarrow{q'} S^3 \times [0,1] \supset S'.$$

The *horizontal composition* of 2-morphisms is given by the fibered product of branched cover cobordisms

$$W_1 \circ W_2 = W_1 \times_{S^3 \times [0,1]} W_2.$$

The *vertical composition* of 2-morphisms is given by gluing of cobordisms

$$W_1 \bullet W_2 = W_1 \cup_M W_2.$$

The convolution algebra $\mathbb{C}[\mathcal{G}^2]$ associated to this 2-category has two product structures that correspond to the horizontal and vertical composition:

$$(f_1 \bullet f_2)(W) = \sum_{W = W_1 \bullet W_2} f_1(W_1) f_2(W_2)$$

$$(f_1 \circ f_2)(W) = \sum_{W = W_1 \circ W_2} f_1(W_1) f_2(W_2).$$

There are two involutions, combined into

$$f^\dagger(W) = \overline{f(\bar{W}^\vee)}$$

where $W \mapsto \bar{W}$ is orientation reversal, which maps $W = W_1 \bullet W_2$ to $\bar{W} = \bar{W}_2 \bullet \bar{W}_1$ and $W = W_1 \circ W_2$ to $\bar{W} = \bar{W}_1 \circ \bar{W}_2$, and $W \mapsto W^\vee$, which is the exchange of covering maps as before, that maps $W = W_1 \circ W_2$ to $W^\vee = W_2^\vee \circ W_1^\vee$ and $W = W_1 \bullet W_2$ to $W^\vee = W_1^\vee \bullet W_2^\vee$.

9.2.2 *Vertical composition and Hartle-Hawking gravity*

The classical (Euclidean) action for gravity is of the form

$$S(g) = -\frac{1}{16\pi} \int_W R \, dv - \frac{1}{8\pi} \int_{\partial W} K \, dv$$

where K is the trace of the II-fundamental form.

The transition amplitude between two 3-dimensional geometries is given by

$$\langle (M_1, g_1), (M_2, g_2) \rangle = \int e^{iS(g)} D[g]$$

a path integral that sums over cobordisms W.

Thus, one can consider an associated time evolution on an algebra of functions of cobordisms, like the convolution algebra we just discussed, given by

$$\sigma_t(f)(W, g) = e^{itS(g)} f(W, g).$$

This time evolution is compatible with the vertical composition of 2-morphisms, provided that, in order to avoid the problem of discontinuity at the gluing $W = W_1 \cup_M W_2$ one imposes conditions on the metric near the boundary ∂W.

Formally, the KMS$_\beta$ states would then be path integrals

$$\varphi_\beta(f) = \frac{\int f(W, g) e^{-\beta S(g)} D[g]}{\int e^{-\beta S(g)} D[g]}.$$

In fact, more generally, any kind of geometric and physics related invariants of four-manifolds that satisfy a suitable gluing property can give rise to a time evolution on our convolution algebra of geometries that is compatible with the vertical composition of morphisms.

For example, moduli spaces of solutions of some classical field equations

$$\mathcal{M} = \{\text{solutions of nonlinear elliptic equations}\}/\text{gauge}.$$

The linearized theory determines a deformation complex

$$\Omega^0 \to \Omega^1 \to \Omega^2$$

where Ω^0 are the infinitesimal gauge transformations, Ω^1 is the tangent space to the configuration space, and Ω^2 are the obstructions. The assembled complex

$$\mathcal{D} : \Omega^{odd} \to \Omega^{ev}$$

has index Ind$\mathcal{D}$ that computes the virtual dimension of the moduli space $\mathcal{M}$.

Gluing theorems for moduli spaces are typically of the form

$$\mathcal{M}(X) = \mathcal{M}(X_+) \times_{\mathcal{M}(M)} \mathcal{M}(X_-)$$

under a decomposition of the manifold

$$X = X_+ \cup_M X_-,$$

with virtual dimensions

$$\dim \mathcal{M}(X) = \dim \mathcal{M}(X_+) + \dim \mathcal{M}(X_-) - \dim \mathcal{M}(M).$$

The case of manifolds with boundary requires an index theory with boundary (Atiyah–Patodi–Singer, L^2-moduli spaces, etc.). Under a gluing decomposition of the cobordisms

$$W = W_1 \cup_M W_2$$

with $\partial W_1 = M_1 \cup -M$ and $\partial W_2 = M \cup -M_2$, the moduli spaces should satisfy

$$\mathcal{M}(W) = \mathcal{M}(W_1) \times_{\mathcal{M}(M)} \mathcal{M}(W_2).$$

Then one can consider a time evolution

$$\sigma_t(f)(W) = \exp(it\delta(W))\, f(W)$$

where $\delta(W) := \dim \mathcal{M}(W) - \dim \mathcal{M}(M_2)$. This satisfies

$$\sigma_t(f_1 \bullet f_2)(W) = \sum_{W = W_1 \bullet W_2} e^{it\delta(W)} f_1(W_1) f_2(W_2)$$

$$= \sum_{W = W_1 \bullet W_2} e^{it\delta(W_1)} f_1(W_1) e^{it\delta(W_2)} f_2(W_2) = \sigma_t(f_1) \bullet \sigma_t(f_2)(W).$$

9.2.3 *Horizontal composition and Connes–Chern character*

The key point in the previous discussion is the additivity of the index

$$\mathrm{Ind}\mathcal{D}_W = \mathrm{Ind}\mathcal{D}_{W_1} + \mathrm{Ind}\mathcal{D}_{W_2}$$

under the vertical composition $W = W_1 \cup_M W_2$.

If one also wants a time evolution compatible with the horizontal composition that is coming from the branched cover structure, one is looking at fibered product operations $X = X_1 \times_Z X_2$. Instead of the index $[\mathcal{D}_X] \in K(X)$, one can consider correspondences

$$U \leftarrow X \rightarrow V,$$

and associated elements in the bivariant KK-theory,

$$[\mathcal{D}_X] \in KK(U, V).$$

This was done, for geometric correspondences of the type we are considering here in [Connes, Skandalis (1984)], where the fibered product of correspondences

$$U \leftarrow X = X_1 \times_Z X_2 \rightarrow V$$

gives rise to the Kasparov product in KK-theory

$$[\mathcal{D}_X] = [\mathcal{D}_{X_1}] \circ [\mathcal{D}_{X_2}]$$

$$KK(U, Z) \times KK(Z, V) \rightarrow KK(U, V).$$

We will not be discussing this in detail here and we refer the reader to [Connes, Skandalis (1984)].

To avoid momentarily the complication caused by working with manifolds with boundary, we consider the simpler situation where W is a 4-manifold endowed with branched covering maps to a closed 4-manifold X (for instance $S^3 \times S^1$ or S^4) instead of $S^3 \times [0, 1]$,

$$S \subset X \xleftarrow{q} W \xrightarrow{q'} X \supset S' \tag{9.1}$$

branched along surfaces S and S' in X.

We can then think of an elliptic operator $\mathcal{D}_W$ on a 4-manifold W, which has branched covering maps as defining an unbounded Kasparov bimodule, that is, as defining a KK-class $[\mathcal{D}_W] \in KK(X, X)$. We can think of this class as being realized by a geometric correspondence in the sense of [Connes, Skandalis (1984)]

$$[\mathcal{D}_W] = kk(W, E_W),$$

with the property that, for the horizontal composition $W = W_1 \circ W_2 = W_1 \times_X W_2$ we have

$$[\mathcal{D}_{W_1}] \circ [\mathcal{D}_{W_2}] = kk(W_1, E_{W_1}) \circ kk(W_2, E_{W_2}) = kk(W, E_W) = [\mathcal{D}_W].$$

The Connes–Chern character pairing K-theory and cyclic homology

$$ch_n : K_i(\mathcal{A}) \to HC_{2n+i}(\mathcal{A})$$

$$ch_n : K^i(\mathcal{A}) \to HC^{2n+i}(\mathcal{A})$$

$$\langle ch(e), ch(x) \rangle = \text{index}$$

Cyclic homology is constructed in terms of a *cyclic category* $\mathcal{A}^\natural$, with

$$HC^n(\mathcal{A}) = \text{Ext}^n_\Lambda(\mathcal{A}^\natural, \mathbb{C}^\natural),$$

see [Connes (1983)]. There is a bivariant version of this character, [Nistor (1993)]. Under this bivariant Chern character that is compatible with the composition products these classes will map to elements in the Yoneda algebra

$$ch_n([\mathcal{D}_W]) \in \mathcal{Y} := \oplus_j \text{Ext}^{2n+j}(\mathcal{A}^\natural, \mathcal{A}^\natural) \qquad (9.2)$$

$$ch_n([\mathcal{D}_{W_1}]) ch_m([\mathcal{D}_{W_2}]) = ch_{n+m}([\mathcal{D}_{W_1}] \circ [\mathcal{D}_{W_2}]).$$

Let $\chi : \mathcal{Y} \to \mathbb{C}$ be a character of the Yoneda algebra. Then by composing $\chi \circ ch$ we obtain

$$\chi ch([\mathcal{D}_{W_1}] \circ [\mathcal{D}_{W_2}]) = \chi ch([\mathcal{D}_{W_1}]) \chi ch([\mathcal{D}_{W_2}]) \in \mathbb{C}.$$

This can be used to define a time evolution for the horizontal product of the form

$$\sigma_t(f)(W) = |\chi ch([\mathcal{D}_W])|^{it} f(W).$$

9.2.4   *Almost-commutative cobordisms*

In [Marcolli, Zainy al-Yasry (2008)] it is also discussed how to extend this setting to almost-commutative geometries, following the viewpoint of [Chamseddine, Connes (2011)] on noncommutative geometries with boundary, in the form presented in [Connes (2007)]. According to this setting, a boundary-even spectral triple with boundary consists of two sets of data $(\mathcal{A}, \mathcal{H}, D, \gamma)$ and $(\mathcal{A}_\partial, \mathcal{H}_\partial, D_\partial)$, where the $\mathbb{Z}/2\mathbb{Z}$-grading γ on $\mathcal{H}$ satisfies $[a, \gamma] = 0$ for all $a \in \mathcal{A}$, with $\text{Dom}(D) \cap \gamma \text{Dom}(D)$ dense in $\mathcal{H}$. The boundary algebra is the quotient $\mathcal{A}_\partial = \mathcal{A}/(\mathcal{J} \cap \mathcal{J}^*)$, where $\mathcal{J}$ is the two-sided ideal $\mathcal{J} = \{a \in \mathcal{A} : a \text{Dom}(D) \subset \gamma \text{Dom}(D)\}$. The boundary Hilbert space $\mathcal{H}_\partial$ is the closure in $\mathcal{H}$ of $D^{-1}\text{Ker}(D_0^*)$ where $D_0 = D|_{\text{Dom}(D) \cap \gamma \text{Dom}(D)}$. The boundary algebra acts on the boundary

Hilbert space by $a - D^{-2}[D^2, a]$. The boundary Dirac operator D_∂ is defined on $D^{-1}\mathrm{Ker}(D_0^*)$ by $\langle \xi, D_\partial\, \eta \rangle = \langle \xi, D\eta \rangle$, for all $\xi \in \mathcal{H}_\partial$ and all $\eta \in D^{-1}\mathrm{Ker}(D_0^*)$. The operator D_∂ has bounded commutators with $\mathcal{A}_\partial$.

We use this approach to define cobordisms between almost commutative correspondences. Note that, since the product of manifolds with boundary is a manifold with corners, in order to remain within the class of spectral triples with boundary, we consider only the case where the finite geometry is a finite spectral triple in the ordinary sense (without boundary) while the manifold part of the almost commutative geometry has boundary. We also consider here, for simplicity, almost-commutative geometries that are actual products of a manifold and a finite spectral triple, rather than the more general form of almost-commutative geometries of [Ćaćić (2012)]. However, the same approach can be adapted to the more general case.

We consider a notion of spectral correspondence given by data $(\mathcal{A}_1, \mathcal{A}_2, \mathcal{H}, D)$, with two associative unital involutive algebras with commuting actions on the same Hilbert space $\mathcal{H}$ and such that the operator D is self-adjoint $D^* = D$, with compact resolvent, with $[D, a]$ and $[D, b]$ bounded, for all $a \in \mathcal{A}_1$ and $b \in \mathcal{A}_2$, and satisfying $[[D, a], b] = 0$, for all $a \in \mathcal{A}_1$ and $b \in \mathcal{A}_2$. The correspondence is even if there is a $\mathbb{Z}/2\mathbb{Z}$-grading γ on $\mathcal{H}$ that anticommutes with D and commutes with the actions of $\mathcal{A}_1$ and $\mathcal{A}_2$. This is, in fact, not all that is needed to obtain a correspondence: we are presently neglecting the crucial additional datum of a *connection* ∇ on the bimodule that is needed in order to be able to combine Dirac operators when taking tensor products of bimodules (fibered products of geometric correspondences). A complete and more general notion of correspondences of spectral triples and a suitable categorical framework were developed in [Mesland (2014)]. For an approach to gauge theory on spectral triples based on the correspondences and the categorical setting of [Mesland (2014)] see also [Brain, Mesland, van Suijlekom (2016)]. Other forms of categorification of noncommutative spaces where considered in [Bertozzini, Conti, Lewkeeratiyutkul (2011)], [Dawe Martins (2008)].

Consider again the correspondences defined above by branched coverings

$$L_1 \subset S^3 \xleftarrow{\pi_1} M \xrightarrow{\pi_2} S^3 \supset L_2.$$

We can associate to the manifold M the corresponding spectral triple

$$S_M := (C^\infty(S^3), C^\infty(S^3), L^2(M, \mathbb{S}), \partial\!\!\!/_M).$$

Consider also a pair $\mathcal{A}$, $\mathcal{B}$ of finite dimensional unital (in general noncommutative) involutive algebras, and a finite dimensional vector space V, endowed with commuting $\mathcal{A}$ and $\mathcal{B}$ actions. Let $T \in \mathrm{End}(V)$ by an operator such that $[[T, a], b] = 0$, for all $a \in \mathcal{A}, b \in \mathcal{B}$, and set

$$S_F := (\mathcal{A}, \mathcal{B}, V, T)$$

and

$$S_M \cup S_F = (C^\infty(S^3) \otimes \mathcal{A}, C^\infty(S^3) \otimes \mathcal{B}, \mathcal{H}, D).$$

If the triple S_F is even, with $\mathbb{Z}/2\mathbb{Z}$-grading γ, on the Hilbert space $\mathcal{H} = L^2(M, \mathbb{S}) \otimes V$ consider the Dirac operator

$$D = T \otimes 1 + \gamma \otimes \partial\!\!\!/_M.$$

If S_F is odd, then on the Hilbert space $\mathcal{H} = L^2(M, \mathbb{S}) \otimes V \oplus L^2(M, \mathbb{S}) \otimes V$ consider the Dirac operator

$$D = \begin{pmatrix} 0 & \delta^* \\ \delta & 0 \end{pmatrix}$$

with $\delta = T \otimes 1 + i \otimes \partial\!\!\!/_M$

$$\gamma = \begin{pmatrix} 1 & 0 \\ 0 & -1 \end{pmatrix}.$$

The notion of a spectral triple with boundary mentioned above can be adapted to a notion of spectral correspondences with boundary

$$(\mathcal{A}_1, \mathcal{A}_2, \mathcal{H}, D), \quad \text{with boundary} \quad (\mathcal{A}_{\partial,1}, \mathcal{A}_{\partial,2}, \mathcal{H}_\partial, D_\partial).$$

We can then say that two spectral correspondences $S_{M_i} \cup S_{F_i}$ with $S_{F_i} = (A_i, B_i, V_i, T_i)$ are cobordant if there is a finite spectral correspondence $S_F = (A, B, V_F, D_F)$ without boundar and a spectral triple with boundary associated to a cobordism manifold W,

$$S_W = (C^\infty(S^3 \times I), C^\infty(S^3 \times I), \mathcal{H}_W, \partial\!\!\!/_W),$$

such that $S_W \cup S_F = (\mathcal{A}, \mathcal{B}, \mathcal{H}, \mathcal{D})$ is a spectral correspondence with boundary

$$(\mathcal{A}_\partial, \mathcal{B}_\partial, \mathcal{H}_\partial, \mathcal{D}_\partial) = (\partial S_W) \cup S_F,$$

with Dirac operator $\mathcal{D}_\partial$ obtained from $\partial\!\!\!/_{M_i}$ and an operator T obtained from the operators T_i by considering $\mathcal{A}$–$\mathcal{A}_i$ bimodules $\mathcal{E}_i$ and $\mathcal{B}_i$–$\mathcal{B}$ bimodules $\mathcal{F}_i$ with connections, and $V = \mathcal{E}_i \otimes_{\mathcal{A}} V_i \otimes_{\mathcal{B}} \mathcal{F}_i$.

The fibered product of geometric correspondences becomes a Kasparov product in the noncommutative settings,

$$(S_{W_1} \cup S_{F_1}) \circ (S_{W_1} \cup S_{F_2}).$$

The formalism of correspondences of spectral triples developed in [Mesland (2014)] shows how to handle this composition of morphisms under very general assumptions. By analogy with the different orders of coverings in the commutative case, one can make correspondence dynamical using the ranks of $\mathcal{E}$ and $\mathcal{F}$. However, the other composition that is coming from the gluing of cobordism is more delicate in this setting, because in the framework of spectral triples with boundary used in [Chamseddine, Connes (2011)] one is using non-APS boundary conditions for the Dirac operator. A suitable formalism of gluing of spectral triples with boundary along a common boundary is needed for this composition.

9.3 Topological spin networks and foams

This section is based on [Denicola, Marcolli, Zainy al-Yasry (2010)]. We use the point of view discussed in the previous section, adapted to the setting of spin networks and spin foams models of quantum gravity, where 3-dimensional geometries are discretized in the form of graphs with representation theoretic data at the edges and vertices (spin networks), and 4-dimensional cobordisms are discretized by 2-dimensional simplicial complexes with analogous data (spin foams). We show that one can use branched cover data to enrich the geometry of spin networks and spin foams so that not only the spacetime geometry, but also its topology can be made dynamical.

9.3.1 *Spin networks and monodromies*

A *spin network* in a 3-manifold M consists of a triple (Γ, ρ, ι) with

(1) an oriented graph $\Gamma \subset M$ embedded in the 3-manifold;
(2) a labeling ρ of each edge e of Γ by a representation ρ_e of a given compact Lie group G;
(3) a labeling ι of each vertex v of Γ by an intertwiner of the representations,

$$\iota_v : \rho_{e_1} \otimes \cdots \otimes \rho_{e_n} \to \rho_{e'_1} \otimes \cdots \otimes \rho_{e'_m},$$

where $e_1, \ldots, e_n$ are the incoming edges at v and $e'_1, \ldots, e'_m$ are the outgoing edges.

The main idea is that a "quantum three-geometry" is represented by a graph where vertices account for quanta of volume and egdes describe quanta of area separating them. The representation data encode *holonomies* of the gravitational field. In this viewpoint the ambient topology M is fixed. Spin networks have been used in the construction of invariants of 3-manifolds, such as the Turaev–Viro invariants.

In [Denicola, Marcolli, Zainy al-Yasry (2010)] this point of view is enriched with additional topological data, which determine the notion of a *topspin network*. The ambient topology is now variable and dynamical and encoded in spin network data, by representing the manifold M as a branched covering of S^3 and encoding the branched covering data in terms of the monodromy representation of the fundamental group of the complement of the branch locus. Thus, a topspin network contains data of *monodromies* in addition to the *holonomies*.

Consider embedded graphs in S^3 up to ambient isotopy. The ambient isotopy condition is equivalent to considering planar projections of embedded graphs up to Reidemeister moves. This is similar to the usual case of knots and links, except that the set of Reidemeister moves is slightly more general as illustrated in the figure.

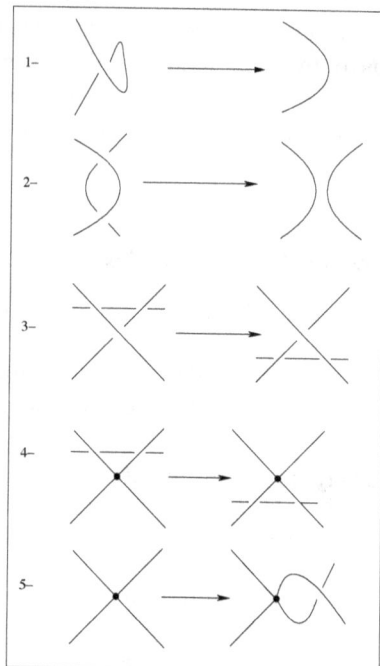

As in the previous section, we consider compact smooth 3-manifolds as branched coverings $p : M \to S^3$ of the 3-sphere, branched along an embedded graph $\Gamma \subseteq S^3$. The restriction $p_| : M \setminus p^{-1}(\Gamma) \to S^3 \setminus \Gamma$ of the projection map to the complement of the embedded graph is an ordinary covering of some degree n.

We consider 4-dimensional cobordisms between 3-manifolds given by PL or smooth 4-manifolds. It is known by [Piergallini (1995)] that all PL 4-manifolds are branched coverings of the four-sphere S^4, branched along an embedded 2-surface. We will more generally allow the branch locus to an embedded simplicial 2-complex. We consider branched cover cobordisms, defined as in the previous section.

Branched coverings $p : M \to S^3$ are determined by representations

$$\sigma : \pi_1(S^3 \setminus \Gamma) \to S_n.$$

The fundamental group $\pi_1(S^3 \setminus \Gamma)$ can be computed explicitly through a *Wirtinger presentation* obtained from a planar projection $D(\Gamma)$ (planar diagram) of the embedded graph. This generalizes the usual Wirtinger presentations of the fundamental groups of knot complements.

The Wirtinger presentation is generated by permutations $\sigma_i \in S_n$ assigned to arcs of $D(\Gamma)$ with relations

$$\sigma_j = \sigma_k \sigma_i \sigma_k^{-1}$$

at crossings and relations

$$\prod_i \sigma_i \prod_j \sigma_j^{-1} = 1$$

at vertices, where σ_i are incoming and σ_j are outgoing arcs. Thus, the presentation is given by monodromies around the edges of the embedded graph with consistency constraints at the vertices.

A result of [Bobtcheva, Piergallini (2012)] shows that different coverings that determine the same 3-manifold are related by a sequence of combinatorial moves, the *covering moves*. The covering moves *at vertices* are given by

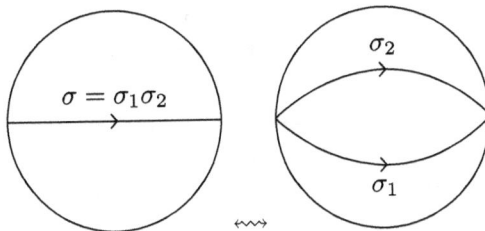

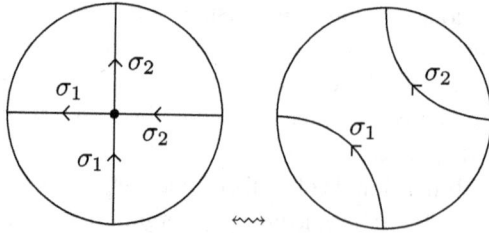

while the covering moves *at crossings* are given by

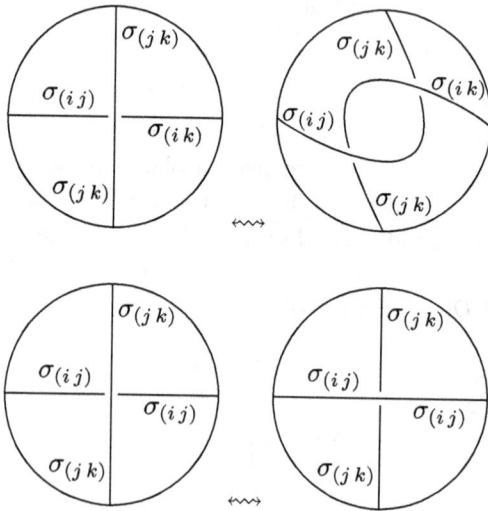

Topspin networks (topologically enriched spin networks) are given by

(1) a spin network (Γ, ρ, ι) with $\Gamma \subset S^3$,
(2) a representation $\sigma : \pi_1(S^3 \smallsetminus \Gamma) \to S_n$.

This gives a spin network in M, a branched covering of S^3, where the topology of M is encoded in the spin network data through the monodromy representation.

Spin network data are compatible with covering moves. This gives a way to extend the holonomy data ρ, ι of spin networks in a way that is compatible with covering moves. The compatibility constraints are given by the diagrams

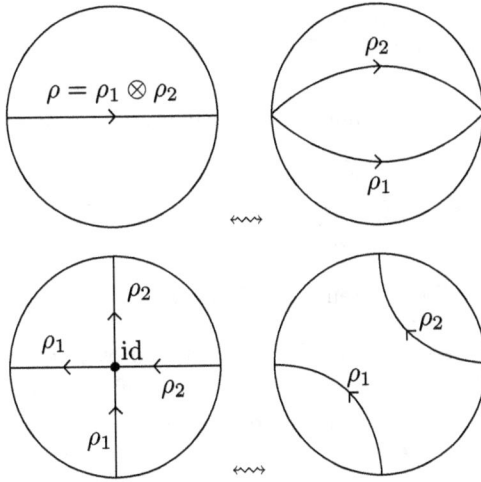

for the moves at vertices, while at crossings there are no conditions (the holonomies assigned to edges not to strands)

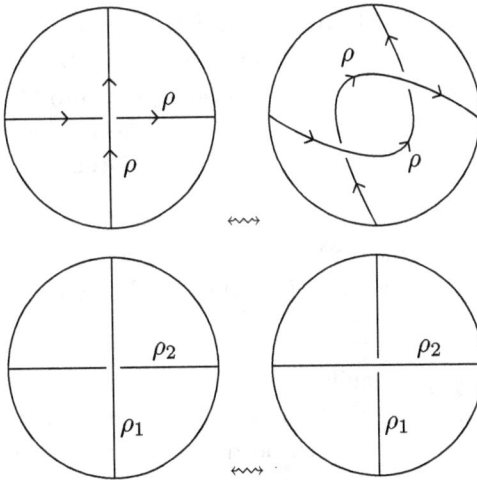

9.3.2 *Spin foams and monodromies*

Spin foams are spin network cobordisms. Given two spin networks $\psi = (\Gamma, \rho, \iota)$ and $\psi' = (\Gamma', \rho', \iota')$ spin networks, with graphs Γ and Γ' embedded in M and M', respectively, a spin foam $\Psi : \psi \to \psi'$ in a cobordism W with $\partial W = M \cup \bar{M}'$ is a triple $\Psi = (\Sigma, \tilde{\rho}, \tilde{\iota})$ of data:

(1) an oriented two-complex $\Sigma \subseteq W$, with $\partial \Sigma = \Gamma \cup \bar{\Gamma}'$
(2) a labeling $\tilde{\rho}$ of each face f of Σ by a representation $\tilde{\rho}_f$ of a given compact Lie group G;
(3) a labeling $\tilde{\iota}$ of each edge e of Σ that does not lie in Γ or Γ' by an intertwiner

$$\tilde{\iota}_e : \bigotimes_{f : e \in \partial(f)} \tilde{\rho}_f \to \bigotimes_{f' : \bar{e} \in \partial(f')} \tilde{\rho}_{f'}.$$

There are additional consistency conditions:

(1) Given an edge e in Γ and a face f_e bordered by e then $\tilde{\rho}_{f_e} = \rho_e$ (or the dual depending on the relative orientation of the face and the edge).
(2) Given a vertex v of Γ and an edge e_v adjacent to v in Σ the intertwiners satisfy $\tilde{\iota}_{e_v} = \iota_v$
   (or the dual depending on the orientations).
(3) The same compatibility conditions hold for Γ'.

Similarly to what we described before in the case of spin network, we can enrich spin foams with topological data of monodromies that encode the background topology of the cobordism W as dynamical variables.

A *Topspin foam* (topologically enriched spin foam) $\Psi : \psi \to \psi'$, where $\psi = (\Gamma, \rho, \iota, \sigma)$ and $\psi' = (\Gamma', \rho', \iota', \sigma')$ are topspin networks with monodromy reps in same S_n (and $\Gamma, \Gamma' \subset S^3$), is a datum $\Psi = (\Sigma, \tilde{\rho}, \tilde{\iota}, \tilde{\sigma})$ with

(1) a spin foam $(\Sigma, \tilde{\rho}, \tilde{\iota})$ between ψ and ψ' with $\Sigma \subset S^3 \times [0, 1]$;
(2) a representation $\tilde{\sigma} : \pi_1((S^3 \times [0, 1]) \setminus \Sigma) \to S_n$, defining branched cover cobordism W between M and M' (branched coverings defined by (Γ, σ) and (Γ', σ')).

The PL (smooth) 4-manifold W that gives the cobordism between M and M' is encoded in the spin foam data in the same way that the topology of the 3-manifolds M and M' is encoded in the data of the topspin networks.

Note that in a path integral formulation, the sum over geometries is now also a sum over topologies, through these monodromy data.

We can describe the topspin foam data in diagrammatic terms. Consider a 3-dimensional projection diagram $D(\Sigma)$ of the embedded 2-complex Σ.

(1) Assign to each one-dimensional strand e_i of $D(\Sigma)$ the same intertwiner $\tilde{\iota}_e$ assigned to the edge e;

(2) Assign to each two-dimensional strand f_α of $D(\Sigma)$ the same representation $\tilde{\rho}_f$ of G assigned to the face f;

(3) Assign to each two-dimensional strand f_α of $D(\Sigma)$ a topological label $\tilde{\sigma}_\alpha \in S_n$ such that, taken in total, such assignments satisfy the Wirtinger relations

$$\tilde{\sigma}_\alpha = \tilde{\sigma}_\beta \tilde{\sigma}_{\alpha'} \tilde{\sigma}_\beta^{-1}$$

at crossings of faces, and along edges

$$\prod_{\alpha:e\in\partial(f_\alpha)} \tilde{\sigma}_\alpha \quad \prod_{\alpha':\bar{e}\in\partial(f_{\alpha'})} \tilde{\sigma}_{\alpha'}^{-1} = 1.$$

As we discussed in the previous section, we can form categories and algebras of topspin networks and topspin foams, which can be regarded as a discretized version of the categories and algebras of geometries discussed before.

9.3.3 2-categories and convolution algebras

A 2-category (or 2-semigroupoid) has an associated *2-semigroupoid algebra*, as we have seen in the previous section, that has two associative multiplications ($\circ$ and $\bullet$) with

$$(a_1 \circ b_1) \bullet (a_2 \circ b_2) = (a_1 \bullet a_2) \circ (b_1 \bullet b_2).$$

The elements of the 2-semigroupoid algebra are finitely supported functions of 2-morphisms $f : \mathcal{C}^{(2)} \to \mathbb{C}$ with the two convolution products given by

$$(f_1 \bullet f_2)(\Phi) = \sum_{\Phi=\Phi_1\bullet\Phi_2} f_1(\Phi_1)f_2(\Phi_2)$$

$$(f_1 \circ f_2)(\Phi) = \sum_{\Phi=\Psi\circ\Upsilon} f_1(\Psi)f_2(\Upsilon).$$

In the 2-semigroupoid algebra of topspin foams we have elements $f = \sum_\Phi c_\Phi \delta_\Phi$ with the two associative products coming from the vertical and horizontal composition of 2-morphisms

$$(f_1 \bullet f_2)(\Phi) = \sum_{\Phi=\Phi_1\bullet\Phi_2} f_1(\Phi_1)f_2(\Phi_2)$$

$$(f_2 \circ f_2)(\Phi) = \sum_{\Phi=\Phi_1\circ\Phi_2} f_1(\Phi_1)f_2(\Phi_2).$$

As we have done in the previous section, one can define vertical and horizontal time evolutions:

$$\sigma_t(f_1 \bullet f_2) = \sigma_t(f_1) \bullet \sigma_t(f_2)$$

$$\sigma_t(f_2 \circ f_2) = \sigma_t(f_1) \circ \sigma_t(f_2).$$

In the case of topspin networks and topspin foams, we would like to consider time evolutions that are compatible with *both* vertical and horizontal composition at the same time. We write

$$(f_1 \circ f_2)(_\Psi W_{\Psi''}) = \sum_{\text{fiber product}} f_1(_\Psi W_{\Psi'}) f_2(_{\Psi'} W'_{\Psi''})$$

$$(f_1 \bullet f_2)(_\Psi W_{\Psi'}) = \sum_{\text{gluing}} f_1(_{\Psi_1} W_{1\Psi'_1}) f_2(_{\Psi_2} W_{2\Psi'_2}).$$

A simple possibility is a variant on the "order of covering" idea used in the previous section. We set

$$\sigma_t(f(_\Psi W_{\Psi'})) = n^{itF} f(_\Psi W_{\Psi'})$$

where F is the number of faces of the source Σ and n is the order of the covering W.

9.3.4 *Quantized area operator and dynamics*

A better example, which is more significant physically, is a dynamical evolution of the algebra of the 2-category determined by the *quantized area operator*.

For spin networks, the quantized area operator is obtained as follows. Let $S \subset S^3$ be a closed embedded smooth (or PL) surface. It generically intersects Γ transversely in a finite number of points. Set

$$A_S f(_\psi M_{\psi'}) = \hbar \left(\sum_{x \in S \cap \Gamma} (j_x(j_x + 1))^{1/2} \right) f(_\psi M_{\psi'})$$

for $f(_\psi M_{\psi'})$ in the convolution algebra of topspin networks with fibered product, with $j_x = j_e$ the spin of the $SU(2)$ representation ρ_e assigned to the edge containing the point x. More generally let $N : \bigcup_\Gamma E(\Gamma) \to \mathbb{Z}$ and set

$$A f(_\psi M_{\psi'}) = \hbar \left(\sum_{e \in E(\Gamma)} N(e) (j_e(j_e + 1))^{1/2} \right) f(_\psi M_{\psi'}).$$

This generalizes the counting of the multiplicity of intersection with a surface S. In particular, for

$$\Gamma \subset \hat{\Gamma} \subset S^3 \xleftarrow{p} M \xrightarrow{p'} S^3 \supset \hat{\Gamma}' \supset \Gamma'$$

let the multiplicity function $\chi_\Gamma : E(\hat{\Gamma}) \to \{0, 1\}$ be the characteristic function of Γ, that is, the function that checks whether a given edge of $\hat{\Gamma}$ is also an edge of Γ. Then set

$$A_{\mathsf{s}} f({}_\psi M_{\psi'}) = \hbar \left(\sum_{e \in E(\hat{\Gamma})} \chi_\Gamma(e) \, (j_e(j_e + 1))^{1/2} \right) f({}_\psi M_{\psi'})$$

and the same for A_{t}, the target. The time evolution

$$\sigma_t(f) = \exp(it(A_{\mathsf{s}} - A_{\mathsf{t}}))f$$

determines a dynamics on the convolution algebra of 3-dim topspin networks with fibered product given by the quantized area difference for the two different realizations of M as branched cover.

For the 2-semigroupoid algebra of topspin foams we similarly set

$$A_{\mathsf{s}} f(W) = \hbar \left(\sum_{\mathfrak{f} \in F(\hat{\Sigma})} \chi_\Sigma(\mathfrak{f}) \, (j_{\mathfrak{f}}(j_{\mathfrak{f}} + 1))^{1/2} \right) f(W),$$

and we consider a time evolution (up to a possible topological factor $e^{it\chi(\Sigma, W)}$)

$$\sigma_t(f) = e^{it(A_{\mathsf{s}} - A_{\mathsf{t}})} f.$$

There is again a problem with the Hamiltonian and the spectral multiplicities of the kind that we encountered in the previous section, and one can use the possible inclusion of a topological factor (a topological invariant of the cobordisms W and the embedded surface Σ) to break the multiplicities of the eigenspaces. This would require the construction of invariants $\chi(\Sigma, W)$ with the properties:

(1) The values of $\chi(\Gamma, W)$ are a discrete set in $\mathbb{R}_+^*$, growing at least linearly $c_1 n + c_0$ for large n, $c_i > 0$, where n is the order of the covering;

(2) for a fixed branched cover, the number of embedded Σ with $\chi(\Sigma, W)$ fixed grows at most like $e^{\kappa n}$ some $\kappa > 0$ (independent of W);

(3) on the fibered product $\tilde{W} = W \times_{S^3 \times I} W'$ the invariant satisfies $\chi(\Sigma \cup q q_1^{-1}(\Sigma_2), \tilde{W}) = \chi(\Sigma, W) + \chi(\Sigma_2, W')$

where

$$\Sigma \subset S^3 \times I \xleftarrow{q} W \xrightarrow{q_1} S^3 \times I \supset \Sigma_1 \quad \text{and} \quad \Sigma_2 \subset S^3 \times I \xleftarrow{q_2} W' \xrightarrow{q'} S^3 \times I \supset \Sigma'.$$

We refer the reader to [Denicola, Marcolli, Zainy al-Yasry (2010)] for a more detailed discussion of spin network and spin foam amplitudes, time evolutions, and topological invariants.

This theory of topologically enriched spin networks and spin foams was further developed in [Duston (2012)], where it was shown that the additional datum of the monodromies can be accommodated with only relatively minor changes to the usual setting of the phase space, the C^*-algebra and the Hilbert space of cylindrical functions used in loop quantum gravity. This leads to a formulation as a $SU(N) \times S_n$ gauge theory, where the Lie group part accounts for the holonomies (the metric structure) and the finite permutation group S_n accounts for the monodromies (the underlying manifold topology). The quantized area operator and the Hamiltonian for topspin networks and foams are also discussed in greater detail. At the level of kinematics, it is shown that this approach allows topology changes, namely changes to the topspin networks produced by the Hamiltonian that cannot be realized via a sequence of the covering moves of [Bobtcheva, Piergallini (2012)] relating equivalent topologies.

A different approach to quantum gravity as quantization of geometry was recently proposed in [Chamseddine, Connes, Mukhanov (2015)], [Chamseddine, Connes, Mukhanov (2014)], also based on the realization of PL 4-manifolds as branched coverings of the 4-spheres, branched along an embedded surface of [Piergallini (1995)].

9.4 Discretized almost-commutative geometries

This section describes an approach developed in [Marcolli, van Suijlekom (2014)], which can be thought of as a discretization of the almost commutative geometries used to construct models of gravity coupled to matter. This approach combines the idea of gauge theory on a lattice with the idea of discretization of spacetime as an approach to quantum gravity, in the form of spin networks and spin foams. The resulting geometries are built out of graphs decorated with data of finite spectral triples. An action functional is constructed in terms of Dirac operators, like the spectral action, which recovers the Wilson action (which in continuum limit gives Yang–Mills) with additional terms for a Higgs field in the adjoint representation.

The general construction can be formulated in categorical terms, with a category of finite spectral triples with morphisms built from algebra morphisms and unitary operators, and representations of quivers (oriented graphs) in this category of finite spectral triples. The main configuration space for the theory is then the configuration space of such representations modulo gauge action. Morphisms between gauge networks are by correspondences determined by bimodules.

9.4.1 *Categorical data and finite spectral triples*

Consider first a category C_0 where the objects $(\mathcal{A}, \pi, \mathcal{H})$ consist of a finite dimensional algebra $\mathcal{A}$ and a finite dimensional Hilbert space representation $\pi : \mathcal{A} \to \mathcal{L}(\mathcal{H})$. These are finite spectral triples with trivial Dirac operator $D = 0$ (as used in the case of gauge theory models ilke Yang–Mills). Morphisms $\Phi : (\mathcal{A}_1, \pi_1, \mathcal{H}_1) \to (\mathcal{A}_2, \pi_2, \mathcal{H}_2)$ are given by pairs $\Phi = (\phi, L)$ with $\phi : \mathcal{A}_1 \to \mathcal{A}_2$ a morphism of unital $\star$-algebras and $L : \mathcal{H}_1 \to \mathcal{H}_2$ a unitary operator with $L\pi_1(a)L^* = \pi_2(\phi(a))$.

More generally, we can consider a category C of finite dimensional spectral triples with objects $(\mathcal{A}, \pi, \mathcal{H}, D)$ and morphisms

$$\Phi : (\mathcal{A}_1, \pi_1, \mathcal{H}_1, D_1) \to (\mathcal{A}_2, \pi_2, \mathcal{H}_2, D_2)$$

as above with the additional condition $LD_1L^* = D_2$.

By Wedderburn theorem the algebras are of the form

$$\mathcal{A}_1 = \bigoplus_{i=1}^{k} M_{N_i}(\mathbb{C}), \quad \mathcal{A}_2 = \bigoplus_{j=1}^{k'} M_{N_j'}(\mathbb{C})$$

and a unital $*$-algebra morphism $\phi : \mathcal{A}_1 \to \mathcal{A}_2$ is a direct sum

$$\phi_j : \bigoplus_{i=1}^{k} M_{N_i}(\mathbb{C}) \to M_{N_j'}(\mathbb{C})$$

where ϕ_j splits as a direct sum of representations $\phi_{ij} : M_{N_i}(\mathbb{C}) \to M_{N_j'}(\mathbb{C})$ with multiplicity $d_{ij} \geq 0$, with $N_j' = \sum_i d_{ij} N_i$. Graphically, these can be represented using Bratteli diagrams. Namely, (multi)graphs with two rows of vertices, where the top k vertices are labeled $N_1, \ldots, N_k$, the bottom k' vertices are labeled by $N_1', \ldots, N_{k'}'$, and there are d_{ij} edges between vertex i of the top row and vertex j of the bottom row. Since $\phi : \mathcal{A}_1 \to \mathcal{A}_2$ is unital, all vertices in the bottom row are reached by an edge, but the top row can have vertices with no incident edge.

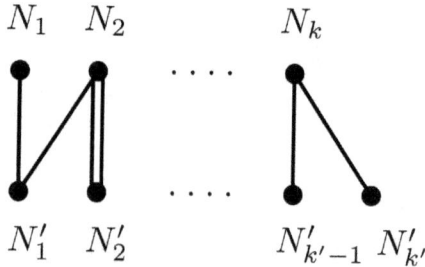

One can refine this usual notion of Bratteli diagrams, in order to also record graphically the possible permutations of identical blocks in the Wedderburn decomposition of the algebra. This can be done by using a *braided* version of the Bratteli diagram, for example as in the figure, for a permutation of matrix blocks of same rank in the algebra $M_2(\mathbb{C})^{\oplus 3} \oplus M_4(\mathbb{C})^{\oplus 2} \oplus M_5(\mathbb{C})$.

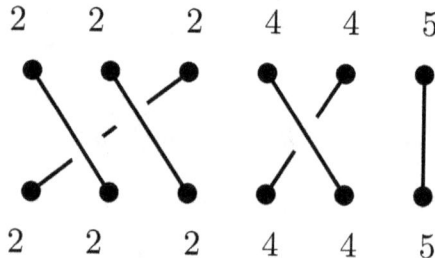

We refer the reader to [Marcolli, van Suijlekom (2014)] for specific examples and a more detailed discussion.

A *quiver* Γ is a (finite) directed graph. A representation π of a quiver Γ in a category $\mathcal{C}$ is the datum of an object π_v for each vertex v and a morphism π_e in $\mathrm{Hom}(\pi_{s(e)}, \pi_{t(e)})$ for each directed edge e.

Two representations π, π' of Γ in same category are equivalent if $\pi_v = \pi'_v$, for all $v \in V(\Gamma)$ and if there is a family of invertible morphisms $\phi_v \in \mathrm{Hom}(\pi(v), \pi(v))$ for $v \in V(\Gamma)$ such that

$$\pi_e = \phi_{t(e)} \circ \pi'_e \circ \phi_{s(e)}^{-1}.$$

For the categories $\mathcal{C}$ or $\mathcal{C}_0$ of finite spectral triples, a representation π of a quiver Γ assigns spectral triples $(\mathcal{A}_v, \mathcal{H}_v, D_v)$ (with $D_v = 0$ in the case of $\mathcal{C}_0$) to the vertices $v \in V(\Gamma)$ and pairs $(\phi, L) \in \mathrm{Hom}((\mathcal{A}_{s(e)}, \mathcal{H}_{s(e)}, D_{s(e)}), (\mathcal{A}_{t(e)}, \mathcal{H}_{t(e)}, D_{t(e)}))$ to the edges $e \in E(\Gamma)$.

For example one recovers with this setting the case of $U(N)$ spin networks of [Baez (1996)], with $(\mathcal{A}_v, \mathcal{H}_v) = (M_N(\mathbb{C}), \mathbb{C}^N)$ and $D = 0$, and with a unitary $u_e \in U(N)$ along each edge and the gauge action $g_v \in U(N)$ at each vertex given by

$$u_e \mapsto g_{t(e)} u_e g_{s(e)}^*.$$

Then only possible Bratteli diagrams in this case, for $\phi : M_N(\mathbb{C}) \to M_N(\mathbb{C})$, consist of a single edge between one upper row vertex and one lower row vertex.

9.4.2 Gauge networks

The data of a *gauge network* defined in [Marcolli, van Suijlekom (2014)] consist of

$$\{\Gamma, (A_v, \lambda_v, H_v; \iota_v)_v, (\rho_e, \mathbb{B}_e)_e\}$$

where Γ is a directed graph, (A_v, λ_v, H_v) is an object in the category $\mathcal{C}_0^s$ for each vertex $v \in V(\Gamma)$, and to every $e \in E(\Gamma)$ one assigns a representation ρ_e of the unitary group $G_e = \mathrm{Aut}_{\tilde{A}_{t(e)}}(H_{t(e)}) \times U(\mathrm{Ker}\lambda_{t(e)})$. At the edges $e \in E(\Gamma)$, one also assigns the Bratteli diagrams $\mathbb{B}_e$ for the $*$-algebra maps $A_{s(e)} \to A_{t(e)}$. At the vertices $v \in V(\Gamma)$ one also considers the intertwiners ι_v for the group $\mathcal{G}_v = U(\mathcal{A}_v) \rtimes S(\mathcal{A}_v)$,

$$\iota_v : \rho_{e'_1} \otimes \cdots \otimes \rho_{e'_k} \to \rho_{e_1}^{K_{\mathbb{B}_{e_1}}} \circ \phi_{\mathbb{B}} \otimes \cdots \otimes \rho_{e_l}^{K_{\mathbb{B}_{e_l}}} \circ \phi_{\mathbb{B}}$$

where $e'_1, \ldots, e'_k$ are the incoming edges and $e_1, \ldots, e_l$ are the outgoing edges at v, with isotropy group $K_{\mathbb{B}_e} = U(\mathrm{Ker}\lambda_{t(e)})_{\mathbb{B}_{e0}}$.

Correspondences between gauge networks are obtained in the following way. Given two quiver representations π, π' of Γ, consider $\mathcal{A}_v$–$\mathcal{A}'_v$ bimodules $\mathcal{E}_v$, such that the representations are related by

$$H_v = \mathcal{E} \otimes_{\mathcal{A}'_v} H'_v$$

with morphisms $T_e : \mathcal{E}_{s(e)} \to \mathcal{E}_{t(e)}$ compatible with the algebra maps ϕ_e, ϕ'_e

$$T_e(a\eta b) = \phi_e(a)T_e(\eta)\phi'_e(b), \quad a \in A_{s(e)}, \eta \in E_{s(e)}, b \in A'_{s(e)}.$$

We think of this datum as a diagram of compatible correspondences

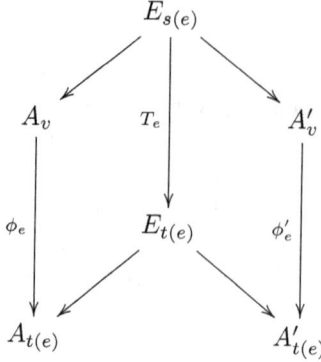

Given gauge networks

$$\psi = (\Gamma, (A_v, H_v, \iota_v)_v, (\rho_e, \mathbb{B}_e)_e), \qquad \psi' = (\Gamma, (A'_v, H'_v, \iota_v)_v, (\rho'_e, \mathbb{B}'_e)_e)$$

and correspondences $_\psi\Psi_{\psi'}$

$$\Psi = \{\Gamma, (_{A_v}E_{A'_v}, \iota_v \otimes \iota'_v)_v, (\rho_e \otimes \rho'_e, \mathbb{B}_e \times \mathbb{B}'_e)_e\},$$

the composition of correspondences is given by the tensor product of bi-modules,

$$\Psi_1 = \{\Gamma, (_{A_v}E_{A'_v}, \iota_v \otimes \iota'_v)_v, (\rho_e \otimes \rho'_e, \mathbb{B}_e \times \mathbb{B}'_e)_e\}$$

$$\Psi_2 = \{\Gamma, (_{A'_v}F_{A''_v}, \iota'_v \otimes \iota''_v)_v, (\rho'_e \otimes \rho''_e, \mathbb{B}'_e \times \mathbb{B}''_e)_e\}$$

$$\Psi_1 \circ \Psi_2 = \{\Gamma, (_{A_v}E \otimes_{A'_v} F_{A''_v}, \iota_v \otimes \iota''_v)_v, (\rho_e \otimes \rho''_e, \mathbb{B}_e \times \mathbb{B}''_e)_e\}.$$

We denote by $\mathcal{S}$ the category of gauge networks with correspondences as morphisms. The algebra $\mathbb{C}[\mathcal{S}]$ given by the semigroupoid ring of the category has elements $a = \sum_\Psi a_\Psi \Psi$ with the convolution product

$$(a * b)_\Psi = \sum_{\Psi = \Psi_1 \circ \Psi_2} a_{\Psi_1} b_{\Psi_2}.$$

It is shown in [Marcolli, van Suijlekom (2014)] that this algebra can be completed to a C^*-algebra represented on a Hilbert space, and the resulting operator algebra can be made dynamical by the construction of a Hamiltonian and a time evolution, built using the quadratic Casimir operator on $U(A_{t(e)})$. This should be thought of as a discretization of the algebra of geometries and correspondences constructed in [Marcolli, Zainy al-Yasry (2008)].

9.4.3 *Spectral action on a lattice*

Consider a graph Γ embedded in a compact Riemannian spin manifold M. It is possible to pull back the spin geometry of M to the graph Γ. Indeed, let $\mathbb{S}$ be the fiber of the spinor bundle on M and take $\mathbb{S}^{V(\Gamma)}$ to be the space of spinors on Γ. Consider the holonomy $\text{Hol}(e, \nabla^S)$ of the spin connection along the edges e of Γ. This satisfies

$$\text{Hol}(e, \nabla^S) = \mathcal{P}e^{\int_e \omega \cdot dx} \sim 1 + \ell_e \omega_e(s(e)) + \mathcal{O}(\ell_e^2)$$

where $\omega_e(v)$ is the pairing of the 1-form ω and a vector $\dot{e}$ at the vertex v, and ℓ_e is the geodesic length of the edge e, which we regard as a continuous variable that tends to zero in the continuum limit of the model.

The Dirac operator on Γ is constructed by setting

$$(D_\Gamma \psi)_v = \sum_{t(e)=v} \frac{1}{2l_e} \gamma_e \text{Hol}(e, \nabla^S) \psi_{s(e)} + \sum_{s(e)=v} \frac{1}{2l_{\bar{e}}} \gamma_{\bar{e}} \text{Hol}(\bar{e}, \nabla^S) \psi_{t(e)},$$

for $\psi \in \mathbb{S}^{V(\Gamma)}$, with ℓ_e the geodesic length of the embedded edge e and $\bar{e}$ the same edge taken with the opposite orientation. The gamma matrices γ_e are defined so that

$$\sum_{e \in S(v)} \gamma_e \omega_e = \gamma^\mu \omega_\mu,$$

where γ^μ are the gamma matrices on the manifold M. This compatibility condition will be needed to ensure that the continuum limit delivers the correct Dirac operator.

In order to consider a "continuum limit" case it is best to restrict to the situation where Γ is a lattice. Since the spectral action setting requires the manifold M to be Riemannian and compact, we can consider the case of tori with a lattice periodically wrapping around the torus.

We look at a setting where the lattice spacing ℓ_e goes to zero. Assume for simplicity that all the edges have the same length $\ell_e = \ell$, the square lattice case, so that

$$(D_\Gamma \psi)_v = \sum_{v_1, v_2} \frac{1}{2\ell} \gamma_e (\psi_{v_1} - \psi_{v_2}) + \frac{1}{2} \gamma_e \omega_e(v)(\psi_{v_1} + \psi_{v_2}) + \mathcal{O}(\ell)$$

where the sum is over all collinear $v_1 \xrightarrow{e'} v \xrightarrow{e} v_2$. Formally, when $\ell \to 0$ one then finds

$$(D_\Gamma \psi)_v \longrightarrow \gamma^\mu (\partial_\mu + \omega_\mu) \psi(v).$$

Suppose given also a quiver representation of Γ in the category of finite spectral triples. Then we can consider Dirac operators on the discretized almost-commutative geometry, of the form

$$(D_{\Gamma,L}\psi)_v = \sum_{t(e)=v} \frac{1}{2l_e}\gamma_e \left(\text{Hol}(e, \nabla^S) \otimes L_e\right) \psi_{s(e)}$$

$$+ \sum_{s(e)=v} \frac{1}{2l_{\bar{e}}}\gamma_{\bar{e}} \left(\text{Hol}(\bar{e}, \nabla^S) \otimes L_{\bar{e}}\right) \psi_{t(e)} + \gamma D_v \psi_v,$$

where D_v is the Dirac operator of the finite spectral triple at the vertex v, and where $L_{\bar{e}} = L_e^*$ and γ are the grading on the spinor bundle of M if the manifold is even dimensional.

If $(A_v, H_v) = (M_N(\mathbb{C}), \mathbb{C}^N)$ at all vertices v, then the morphism (ϕ, L) is a unitary in $U(N)$, which is the holonomy of some gauge connection 1-form A_μ. Then the Dirac operator on Γ recovers the Dirac operator on M twisted by gauge field, in the continuum limit.

In order to define a spectral action functional in this setting with finite spectral triples, consider an action functional of the form

$$S[\{L_e\}, \{D_v\}] = \text{Tr} f(D_{\Gamma,L}),$$

for some function f on the real line.

For a lattice gauge field theory, assuming the cutoff energy scales like $\Lambda \propto \ell^{-1}$ and f is quartic, when we consider the small ℓ limit, we have

$$S_\Lambda[\{L_e\}, \{D_v\}] := \text{Tr} f(D_{\Gamma,L}/\Lambda) \equiv \ell^4 \text{Tr}((D_{\Gamma,L})^4).$$

On a square lattice we then, from counting contributions of different cycles in the lattice, we find

$$S_\Lambda[\{L_e\}, \{D_v\}] = -\frac{1}{4} \sum_{\partial p = e_4 \cdots e_1} \left(\text{Tr}\left(L_{\bar{e}_4} L_{\bar{e}_3} L_{e_2} L_{e_1}\right) + \text{Tr}\left(L_{\bar{e}_1} L_{\bar{e}_2} L_{e_3} L_{e_4}\right)\right) + C$$

$$+ \sum_v \ell^4 \text{Tr} D_v^4 + 4\ell^2 \sum_e \left(\text{Tr} D_{s(e)}^2 + \text{Tr} D_{t(e)}^2 - \text{Tr} L_e^* D_{t(e)} L_e D_v\right).$$

for some constant C. We refer the reader to [Marcolli, van Suijlekom (2014)] for a more detailed derivation.

In the flat case, the holonomy of the spin connection is trivial, and $S_\Lambda[\{L_e\}]$ is

$$= 4\ell^4 \sum_{\partial p = \bar{e}_4 \bar{e}_3 e_2 e_1} \frac{1}{(2\ell)^4} \text{Tr}(\gamma_\nu \gamma_\mu)^2 \left(\text{Tr}\left(L_{\bar{e}_4} L_{\bar{e}_3} L_{e_2} L_{e_1}\right) + \text{Tr}\left(L_{\bar{e}_1} L_{\bar{e}_2} L_{e_3} L_{e_4}\right)\right)$$

plus constant terms, hence it is

$$= -\frac{1}{4} \sum_{\partial p = \bar{e}_4 \bar{e}_3 e_2 e_1} \left(\text{Tr}\left(L_{\bar{e}_4} L_{\bar{e}_3} L_{e_2} L_{e_1}\right) + \text{Tr}\left(L_{\bar{e}_1} L_{\bar{e}_2} L_{e_3} L_{e_4}\right)\right) + \text{const}.$$

The argument for the other terms is similar, see [Marcolli, van Suijlekom (2014)].

9.4.4 *Continuum limit and the Wilson action*

When passing to the continuum limit, one finds in the μ direction of e the continuous gauge field A_μ at $s(e)$,
$$L_e = \mathcal{P}e^{i\int_e A\cdot dx} \sim e^{iA_\mu \ell} \quad \text{for } \ell \to 0.$$
With $(A_v, H_v) = (M_N(\mathbb{C}), \mathbb{C}^N)$ at all vertices v, in the limit $\ell \to 0$ with $\Lambda \propto \ell^{-1}$, the spectral action S_Λ becomes
$$\frac{1}{4}\int_M \mathrm{Tr}F_{\mu\nu}F^{\mu\nu} + 2\int_M \mathrm{Tr}(\partial_\mu\Phi - [iA_\mu, \Phi])(\partial^\mu\Phi - [iA^\mu, \Phi])$$
$$+ 8\Lambda^2 \int_M \mathrm{Tr}\Phi^2 + \int_M \mathrm{Tr}\Phi^4,$$
which is the action for Yang–Mills coupled to a Higgs field with a quartic potential.

For a plaquette
$$\mathrm{Tr}\left(L_{\bar{e}_4}L_{\bar{e}_3}L_{e_2}L_{e_1}\right) = \mathrm{Tr}e^{-i\ell A_\nu(x)}e^{-i\ell A_\mu(x+l\hat{\nu})}e^{i\ell A_\nu(x+\ell\hat{\mu})}e^{i\ell A_\mu(x)}$$
$$\sim \mathrm{Tr}e^{i\ell^2 F_{\mu\nu}} \quad \text{for } \ell \to 0$$
and similarly for $\mathrm{Tr}\left(L_{\bar{e}_1}L_{\bar{e}_2}L_{e_3}L_{e_4}\right)$. Thus, for $\ell \to 0$ (and $\Lambda \to \infty$)
$$S_\Lambda \sim \frac{1}{4}\int_M \mathrm{tr}F_{\mu\nu}F^{\mu\nu}$$
To see the source of the Higgs terms, a vertex v at the position x determines
$$\mathrm{Tr}e^{-iA_\mu\ell}\Phi(x+\ell\hat{\mu})e^{iA_\mu\ell}\Phi(x) \sim$$
$$\mathrm{Tr}\bigg(\Phi(x)\Phi(x+l\hat{\mu}) + l\Phi(x+l\hat{\mu})[iA_\mu, \Phi(x)]$$
$$-\frac{1}{2}\ell^2[iA_\mu, \Phi(x+l\hat{\mu})][iA_\mu, \Phi(x)]\bigg) + \mathcal{O}(\ell^3),$$
where $\Phi(x)$ is a continuous (hermitian) Higgs field corresponding to D_x, and L_e is expanded in the fields A_μ.

Modulo $\mathcal{O}(\ell^3)$, one then finds in S_Λ
$$S_\Lambda = -\frac{1}{4}\sum_{\partial p = \bar{e}_4\bar{e}_3 e_2 e_1}\left(\mathrm{Tr}\left(L_{\bar{e}_4}L_{\bar{e}_3}L_{e_2}L_{e_1}\right) + \mathrm{Tr}\left(L_{\bar{e}_1}L_{\bar{e}_2}L_{e_3}L_{e_4}\right)\right)$$
$$+ \sum_v \ell^4 \mathrm{Tr}D_v^4 + 4\ell^2 \sum_e \left(\mathrm{Tr}D_{s(e)}^2 + \mathrm{Tr}D_{t(e)}^2 - \mathrm{Tr}L_e^* D_{t(e)} L_e D_{s(e)}\right)$$
$$\sim \frac{1}{2}\mathrm{Tr}e^{i\ell^2 F_{\mu\nu}} + \ell^4 \mathrm{Tr}\Phi^4(x) + 2\ell^2 \sum_\mu \mathrm{tr}\Phi^2(x) + \mathrm{tr}\Phi^2(x+\ell\hat{\mu})$$
$$+ 2\ell^4 \sum_\mu \frac{1}{\ell^2}\mathrm{Tr}(\Phi(x+\ell\hat{\mu}) - \Phi(x))^2$$
$$-\frac{2}{\ell}\mathrm{Tr}\Phi(x+\ell\hat{\mu})[iA_\mu(x), \Phi(x)] + \mathrm{Tr}([iA_\mu(x), \Phi(x)])^2.$$

9.5 Random finite noncommutative geometries

In this section we discuss a quantum gravity approach based on finite spectral triples, developed in [Barrett (2015)] and [Barrett, Glaser (2015)].

Consider a geometry given by a finite spectral triple $(\mathcal{A}, \mathcal{H}, D, J, \gamma)$ with real structure. A *random geometry* has a fixed fermion space $(\mathcal{A}, \mathcal{H}, J, \gamma)$ and varying Dirac operator D up to unitary equivalences. Thus, we can view a random geometry as a "random" (in a suitable probability distribution) choice of a point in a moduli space $\mathcal{M}$ (in the sense of [Ćaćić (2011)]) of Dirac operators of a finite spectral triple.

We want the measure to reflect some action functional, as in the usual path integral setting,

$$e^{-S(D)}\, dD.$$

One can view this as a *random matrix model*, where the matrices D are constrained by the properties of Dirac operators of finite spectral triples. In particular, one would like the action functional to be a version of the spectral action,

$$S(D) = \mathrm{Tr}(f(D)) = \sum_{\lambda \in Spec(D)} f(\lambda).$$

Here we drop the explicit reference to the energy scale Λ in the spectral action, since we are considering only finite dimensional settings, where an (energy dependent) cutoff on the spectrum is not needed. For convergence of the integral

$$Z = \int_{\mathcal{M}} e^{-S(D)}\, dD$$

we want some function $f(x)$ with $f(x) \to \infty$ for $|x| \to \infty$. This is different from the usual case of the spectral action in the manifold setting, where f is a rapidly decaying test function that approximates a cutoff function. Note, however, that this is similar to the discretized case of [Marcolli, van Suijlekom (2014)] discussed in the previous section, where one also considers polynomial functions. The simplest choice is a quartic polynomial

$$f(D) = g_2 D^2 + g_4 D^4,$$

with $g_4 > 0$, or with $g_4 = 0$ and $g_2 > 0$.

Observables $\mathcal{O}(D)$ are functions of the Dirac operator D,

$$\langle \mathcal{O} \rangle = \frac{1}{Z} \int_{\mathcal{M}} \mathcal{O}(D) e^{-S(D)}\, dD$$

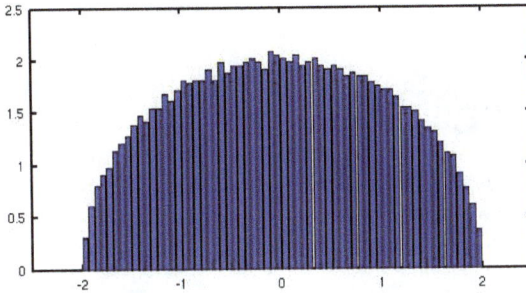

Fig. 9.1   The Wigner semi-circle law.

and one is interested in the behavior in the limit $N \to \infty$ of large matrices.

In the approach of [Barrett (2015)] and [Barrett, Glaser (2015)], one considers the moduli space of all possible Dirac operators on a finite spectral triple, in the same sense that we discussed in Chapter 1, see [Ćaćić (2011)]. A classification of the possible Dirac operators of finite spectral triples is given in [Barrett (2015)], where it is shown that they can be classified in terms of different signatures and Clifford modules. Let $\mathcal{H} = V \otimes M_n(\mathbb{C})$, with $V = \mathbb{C}^k$ a Clifford module of signature (p, q), with $k = 2^{d/2}$ or $k = 2^{(d-1)/2}$. All the possibilities for (p, q) are analyzed by writing the Dirac operators in terms of gamma matrices and commutators or anticommutators with given hermitian matrices H and anti-hermitian matrices L.

For example, in the case $(1, 0)$ one has $D = \{H, \cdot\}$ and in the case of $(0, 1)$ one has $D = -i[L, \cdot]$. Similarly, in the case of $(1, 1)$, one has $(\gamma^1)^2 = 1$ and $(\gamma^2)^2 = -1$ and the form of the Dirac operator is

$$D = \gamma^1 \otimes \{H, \cdot\} + \gamma^2 \otimes [L, \cdot].$$

We refer the reader to [Barrett (2015)] for a detailed analysis of the various cases.

In [Barrett, Glaser (2015)] a Monte Carlo simulation is used to analyze the behavior of these random geometries. One starts with a random D, described in terms of the hermitian and anti-hermitian H_i and L_i, and constructs a variation $D + \delta D$ by introducing variations δH_i and δL_i. A candidate new Dirac operator is accepted if

$$\Delta S(D) = S(D_{new}) - S(D_{old}) < 0$$

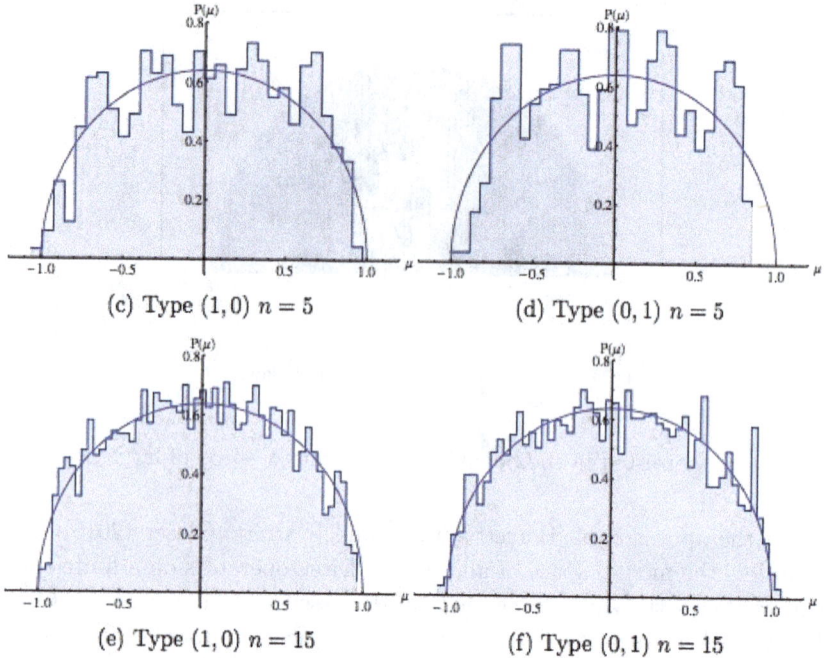

(c) Type $(1,0)$ $n = 5$ (d) Type $(0,1)$ $n = 5$

(e) Type $(1,0)$ $n = 15$ (f) Type $(0,1)$ $n = 15$

Fig. 9.2   Density of states from [Barrett, Glaser (2015)].

or, in order to escape the local minima, if

$$\exp(S(D_{old}) - S(D_{new})) > p,$$

for a uniformly distributed random number on $[0, 1]$. Otherwise one keeps D_{old}. The results are compared with Wigner's semicircle law (shown in Figure 9.1) for random matrix model with real symmetric matrices of large order N. In the Gaussian case, with $\text{Tr}(D^2)$ action, the density of states obtained in [Barrett, Glaser (2015)] for H and L in the type $(1, 0)$ and type $(0, 1)$ cases is shown in the figure, reproduced from their paper.

The distribution for the Dirac operator, described as a combination of H and L as above, correspondingly gives a behavior as in Figure 9.3. In the case of the quartic action $\text{Tr}(g_2 D^2 + D^4)$, for varying values of g_2, ranging from $g_2 = -1$ to $g_2 = -5$, the distribution for the eigenvalues for the action takes the form shown in Figure 9.4. In three of the four cases in Figure 9.4, the graphs show a phase transition. Moreover, the eigenvalue distribution at the critical value of g_2 resembles the eigenvalue distribution on a manifold, namely a power law $|\lambda|^{d-1}$ for dimension d.

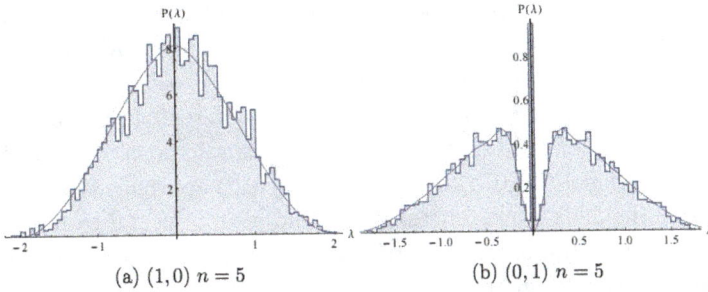

Fig. 9.3   Eigenvalue density for the Dirac operator, Gaussian case, [Barrett, Glaser (2015)].

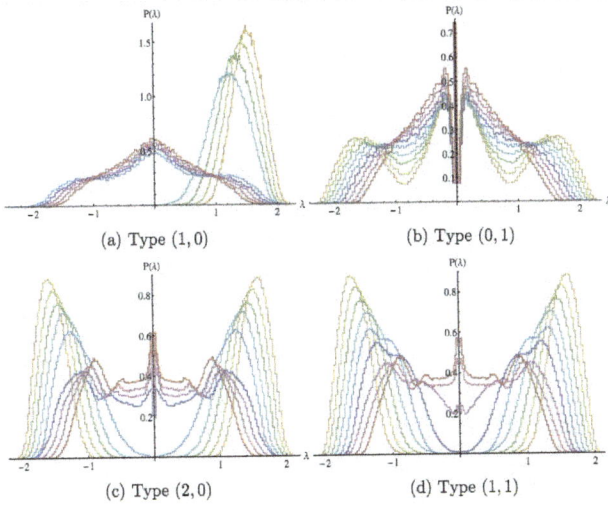

Fig. 9.4   Distribution for the eigenvalues for the quartic action $\mathrm{Tr}(g_2 D^2 + g_4 D^4)$, with g_2 ranging from -5 to -1, from [Barrett, Glaser (2015)].

Following this point of view, one should consider approximations of smooth (compact Riemannian spin) manifolds by finite spectral triples, and replace the integral over smooth geometries of the Hartle–Hawking wave function with an integral on random finite spectral triples. This requires a precise description of the approximation of smooth geometries by finite spectral triple. A result in this direction was obtained in [Rieffel (2004)] using convergence in a Gromov–Hausdorff distance as quantum metric spaces, and in [Rieffel (2016)] where the additional data of vector bundles is taken into account. Another approach to the approximation by finite spectral triples is being developed by Barrett for the case of coadjoint orbits of Lie groups. While the two approaches appear different, they may be compatible: the case of coadjoint orbits was in fact also considered in [Rieffel (2010)] and [Rieffel (2009)]. In order to apply this approach to a version of the Hartle–Hawking wave function, it would also be useful to introduce a suitable notion of finite spectral triples with boundary as approximations to manifolds with boundary.

Bibliography

T. Ackermann, *A note on the Wodzicki residue*, J. Geom. Phys. 20 (1996) no. 4, 404–406.

Y. André, *Une introduction aux motifs (motifs purs, motifs mixtes, périodes)*. Panoramas et Synthèses **17**. Société Mathématique de France, Paris, 2004.

S. Antusch, J. Kersten, M. Lindner, M. Ratz, M.A. Schmidt *Running neutrino mass parameters in see-saw scenarios*, JHEP 03 (2005) 024, hep-ph/0501272v3.

S. Antusch, J. Kersten, M. Lindner, M. Ratz, *Neutrino mass matrix running for non-degenerate see-saw scales*, Phys. Lett. B538 (2002) 87–95.

H. Arason, D.J. Castano, B. Kesthlyi, E.J. Piard, P. Ramond, B.D. Wright, *Renormalization-group study of the standard model and its extensions: the standard model*, Phys. Rev. D, 46 (1992) N.9, 3945–3965.

T. Asselmeyer-Maluga, C.H. Brans, *Exotic Smoothness and Physics*, World Scientific (2007).

M.F. Atiyah, *Global theory of elliptic operators*, Proc. Internat. Conf. on Functional Analysis and Related Topics (Tokyo, 1969) pp. 21–30, Univ. of Tokyo Press, 1970.

R. Aurich, S. Lustig, F. Steiner, H. Then, *Cosmic microwave background alignment in multi-connected universes*, Class. Quantum Grav. 24 (2007) 1879–1894.

I.G. Avramidi, *Covariant methods for the calculation of the effective action in quantum field theory and investigation of higher-derivative quantum gravity*, PhD Thesis, Moscow University, 1986, hep-th/9510140.

M.V. Babich, D.A. Korotkin, *Self–dual SU(2)–Invariant Einstein Metrics and Modular Dependence of Theta–Functions*. Lett. Math. Phys. 46 (1998) 323–337.

K.S. Babu, C.N. Leung, J. Pantaleone, *Renormalization of the neutrino mass operator*, Phys. Lett. B, Vol. 319 (1993) 191–198.

J.C. Baez, *Spin network states in gauge theory*, Adv. Math. 117 (1996) 253–272.

I. Bakas, F. Bourliot, D. Lüst, M. Petropoulos, *The mixmaster universe in Hořava–Lifshitz gravity*, Classical Quantum Gravity 27 (2010) no. 4, 045013, 25 pp.

A. Ball, M. Marcolli, *Spectral action models of gravity on packed swiss cheese cosmology*, Class. Quantum Grav. 33 (2016) 115018, 39 pp.

C. Bär, *The Dirac operator on space forms of positive curvature*, J. Math. Soc. Japan 48 (1996) N.1, 69–83.

C. Bär, *Dependence of Dirac spectrum on the spin structure*, in "Séminaires Congrès, 4. Global Anal. and Harmonic Anal. (Luminy, 2000)", French Math. Soc., 2000, pp. 17–33.

C. Bär, *The Dirac operator on homogeneous spaces and its spectrum on 3-dimensional lens spaces*, Arch. Math. 59 (1992) 65–79.

A.M. Baranov, Yu.I. Manin, I.V. Frolov, A.S. Schwarz, *A superanalog of the Selberg trace formula and multiloop contributions for fermionic strings*, Commun. Math. Phys. Vol. 111 (1987) N.3, 373–392.

J.W. Barrett, *Lorentzian version of the noncommutative geometry of the Standard Model of particle physics.* J. Math. Phys. 48 (2007) no. 1, 012303, 7 pp.

J.W. Barrett, *Matrix geometries and fuzzy spaces as finite spectral triples*, arXiv:1502.05383.

J.W. Barrett, L. Glaser, *Monte Carlo simulations of random non-commutative geometries*, arXiv:1510.01377.

J.W. Barrett, R.A. Dawe Martins, *Noncommutative geometry and the Standard Model vacuum*, J. Math. Phys. Vol. 47 (2006), no. 5, 052305, 14 pp.

J.D. Barrow, *Gravitational memory?* Phys. Rev. D Vol. 46 (1992) N.8 R3227, 4 pp.

D. Baumann, *TASI Lectures on inflation*, Lectures from the 2009 Theoretical Advanced Study Institute at Univ. of Colorado, Boulder, arXiv:0907.5424, 160 pp.

E. Bédos, *An introduction to 3D discrete magnetic Laplacians and noncommutative 3-tori*, J. Geom. Phys. 30 (1999) no. 3, 204–232.

W. Beenakker, T. van den Broek, W.D. van Suijlekom, *Supersymmetry and Noncommutative Geometry*, Springer Briefs in Mathematical Physics (2015).

V. Belinskii, I.M. Khalatnikov, E.M. Lifshitz, *Oscillatory approach to singular point in Relativistic cosmology*, Adv. Phys. 19 (1970) 525–551.

G. Belyi, *On Galois extensions of a maximal cyclotomic field*, Math. USSR Izv. 14 (1980) N.2, 247–256.

J. Bellissard, *Noncommutative geometry and quantum Hall effect*, Proceedings of the International Congress of Mathematicians, Vol. 1, 2 (Zürich, 1994), pp.1238–1246, Birkhäuser (1995).

A.A. Belyanin, V.V. Kocharovsky and VI.V. Kocharovsky, *Gamma-ray bursts from evaporating primordial black holes*, Radiophysics and Quantum Electronics, Vol. 41 (1996) N.1, 22–27.

B.K. Berger, *Numerical approaches to spacetime singularities*, arXiv:gr-qc/0201056.

P. de Bernardis, P.A.R. Ade, J.J. Bock, J.R. Bond, J. Borrill, A. Boscaleri, K. Coble, B.P. Crill, G.De Gasperis, P.C. Farese, P.G. Ferreira, K. Ganga, M. Giacometti, E. Hivon, V.V. Hristov, A. Iacoangeli, A.H. Jaffe, A.E. Lange, L. Martinis, S. Masi, P.V. Mason, P.D. Mauskopf, A. Melchiorri, L. Miglio, T. Montroy, C.B. Netterfield, E. Pascale, F. Piacentini,

D. Pogosyan, S. Prunet, S. Rao, G. Romeo, J.E. Ruhl, F. Scaramuzzi, D. Sforna, N. Vittorio, *A flat Universe from high-resolution maps of the cosmic microwave background radiation*, Nature 404 (2000) 955–959.

P. Bertozzini, R. Conti, W. Lewkeeratiyutkul, *A horizontal categorification of Gelfand duality*, Adv. Math. 226 (2011) no. 1, 584–607.

M. Birkel, S. Sarkar, *Nucleosynthesis bounds on a time-varying cosmological 'constant'*, Astroparticle Physics, Vol. 6 (1997) 197–203.

F. Bittner, *The universal Euler characteristic for varieties of characteristic zero*, Compos. Math. 140 (2004) no. 4, 1011–1032.

S. Bloch, H. Esnault, D. Kreimer, *On motives associated to graph polynomials*, Comm. Math. Phys. 267 (2006), no. 1, 181–225.

I. Bobtcheva, R. Piergallini, *On 4-dimensional 2-handlebodies and 3-manifolds*, J. Knot Theory Ramifications 21 (2012) no. 12, 1250110, 230 pp.

J. Boeijink, W.D. van Suijlekom, *The noncommutative geometry of Yang-Mills fields*, J. Geom. Phys. 61 (2011) 1122–1134.

J. Bognár, *Indefinite inner product spaces*, Springer (1974).

J. Bolte, H.M. Stiepan, *The Selberg trace formula for Dirac operators*, J. Math. Phys. 47 (2006) N.11, 112104, 16 pp.

D. Boyd, *The residual set dimension of the Apollonian packing*, Mathematika 20 (1973) 170–174.

L. Boyle, S. Farnsworth, *Non-commutative geometry, non-associative geometry and the standard model of particle physics*, New J. Phys. 16 (2014) 123027, 5 pp.

L. Boyle, S. Farnsworth, *A new algebraic structure in the standard model of particle physics*, arXiv:1604.00847.

L. Boyle, S. Farnsworth, J. Fitzgerald, M. Schade, *The Minimal Dimensionless Standard Model (MDSM) and its Cosmology*, arXiv:1111.0273.

S. Brain, B. Mesland, W.D. van Suijlekom, *Gauge theory for spectral triples and the unbounded Kasparov product*, J. Noncommut. Geom. Vol. 10 (2016) N.1, 135–206.

N. Brand, *Classifying spaces for branched coverings*, Indiana University Mathematical Journal, Vol. 29 (1980) N.2, 229–248.

N. Brand, G. Brumfiel, *Periodicity phenomena for concordance classes of branched coverings*, Topology, Vol. 19 (1980) 255–263.

T.P. Branson, P.B. Gilkey, *Residues of the eta function for an operator of Dirac type with local boundary conditions*, Differential Geom. Appl. 2 (1992) no. 3, 249–267.

V. Braungardt, D. Kotschick, *Einstein metrics and the number of smooth structures on a four-manifold*, Topology, Vol. 44 (2005) N.3, 641–659.

B. Brenken, *A classification of some noncommutative tori*, Rocky Mount. J. Math. 20 (1990) 389–397.

B. Brenken, J. Cuntz, G. Elliott, R. Nest, *On the classification of noncommutative tori. III*, in "Operator algebras and mathematical physics" (Iowa City, Iowa, 1985), 503–526, Contemp. Math., 62, Amer. Math. Soc., Providence, RI (1987).

T. van den Broek, W.D. van Suijlekom, *Supersymmetric QCD and noncommuta-*

tive geometry, Comm. Math. Phys. 303 (2011) no. 1, 149–173.

C. Brouder, N. Bizi, F. Besnard, *The Standard Model as an extension of the noncommutative algebra of forms*, arXiv:1504.03890.

I.L. Buchbinder, D.D. Odintsov, I.L. Shapiro, *Effective action in quantum gravity*, IOP Publishing (1992).

M. Buck, M. Fairbairn, M. Sakellariadou, *Inflation in models with conformally coupled scalar fields: An application to the noncommutative spectral action*, Phys. Rev. D 82 (2010) 043509, 14 pp.

E.V. Bugaev, V.B. Petkov, A.N. Gaponenko, P.A. Klimai, M.V. Andreev, A.B. Chernyaev, I.M. Dzaparova, D.D. Dzhappuev, Zh.Sh. Guliev, N.S. Khaerdinov, N.F. Klimenko, A.U. Kudzhaev, A.V. Sergeev, V.I. Volchenko, G.V. Volchenko, A.F. Yanin, *Experimental search of bursts of gamma rays from primordial black holes using different evaporation models*, arXiv:0906.3182.

C. Burgess, G. Moore, *The Standard Model. A Primer*, Cambridge University Press (2007).

B. Ćaćić, *Moduli spaces of Dirac operators for finite spectral triples*, arXiv:0902.2068, in "Quantum groups and noncommutative spaces", 9–68, Aspects Math., E41, Vieweg (2011).

B. Ćaćić, *A reconstruction theorem for almost-commutative spectral triples*, Lett. Math. Phys. 100 (2012) no. 2, 181–202.

B. Ćaćić, *Real structures on almost-commutative spectral triples*, Lett. Math. Phys. 103 (2013) no. 7, 793–816.

B. Ćaćić, *On reconstruction theorems in noncommutative Riemannian geometry*, Ph.D. Thesis, California Institute of Technology (2013).

B. Ćaćić, *A reconstruction theorem for Connes–Landi deformations of commutative spectral triples*, arXiv:1408.4429v2 [math-ph].

B. Ćaćić, M. Marcolli, K. Teh, *Coupling of gravity to matter, spectral action and cosmic topology*, J. Noncommutative Geometry, Vol. 8 (2014) N.2, 473–504.

S. Caillerie, M. Lachièze-Rey, J.P. Luminet, R. Lehoucq, A. Riazuelo, J. Weeks, *A new analysis of the Poincaré dodecahedral space model*, Astronomy and Astrophysics, Vol. 476 (2007) N.2, 691–696.

L. Carminati, B. Iochum, T. Schücker, *The noncommutative constraints on the standard model à la Connes*, J. Math. Phys. 38 (1997) no. 3, 1269–1280.

L. Carminati, B. Iochum, D. Kastler, T. Schücker, *Relativity, noncommutative geometry, renormalization and particle physics. Coherent states, differential and quantum geometry*, Rep. Math. Phys. 43 (1999) no. 1-2, 53–71.

B.J. Carr, *Primordial black holes as a probe of the early universe and a varying gravitational constant*, arXiv:astro-ph/0102390v2.

J.A. Casas, V. Di Clemente, A. Ibarra, M. Quirós, *Massive neutrinos and the Higgs mass window*, Phys. Rev. D62 (2000) 053005, 4 pp.

A. Chamseddine, *Noncommutative geometry as the key to unlock the secrets of space-time*, arXiv:0901.0577.

A. Chamseddine, A. Connes, *The spectral action principle*, Comm. Math. Phys. 186 (1997) no. 3, 731–750.

A. Chamseddine, A. Connes, *Inner fluctuations of the spectral action*, J. Geom. Phys. 57 (2006) N.1, 1–21.

A. Chamseddine, A. Connes, *Why the Standard Model*, J. Geom. Phys. 58 (2008) 38–47.

A. Chamseddine, A. Connes, *The uncanny precision of the spectral action*, Comm. Math. Phys. 293 (2010) no. 3, 867–897.

A. Chamseddine, A. Connes, *Noncommutative geometric spaces with boundary: spectral action*, J. Geom. Phys. 61 (2011) N.1, 317–332.

A.H. Chamseddine, A. Connes, *Resilience of the Spectral Standard Model*, J. High Energy Phys. 1209 (2012) 104.

A.H. Chamseddine, A. Connes, *Spectral action for Robertson-Walker metrics*, J. High Energy Phys. (2012) no. 10, 101, 29 pp.

A. Chamseddine, A. Connes, M. Marcolli, *Gravity and the standard model with neutrino mixing*, Advances in Theoretical and Mathematical Physics, 11 (2007) 991–1090.

A.H. Chamseddine, A. Connes, V. Mukhanov, *Quanta of geometry: noncommutative aspects*, Phys. Rev. Lett. Vol. 114 (2015) N.9, 091302, 5 pp.

A.H. Chamseddine, A. Connes, V. Mukhanov, *Geometry and the quantum: basics*, J. High Energy Phys. (2014) N.12, 098, 24 pp.

A.H. Chamseddine, A. Connes, W. van Suijlekom, *Beyond the Spectral Standard Model: Emergence of Pati-Salam Unification*, JHEP 1311 (2013) 132.

A.H. Chamseddine, A. Connes, W. van Suijlekom, *Grand Unification in the Spectral Pati-Salam Model*, arXiv:1507.08161.

L. Chen, A. Gibney, D. Krashen. *Pointed trees of projective spaces*. J. Algebraic Geom. Vol. 18 (2009) N. 3, 477–509.

E. Christensen, C. Ivan, M.L. Lapidus, *Dirac operators and spectral triples for some fractal sets built on curves*, Adv. Math. 217 (2008) N.1, 42–78.

E. Christensen, C. Ivan, E. Schrohe, *Spectral triples and the geometry of fractals*, J. Noncommut. Geom. 6 (2012) no. 2, 249–274.

D. Cimasoni, N. Reshetikhin, *Dimers on surface graphs and spin structures, I*, Comm. Math. Phys., Vol. 275 (2007) 187–208.

F. Cipriani, D. Guido, T. Isola, J.L. Sauvageot, *Spectral triples for the Sierpinski gasket*, J. Funct. Anal. 266 (2014) no. 8, 4809–4869.

L.S. Cirio, G. Landi, R.J. Szabo, *Algebraic deformations of toric varieties I. General constructions*, Adv. Math. 246 (2013) 33–88.

L.S. Cirio, G. Landi, R.J. Szabo, *Algebraic deformations of toric varieties II: noncommutative instantons*, Adv. Theor. Math. Phys. 15 (2011) no. 6, 1817–1907.

L.S. Cirio, G. Landi, R.J. Szabo. *Instantons and vortices on noncommutative toric varieties*. arXiv:1212.3469.

J. Cisneros-Molina, *The η-invariant of twisted Dirac operators of S^3/Γ*, Geometriae Dedicata, Vol. 84 (2001) 207–228.

A. Codello, R. Percacci, *Fixed points of higher derivative gravity*, Phys. Rev. Lett. 97 (2006) 221301.

P.B. Cohen, C. Itzykson, J. Wolfart, *Fuchsian triangle groups and Grothendieck dessins. Variations on a theme of Belyi*, Comm. Math. Phys. 163 (1994) N.3, 605–627.

P. Coles, F. Lucchin, *Cosmology. The origin and evolution of cosmic structure*,

Wiley (1995).

A. Connes, C^*-*algèbres et géométrie différentielle*, C. R. Acad. Sci. Paris Sér. A-B 290 (1980) N.13, A599–A604.

A. Connes, *Cohomologie cyclique et foncteurs* Ext^n. C. R. Acad. Sci. Paris Sér. I Math. Vol. 296 (1983), N.23, 953–958.

A. Connes, *Noncommutative geometry*, Academic Press (1994).

A. Connes, *Geometry from the spectral point of view*. Lett. Math. Phys. 34 (1995), no. 3, 203–238.

A. Connes, Lecture at Oberwolfach, September (2007).

A. Connes, *On the spectral characterization of manifolds*, J. Noncommut. Geom. 7 (2013) N.1, 1–82.

A. Connes, M.R. Douglas, A. Schwarz, *Noncommutative geometry and matrix theory: compactification on tori*, J. High Energy Phys. (1998) N.2, Paper 3, 35 pp.

A. Connes, G. Landi, *Noncommutative manifolds, the instanton algebra and isospectral deformations*, Comm. Math. Phys. 221 (2001) N.1, 141–159.

A. Connes, M. Marcolli, *Noncommutative Geometry, Quantum Fields, and Motives*, Colloquium Publications, Vol. 55, American Mathematical Society (2008).

A. Connes, M. Marcolli, *A walk in the noncommutative garden*, An invitation to noncommutative geometry, pp.1–128, World Scientific (2008).

A. Connes, G. Skandalis, *The longitudinal index theorem for foliations*. Publ. Res. Inst. Math. Sci. 20 (1984), no. 6, 1139–1183.

N.J. Cornish, D.N. Spergel, G.D. Starkman, E. Komatsu, *Constraining the topology of the universe*, Phys. Rev. Lett. 92 (2004) 201302.

L. Dabrowski, G. Dossena, *Product of real spectral triples*, Int. J. Geom. Methods Mod. Phys. 8 (2011) N.8, 1833–1848.

M. Dahl, *Prescribing eigenvalues of the Dirac operator*, Manuscripta Math. 118 (2005) 191–199.

M. Dahl, *Dirac eigenvalues for generic metrics on three-manifolds*, Annals of Global Analysis and Geometry, 24 (2003) 95–100.

K. Davidson, C^*-*algebras by example*, Fields Institute Monographs, Vol. 6, American Mathematical Society (1996).

R.A. Dawe Martins, *Noncommutative geometry, topology, and the standard model vacuum*, J. Math. Phys. 47 (2006) no. 11, 113507, 16 pp.

R.A. Dawe Martins, *Categorified noncommutative manifolds*, Internat. J. Modern Phys. A 24 (2009) no. 15, 2802–2819.

P.D. D'Eath, *Supersymmetric Quantum Cosmology*, Cambridge University Press (1996).

D. Denicola, M. Marcolli, A. Zainy al-Yasry, *Spin foams and noncommutative geometry*, Classical Quantum Gravity 27 (2010) no. 20, 205025, 53 pp.

A. De Simone, M.P. Hertzberg, F. Wilczek, *Running inflation in the Standard Model*, Phys. Lett. B, Vol. 678 (2009) N.1, 1–8.

S. Dodelson, *Modern Cosmology*, Academic Press (2003).

S. Donaldson, R. Friedman, *Connected sums of self-dual manifolds and deformations of singular spaces*, Nonlinearity 2 (1989) no. 2, 197–239.

J.F. Donoghue, *General relativity as an effective field theory: the leading quantum corrections*, Phys. Rev. D, Vol. 50 (1994) N.6, 3874–3888.

J.F. Donoghue, E. Golowich, B.R. Holstein, *Dynamics of the Standard Model*, Cambridge University Press (1994).

C.F. Doran, M.G. Faux, S.J. Gates, Jr., T. Hübsch, K.M. Iga, G.D. Landweber, R.L. Miller, *Codes and supersymmetry in one dimension*, Adv. Theor. Math. Phys., Vol. 15 (2011) N.6, 1909–1970.

C.F. Doran, M.G. Faux, S.J. Gates, Jr., T. Hübsch, K.M. Iga, G.D. Landweber, *Relating doubly-even error-correcting codes, graphs, and irreducible representations of N-supersymmetry*, in "Discrete and computational mathematics", pp. 53–71, Nova Sci. Publ. (2008).

C. Doran, K. Iga, J. Kostiuk, G. Landweber, S. Méndez-Diez, *Geometrization of N-extended 1-dimensional supersymmetry algebras, I*, Adv. Theor. Math. Phys. Vol. 19 (2015) N.5, 1043–1113.

J.S Dowker, *Spherical universe topology and the Casimir effect*, Class. Quant. Grav. 21 (2004) 4247–4271.

K. van den Dungen, M. Paschke, A. Rennie, *Pseudo-Riemannian spectral triples and the harmonic oscillator*, J. Geom. Phys. 73 (2013), 37–55.

G.V. Dunne, *Heat kernels and zeta functions on fractals*, J. Phys. A: Math. Theor. 45 (2012) N.37, 374016.

C.L. Duston, *Exotic smoothness in four dimensions and Euclidean quantum gravity*, Int. J. Geom. Methods Mod. Phys. Vol. 8 (2011) N. 3, 459–484.

C.L. Duston, *Topspin networks in loop quantum gravity*, Classical Quantum Gravity, Vol. 29 (2012) N. 20, 205015, 29 pp.

D. Dutkay, P. Jorgensen, *Iterated function systems, Ruelle operators, and invariant projective measures*, Math. Comp. 75 (2006), no. 256, 1931–1970.

T. Eguchi, A.J. Hanson, *Self-dual solutions to Euclidean Gravity*, Annals of Physics, 120 (1979), 82–106.

T. Eguchi, A.J. Hanson, *Gravitational Instantons*, Gen. Relativity Gravitation 11, No 5 (1979) 315–320.

G.A. Elliott, *On the K-theory of the C^*-algebra generated by a projective representation of a torsion–free discrete group*, Operator algebras and group representations, Monogr. Stud. Math. Vol. 17 (1984) 157–184.

C. Estrada, M. Marcolli, *Asymptotic safety, hypergeometric functions, and the Higgs mass in spectral action models*, Int. J. Geom. Methods Mod. Phys. 10 (2013) no. 7, 1350036, 30 pp.

C. Estrada, M. Marcolli, *Noncommutative mixmaster cosmologies*, Int. J. Geom. Methods Mod. Phys. 10 (2013), no. 1, 1250086, 28 pp.

K.J. Falconer, *The geometry of fractal sets*, Cambridge Tracts in Mathematics, Vol. 85. Cambridge University Press (1986).

W. Fan, F. Fathizadeh, M. Marcolli, *Spectral action for Bianchi type-IX cosmological models*, Journal of High Energy Physics, (2015) 85, 29 pp.

W. Fan, F. Fathizadeh, M. Marcolli, *Modular forms in the spectral action of Bianchi IX gravitational instantons*, arXiv:1511.05321.

W. Fan, F. Fathizadeh, M. Marcolli, *Motives and periods in Bianchi IX gravity models*, preprint (2017).

L.Z. Fang, R. Ruffini (Eds.), *Quantum Cosmology*, World Scientific (1987).

S. Farnsworth, *The graded product of real spectral triples*, arXiv:1605.07035.

S. Farnsworth, L. Boyle, *Rethinking Connes' approach to the standard model of particle physics via non-commutative geometry*, New J. Phys. 17 (2015) 023021, 5 pp.

F. Fathizadeh, A. Ghorbanpour, M. Khalkhali, *Rationality of spectral action for Robertson-Walker metrics*, J. High Energy Phys. (2014) N.12, 064, 21 pp.

F. Fathizadeh, M. Marcolli, *Periods and motives in the spectral action of Robertson-Walker spacetimes*, arXiv:1611.01815.

M. Faux, S.J. Gates, Jr., *Adinkras: a graphical technology for supersymmetric representation theory*, Phys. Rev. D 71(3) (2005) 065002.

N. Franco, *Temporal Lorentzian spectral triples*, Rev. Math. Phys. 26 (2014) N.8, 1430007, 23 pp.

E. Gausmann, R. Lehoucq, J.P. Luminet, J.P. Uzan, J. Weeks, *Topological lensing in spherical spaces*, Class. Quant. Grav. 18 (2001) 5155–5186.

L. Gerritzen, M. van der Put, *Schottky groups and Mumford curves*, Lecture Notes in Mathematics, Vol. 817, Springer (1980).

P.B. Gilkey, *Invariance theory, the heat equation, and the Atiyah-Singer index theorem*, Second edition. Studies in Advanced Mathematics. CRC Press (1995).

N. Ginoux, *The spectrum of the Dirac operator on $SU(2)/Q8$*, Manus. Math. 125 (2008) N.3, 383–409.

N. Ginoux, *The Dirac spectrum*, Lecture Notes in Mathematics, Vol. 1976. Springer (2009).

G.I. Gomero, M.J. Reboucas, A.F.F. Teixeira, *Spikes in cosmic crystallography II: topological signature of compact flat universes*. Phys. Lett. A 275 (2000) 355–367.

G.I. Gomero, M.J. Reboucas, R. Tavakol, *Detectability of cosmic topology in almost flat universes*, Class. Quant. Grav. 18 (2001) 4461–4476.

J.M. Gracia-Bondía, J.C. Várilly, H. Figueroa, *Elements of noncommutative geometry*, Birkhäuser (2001).

R.L. Graham, J.C. Lagarias, C.L. Mallows, A.R. Wilks, C.H. Yan, *Apollonian circle packings: number theory*, J. Number Theory 100 (2003) 1–45.

R.L. Graham, J.C. Lagarias, C.L. Mallows, A.R. Wilks, C.H. Yan, *Apollonian Circle Packings: Geometry and Group Theory III. Higher Dimensions*, Discrete Comput. Geom. 35 (2006) 37–72.

R. Gregory, V.A. Rubakov, S.M. Sibiryakov, *Gravity and antigravity in a brane world with metastable gravitons*, Phys. Lett. B Vol. 489 (2000) 203–206.

V.I. Gromak, I. Laine, S. Shimomura, *Painlevé differential equations in the complex plane*, Walter de Gruyter (2002).

C. Grosche, *Selberg supertrace formula for super Riemann surfaces, analytic properties of Selberg super zeta-functions and multiloop contributions for the fermionic string*, Comm. Math. Phys. 133 (1990) N.3, 433–485.

C. Grosche, *Path Integrals, hyperbolic spaces and Selberg trace formulae*, World Scientific (2013).

U. Guenther, A. Zhuk, V. Bezerra, C. Romero, *AdS and stabilized extra dimen-*

sions in multidimensional gravitational models with nonlinear scalar curvature terms $1/R$ *and* R^4, Class. Quant. Grav. 22 (2005) 3135–3167.

V.G. Gurzadyan, R. Penrose. *On CCC–predicted concentric low–variance circles in the CMB sky*, Eur. Phys. J. 128 (2013) 22–38.

A.H. Guth, *Phase transitions in the very early universe*, The very early universe (G.W. Gibbons, S.W. Hawking, S.T.C. Siklos, Eds.) pp. 171–204, Cambridge University Press (1982).

D. Harlow, S. Shenker, D. Stanford, L. Susskind, *Tree-like structure of eternal inflation: A solvable model*, Phys. Rev. D 85 (2021) 063516; on the arXiv as "Eternal Symmetree", arXiv:1110.0496.

J.B. Hartle, S.W. Hawking, *Wave function of the universe*, Phys. Rev. D Vol. 28 (1983) N.12, 2960–2975.

S.W. Hawking, *The quantum state of the universe*, Nuclear Phys. B239 (1984) 257–276.

S.W. Hawking, *Quantum Cosmology*, in "Relativity, groups and topology, II (Les Houches, 1983)", pp. 333–379, North-Holland (1984).

S.W. Hawking, J.C. Luttrell, *Higher derivatives in quantum cosmology (I). The isotropic case*, Nucl. Phys. B247 (1984) 250–260.

D. Hensley, *Continued fraction Cantor sets, Hausdorff dimension, and functional analysis*, J. Number Theory 40 (1992) 336–358.

M. Heydeman, M. Marcolli, I. Saberi, B. Stoica, *Tensor networks, p-adic fields, and algebraic curves: arithmetic and the* AdS_3/CFT_2 *correspondence*, arXiv:1605.07639.

N. Hitchin, *Harmonic spinors*, Advances in Math. 14 (1974) 1–55.

N.J. Hitchin. *Twistor spaces, Einstein metrics and isomonodromic deformations*, J. Diff. Geom., Vol. 42, No. 1 (1995) 30–112.

F. Hoyle, J.V. Narlikar, *A new theory of gravitation*, Proc. Royal Soc. London. Series A, Mathematical and Physical Sciences, Vol. 282 (1964) No. 1389, 191–207.

S. Iso, N. Okada, Y. Orikasa, *Classically conformal B–L extended Standard Model*, Phys. Lett. B Vol. 676 (2009) N.1–3, 81–87.

P. Jordan, *Schwerkraft und Weltall*, Vieweg Braunschweig (1955).

A. Kahle, *Superconnections and index theory*, J. Geom. Phys. 61 (2011) 1601–1624.

W. Kalau, M. Walze, *Gravity, non-commutative geometry and the Wodzicki residue*, J. Geom. Phys. 16 (1995) N.4, 327–344.

M. Kamionkowski, D.N. Spergel, N. Sugiyama, *Small-scale cosmic microwave background anisotropies as a probe of the geometry of the universe*, Astrophysical J. 426 (1994) L 57–60.

D. Kastler, *The Dirac operator and gravitation*, Comm. Math. Phys. 166 (1995) N.3, 633–643.

J. Kellendonk, J. Savinien, *Spectral triples and characterization of aperiodic order*, Proc. Lond. Math. Soc. (3) 104 (2012) no. 1, 123–157.

J. Kellendonk, J. Savinien, *Spectral triples from stationary Bratteli diagrams*, arXiv:1210.7360.

I.M. Khalatnikov, E.M. Lifshitz, K.M. Khanin, L.N. Shchur, and Ya.G. Sinai. *On* the stochasticity in relativistic cosmology. Journ. Stat. Phys., Vol. 38, Nos. 1/2 (1985) 97–114.

M. Khalkhali, *Basic Noncommutative Geometry*, European Mathematical Society (2009).

D. Kolodrubetz, M. Marcolli, *Boundary conditions of the RGE flow in the noncommutative geometry approach to particle physics and cosmology*, Phys. Lett. B, Vol. 693 (2010) 166–174.

A. Kontorovich, H. Oh, *Apollonian circle packings and closed horospheres on hyperbolic 3-manifolds*, Journal of AMS, Vol 24 (2011) 603–648.

T. Krajewski, *Classification of Finite Spectral Triples*, J. Geom. Phys. 28 (1998) 1–30.

M. Lachièze-Rey, J.P. Luminet, *Cosmic topology*, Phys. Rep. 254 (1995) 135–214.

A. Lai, K. Teh, *Spectral action for a one-parameter family of Dirac operators on $SU(2)$ and $SU(3)$*, J. Math. Phys. 54 (2013) no. 2, 022302, 21 pp.

N.P. Landsman, *Notes on Noncommutative Geometry*, manuscript available at http://www.math.ru.nl/ landsman/ncg2010.pdf

M.L. Lapidus, M. van Frankenhuijsen, *Fractal geometry, complex dimensions and zeta functions. Geometry and spectra of fractal strings*, Second edition. Springer (2013).

H.B. Lawson, M.L. Michelsohn, *Spin geometry*, Princeton University Press (1989).

O.M. Lecian, *Reflections on the hyperbolic plane*, arXiv:1303.6343.

R. Lehoucq, J. Weeks, J.P. Uzan, E. Gausmann, J.P. Luminet, *Eigenmodes of three-dimensional spherical spaces and their applications to cosmology*, Class. Quant. Grav. 19 (2002) 4683–4708.

J.E. Lidsey, A.R. Liddle, E.W. Kolb, E.J. Copeland, T. Barreiro, M. Abney, *Reconstructing the inflaton potential – an overview*, Rev. Mod. Phys. 69 (1997) 373–410.

A.D. Linde, *Gauge theories, time-dependence of the gravitational constant and antigravity in the early universe*, Phys. Letters B, Vol. 93 (1980) N.4, 394–396.

J.P. Luminet, J. Weeks, A. Riazuelo, R. Lehoucq, *Dodecahedral space topology as an explanation for weak wide-angle temperature correlations in the cosmic microwave background*, Nature 425 (2003) 593–595.

C. Mallows, *Growing Apollonian packings*, J. Integer Seq. 12 (2009) N.2, Article 09.2.1, 8 pp.

Yu.I. Manin, *p-adic automorphic functions*, Journ. of Soviet Math., 5 (1976) 279–333.

Yu.I. Manin, *Topics in Noncommutative Geometry*, Princeton University Press (1991).

Yu.I. Manin, *Gauge field theory and complex geometry*, Springer Verlag (1988) (Second Edition, 1997.)

Yu.I. Manin, *Sixth Painlevé equation, universal elliptic curve, and mirror of P^2*, in "Geometry of Differential Equations", Amer. Math. Soc. Transl. (2), vol. 186 (1998) 131–151.

Yu.I. Manin, *Real multiplication and noncommutative geometry (ein Alterstraum)*, The legacy of Niels Henrik Abel, pp. 685–727, Springer (2004).

Yu.I. Manin, M. Marcolli, *Holography principle and arithmetic of algebraic curves*, Adv. Theor. Math. Phys. 5 (2001) no. 3, 617–650.

Yu.I. Manin, M. Marcolli, *Continued fractions, modular symbols, and noncommutative geometry*, Selecta Math. (N.S.) 8 (2002) no. 3, 475–521.

Yu.I. Manin, M. Marcolli, *Modular shadows and the Lévy-Mellin ∞-adic transform*, in "Modular forms on Schiermonnikoog", 189–238, Cambridge Univ. Press (2008).

Yu.I. Manin, M. Marcolli, *Error-correcting codes and phase transitions*, Mathematics in Computer Science, Vol. 5 (2011) 133–170.

Yu.I. Manin, M. Marcolli, *Big Bang, blowup, and modular curves: algebraic geometry in cosmology*, SIGMA Symmetry Integrability Geom. Methods Appl. 10 (2014) Paper 073, 20 pp.

Yu.I. Manin, M. Marcolli, *Symbolic Dynamics, Modular Curves, and Bianchi IX Cosmologies*, Annales de la Faculté des Sciences de Toulouse, Vol. XXV (2016) N. 2–3, 313–338.

P.D. Mannheim, *Making the case for conformal gravity*, Foundations of Physics, Vol. 42 (2012) N.3, 388–420.

P.D. Mannheim, *Cosmological perturbations in conformal gravity*, Phys. Rev. D 85 (2012) N.12, 124008, 43 pp, arXiv:1109.4119.

M. Marcolli, *Limiting modular symbols and the Lyapunov spectrum*, J. Number Theory Vol. 98 N.2 (2003) 348–376.

M. Marcolli, *Arithmetic noncommutative geometry*, University Lecture Series, Vol. 36. American Mathematical Society (2005).

M. Marcolli, *Modular curves, C^*-algebras, and chaotic cosmology*, in "Frontiers in number theory, physics, and geometry. II", pp. 361–372, Springer (2007).

M. Marcolli, *Solvmanifolds and noncommutative tori with real multiplication*, Commun. Number Theory Phys. 2 (2008) N.2, 421–476.

M. Marcolli, *Feynman Motives*, World Scientific (2010).

M. Marcolli, A.M. Paolucci, *Cuntz-Krieger algebras and wavelets on fractals*, Complex Anal. Oper. Theory 5 (2011) N.1, 41–81.

M. Marcolli, C. Perez, *Codes as fractals and noncommutative spaces*, Math. Comput. Sci. Vol. 6 (2012) N.3, 199–215.

M. Marcolli, E. Pierpaoli, *Early universe models from Noncommutative Geometry*, Advances in Theoretical and Mathematical Physics, Vol. 14 (2010) 1373–1432.

M. Marcolli, E. Pierpaoli, K. Teh, *The spectral action and cosmic topology*, Comm. Math. Phys. 304 (2011) N.1, 125–174.

M. Marcolli, E. Pierpaoli, K. Teh, *The coupling of topology and inflation in noncommutative cosmology*, Comm. Math. Phys. 309 (2012) N.2, 341–369.

M. Marcolli, W. van Suijlekom, *Gauge Networks in Noncommutative Geometry*, J. Geom. Phys., Vol. 75 (2014) 71–91

M. Marcolli, N. Tedeschi, *Multifractals, Mumford curves and eternal inflation*, p-Adic Numbers Ultrametric Anal. Appl. 6 (2014) N.2, 135–154.

M. Marcolli, A. Zainy al-Yasry, *Coverings, correspondences, and noncommutative geometry*, J. Geom. Phys. 58 (2008) N.12, 1639–1661.

M. Marcolli, N. Zolman, *Adinkras, Dessins, Origami, and Supersymmetry Spectral*

Triples, arXiv:1606.04463.

D.H. Mayer, *Relaxation properties of the Mixmaster Universe*, Phys. Lett. A 121 (1987), no. 8,9, 390–394.

B. Mesland, *Unbounded bivariant K-theory and correspondences in noncommutative geometry*, J. Reine Angew. Math. Vol. 691 (2014) 101–172.

B. McInnes, *APS instability and the topology of the brane-world*, Phys. Lett. B593 (2004) N.1-4, 10–16.

A. Moss, D. Scott, J.P. Zibin, *No evidence for anomalously low variance circles on the sky*, arXiv:1012.1305 [astro-ph.CO].

J.R. Mureika, C.C. Dyer, *Multifractal analysis of Packed Swiss Cheese Cosmologies*, General Relativity and Gravitation, Vol. 36 (2004) N.1, 151–184.

D. Mumford, *An analytic construction of degenerating curves over complete local rings*, Compositio Math. 24 (1972) 129–174.

A. Nautiyal, *Anisotropic non-gaussianity with noncommutative spacetime*, Phys. Lett. B728 (2014) 472–481.

W. Nelson, J. Ochoa, M. Sakellariadou, *Constraining the noncommutative spectral action via astrophysical observations*. Phys. Rev. Lett. 105 (2010) 101602.

W. Nelson, J. Ochoa, M. Sakellariadou, *Gravitational waves in the spectral action of noncommutative geometry*, Phys. Rev. D 82 (2010) 085021.

W. Nelson, M. Sakellariadou, *Cosmology and the noncommutative approach to the standard model*, Phys. Rev. D 81 (2010) 085038.

W. Nelson, M. Sakellariadou, *Inflation mechanism in asymptotic noncommutative geometry*, Phys. Lett. B 680 (2009) 263–266. MR 2574318.

E.T. Newman, *A fundamenal solution to the CCC equations*, arXiv:1309.7271.

A. Niarchou, A. Jaffe, *Imprints of spherical nontrivial topologies on the cosmic microwave background*, Phys. Rev. Lett. 99 (2007) 081302.

V. Nistor, *A bivariant Chern–Connes character*, Annals of Math. Vol. 138 (1993) 555–590.

I.D. Novikov, A.G. Polnarev, A.A. Starobinsky, Ya.B. Zeldovich, *Primordial black holes*, Astron. Astrophys. 80 (1979) 104–109.

E. Newman, L. Tamburino, T. Unti, *Empty-space generalization of the Schwarzschild metric*, Journ. Math. Phys. 4 (1963) 915–923.

S. Okumura. *The self–dual Einstein–Weyl metric and classical solutions of Painlevé VI*, Lett. in Math. Phys., 46 (1998) 219–232.

P. Olczykowski, A. Sitarz, *On spectral action over Bieberbach manifolds*, Acta Phys. Polon. B 42 (2011) N.6, 1189–1198.

J.M. Overduin, F.I. Cooperstock, *Evolution of the scale factor with a variable cosmological term*, Phys. Rev. D 58, 043506, 23 pp.

J. Pearson, J. Bellissard, *Noncommutative Riemannian geometry and diffusion on ultrametric Cantor sets*, J. Noncommut. Geom. 3 (2009) N.3, 447–480.

R. Penrose, *The complex geometry of the natural world*, Proceedings of the International Congress of Mathematicians (Helsinki, 1978) pp. 189–194, Acad. Sci. Fennica, 1980.

R. Penrose. *Cycles of time*, Vintage Books (2010).

R. Piergallini *Four-manifolds as 4-fold branched covers of S^4*, Topology 34 (1995) 497–508.

M. Pimsner, D. Voiculescu, *Imbedding the irrational rotation C*-algebra into an AF-algebra*, J. Operator Theory 4 (1980) N.2, 201–210.

M.D. Pollock, *On the proposed existence of an anti-gravity regime in the early universe*, Phys. Lett. B, Vol. 108 (1982) 386–388.

M.D. Pollock, *Determination of the superstring moduli from the higher-derivative terms*, Phys. Lett. B Vol. 495 (2000) 401–406.

S. Popa, M. Rieffel, *The Ext groups of the C*-algebras associated with irrational rotations*, J. Operator Theory 3 (1980) N.2, 271–274.

M.J. Rees, D.W. Sciama, *Large-scale density inhomogeneities in the universe*, Nature, Vol. 217 (1968) 511–516.

M. Reuter, *Nonperturbative evolution equation for quantum gravity*, Phys Rev D 57 (1998) N.2, 971–985.

A. Riazuelo, J.P. Uzan, R. Lehoucq, J. Weeks, *Simulating cosmic microwave background maps in multi-connected spaces*, Phys. Rev. D 69 (2004) 103514.

A. Riazuelo, J. Weeks, J.-P. Uzan, R. Lehoucq, J. P. Luminet, *Cosmic microwave background anisotropies in multiconnected flat spaces*, Phys. Rev. D 69 (2004) 103518.

M. Rieffel, *C*-algebras associated with irrational rotations*, Pacific J. Math. 93 (1981) N.2, 415–429.

M. Rieffel, *Projective modules over higher-dimensional noncommutative tori*, Can. J. Math. (1988) XL, 257–338.

M. Rieffel, *Matrix algebras converge to the sphere for quantum Gromov-Hausdorff distance*, in "Gromov-Hausdorff distance for quantum metric spaces", Mem. Amer. Math. Soc. 168 (2004) N.796, pp. 67–91.

M. Rieffel, *Dirac operators for coadjoint orbits of compact Lie groups*, Münster J. Math. Vol. 2 (2009) 265–297.

M. Rieffel, *Distances between matrix algebras that converge to coadjoint orbits*, in "Superstrings, geometry, topology, and C*-algebras", pp. 173–180, Proc. Sympos. Pure Math., Vol. 81, Amer. Math. Soc. (2010).

M. Rieffel, *Matricial bridges for "Matrix algebras converge to the sphere"*, in "Operator Algebras and Their Applications: A Tribute to Richard V. Kadison", pp. 209–233, Contemporary Mathematics, Vol. 671, Amer. Math. Soc. (2016).

M. Rost, *The motive of a Pfister form*, Preprint (1998). www.physik.uni-regensburg.de/~rom03516/motive.html

B.F. Roukema, P.T. Rózański, *The residual gravity acceleration effect in the Poincaré dodecahedral space*, Astron. and Astrophy. 502 (2009) 27.

M.V. Safonova, D. Lohiya, *Gravity balls in induced gravity models – 'gravitational lens' effects*, Gravitation and Cosmology (1998) N.1, 1–10.

M. Salvetti, *On the number of nonequivalent differentiable structures on 4-manifolds*, Manuscripta Math., Vol. 63 (1989) N.2, 157–171.

L. Schneps, (Ed.) *The Grothendieck theory of Dessins d'Enfants*, Cambridge University Press (1994).

M.D. Schwartz, *Quantum field theory and the standard model*, Cambridge University Press (2013).

A. Scorpan, *The wild world of 4-manifolds*, American Mathematical Society (2005).

C. Series, *The modular surface and continued fractions*, J. London MS, Vol. 2, no. 31 (1985) 69–80.

M. Shaposhnikov, C. Wetterich, *Asymptotic safety of gravity and the Higgs boson mass*, Phys. Lett. B Vol. 683 (2010) N.2-3, 196–200.

M. Sher, *Electroweak Higgs potential and vacuum stability*, Phys. Rep. Vol. 179 (1989) N.5-6, 273–418.

M. Shiraishi, D.F. Mota, A. Ricciardone, F. Arroja, *CMB statistical anisotropy from noncommutative gravitational waves*, arXiv:1401.7936 [astro-ph].

D. Sloan, *Loop quantum cosmology and the early universe*, PhD Thesis, The Pennsylvania State University (2010).
https://etda.libraries.psu.edu/catalog/10964

T.L. Smith, M. Kamionkowski, A. Cooray, *Direct detection of the inflationary gravitational wave background*, Phys. Rev. D 73, N.2, 023504 (2006), 14 pp.

E.D. Stewart, D.H. Lyth, *A more accurate analytic calculation of the spectrum of cosmological perturbations produced during inflation*, Phys. Lett. B 302 (1993) 171–175.

W. van Suijlekom, *Noncommutative geometry and particle physics*, Springer (2015).

W. van Suijlekom, *Perturbations and operator trace functions*, J. Funct. Anal. 260 (2011) N.8, 2483–2496.

W. van Suijlekom, *Renormalizability conditions for almost-commutative manifolds*, Ann. Henri Poincar 15 (2014) N.5, 985–1011.

W. van Suijlekom, *Renormalization of the asymptotically expanded Yang-Mills spectral action*, Comm. Math. Phys. 312 (2012) N.3, 883–912.

W. van Suijlekom, *The noncommutative Lorentzian cylinder as an isospectral deformation*, J. Math. Phys. 45 (2004) N.1, 537–556.

L. Susskind, *Fractal-flows and time's arrow*, arXiv:1203.6440.

F. Sylos Labini, M. Montuori, L. Pietroneo, *Scale-invariance of galaxy clustering*, Phys. Rep. Vol. 293 (1998) N. 2-4, 61–226.

A.H. Taub, *Empty space-times admitting a three parameter group of motions*, Annals of Mathematics 53 (1951) 472–490.

K. Teh, *Nonperturbative spectral action of round coset spaces of SU(2)*, J. Noncommut. Geom. 7 (2013) N.3, 677–708.

K. Teh, *Dirac spectra, summation formulae, and the spectral action*, Ph.D. Thesis, California Institute of Technology (2013).

K.P. Tod. *Self–dual Einstein metrics from the Painlevé VI equation*, Phys. Lett. A 190 (1994) 221–224.

P. Tod. *Penrose's circles in the CMB and a test of inflation*, Gen. Rel. and Gravitation, vol. 44, issue 11, (2013) pp. 2933–2938.

J.P. Uzan, U. Kirchner, G.F.R. Ellis, *WMAP data and the curvature of space*, Mon. Not. Roy. Astron. Soc. 344, L65 (2003).

A. Vishik, *Motives of quadrics with applications to the theory of quadratic forms*, in "Geometric methods in the algebraic theory of quadratic forms", pp. 25–101, Lecture Notes in Math., Vol.1835, Springer (2004).

V. Voevodsky, *Triangulated categories of motives over a field*, in "Cycles, trans-
fers, and motivic homology theories", pp. 188–238, Ann. of Math. Stud.,
143, Princeton Univ. Press, Princeton, NJ (2000).

J. Weeks, J.Gundermann, *Dodecahedral topology fails to explain quadrupole-
octupole alignment*, Class. Quant. Grav. 24 (2007) 1863–1866.

I.K. Wehus, H.K. Eriksen, *A search for concentric circles in the 7-year WMAP
temperature sky maps*, Astrophysical Journal Letter, 733 (2011) N.2, L29,
6 pp.

S. Weinberg, *Ultraviolet divergences in quantum theories of gravitation*, in "General Relativity: an Einstein centenary survey" Cambridge Univ. Press
(1979) pp. 790–831.

J.A. Wolf, *Spaces of constant curvature*, AMS Chelsea Publishing (2011).

M. Yamashita, *Connes–Landi deformation of spectral triples*, Lett. Math. Phys.
94 (2010) N.3, 263–291.

Index

2-category, 218, 225, 239, 241

adinkras, 51
aeons, 117, 122
algebraic curves, 53
algebraic differential form, 177
algebraic geometry, 171
almost-commutative geometry, 10,
    109, 211, 213, 231, 248
analytic continuation, 193
analytic packing, 199
anomalous dimensions, 84
Apollonian group, 191
Apollonian packing, 188, 189, 197
asymptotic expansion, 20, 28, 39, 157,
    199
asymptotic safety, 84
asymptotics, 150

Berger sphere, 114, 152
Bernoulli process, 204
Bethe tree, 203
Bianchi IX, 127, 139, 146, 150, 152,
    155, 163, 185, 210, 215, 216
Bieberbach manifolds, 90, 102, 108
Big Bang, 117, 123
billiard, 132
bimodule, 231, 243
binary dihedral group, 97
binary icosahedral group, 97
binary octahedral group, 97
binary tetrahedral group, 97

blowup, 117, 118, 120
bosons, 37
branched cover cobordism, 222, 235
branched covering, 217, 218, 231
brane-world scenarios, 92
Bratteli diagram, 244
Bruhat-Tits tree, 120, 202

Cabibbo-Kobayashi-Maskawa, 36
Cantor set, 17, 137, 204
Casimir operator, 59, 246
categorification, 231
chromotopology, 52
coadjoint orbits, 254
cobordism, 222, 226, 227, 231, 233,
    237
coding, 52, 130, 204
complexified spacetime, 116
configuration space, 120, 190
conformal factor, 146, 149
conformal gravity, 25, 42, 76, 78
conformal infinity, 117, 121
conformally cyclic cosmology, 119
Connes-Chern character, 230
continued fraction, 126, 129
continuum limit, 247, 249
contragradient module, 32
correspondence, 218, 229, 231
cosmic crystallography, 96
cosmic topology, 89, 92
cosmological constant, 23, 42, 69, 73,
    74, 149, 164

cosmological horizon, 93
cosmological inflation, 73, 75, 79, 94,
    119–121, 143
cosmological timeline, 71
coupling constants, 62
covering moves, 235
crossover models, 119
cyclic branched covering, 219
cyclic group, 97
cyclic homology, 230

deformation, 123
Descartes configuration, 191, 194
dessins d'enfant, 53
dimension, 6, 18, 137, 189, 192, 196
Dirac spectrum, 98, 104, 109, 140,
    152, 252
distinguished triangle, 180
dodecahedral space, 97, 98, 100, 195,
    197, 218

early universe, 61, 71
effective field theory, 66
eigenvalue distribution, 252
Einstein metric, 164, 217
Einstein-Hilbert action, 22, 24, 217
elliptic curve, 122, 130, 148
embedded graph, 218, 234
eternal inflation, 120, 202
Eternal Symmetree, 187, 205
Euler characteristic, 179
exotic smoothness, 142, 216
extrinsic curvature, 213, 227

Farey tessellation, 122, 130
fermions, 30, 31, 33
Feynman diagrams, 62, 172
fibered product, 226, 232, 241
fine tuning, 69
finite spectral triple, 243, 250
flat tori, 102
fractal, 17, 28, 137, 187, 189, 192,
    195, 196, 201
fractal dodecahedra, 195
Friedmann equation, 143

Friedmann-Robertson-Walker
    spacetime, 116, 122, 143, 171, 173,
    179, 185, 188, 215, 216
Fulton-MacPherson compactification,
    120
fused algebra, 43

gauge network, 243, 245
Gauss-Bonnet gravity, 22, 25, 42, 164,
    214
geodesic, 122, 130, 133
geometric engineering, 113
golden ratio, 97
grand unified theories, 57
Grassmann variables, 31
Grassmannian, 115, 123
gravitational constant, 23, 42, 69, 73,
    78
gravitational instantons, 146, 147,
    149, 155, 163, 185, 210
gravitational memory, 73, 77
gravitational waves, 73
gravity balls, 73, 77
Gromov-Hausdorff distance, 254
Grothendieck ring of varieties, 179,
    181, 183
groupoid, 220
Gysin sequence, 179, 180, 184

half-turn space, 102
Hamiltonian constraint, 213
Hantzsche-Wendt space, 102
harmonic oscillator, 215
Hartle-Hawking quantum cosmology,
    209, 216, 227, 254
Hartle-Hawking wave function, 213
Hawking radiation, 78
heat kernel, 20, 28, 41, 107, 109, 156,
    163
Higgs, 20, 39, 41, 43, 50, 60, 75, 77,
    79–81, 84, 243, 249
Hilden-Montesinos theorem, 218
holonomy, 234, 247
Hopf fibration, 151, 173
horizontal composition, 226, 240
Hoyle-Narlikar cosmology, 25, 73, 78

Hubble parameter, 143
hyperbolic manifolds, 90
hyperbolic quadratic form, 183
hypercharge, 33, 38, 50
hypergeometric function, 88

inclusion-exclusion, 181
inner fluctuations, 37
invariant density, 127

Kasner cycle, 126, 135, 141, 142
Kasner epoch, 122, 126, 127, 136,
    141, 142
Kasner metric, 125, 139, 142, 144
Kasner time, 133
Kasparov bimodule, 229
Kasteleyn orientation, 54
Keane potential, 204, 206
KK-theory, 229
Klein quadric, 116, 124
KMS state, 138, 225, 227
Krajewski diagrams, 13, 49

Laplace transform, 108
lattice property, 194
Lefschetz motive, 182
left-right symmetry, 31
length spectrum, 191
lens spaces, 97
Levi-Civita connection, 157
light cone, 117
Linde antigravity, 73, 75
Lorentzian geometry, 23
Lyapunov exponent, 135
Lyapunov spectrum, 135

Majorana fermions, 31, 64, 66
Markov partition, 137
Markov process, 204
Marshak-Mohapatra model, 60
matching circles, 91
maximal mixing, 68
Mayer-Vietoris sequence, 179, 180,
    184
Mellin transform, 21, 29, 161
Melzak packing constant, 191

mini superspace, 210, 214, 216
mixed motives, 179, 183
mixed Tate motives, 180, 181, 185
mixmaster universe, 122, 125, 128,
    132, 134, 139, 142, 155
modified gravity, 22, 171, 214
modular curve, 121, 122, 129, 130,
    134
modular form, 165
moduli space, 34, 120, 228
monodromy, 148, 233, 234, 236, 238
motives, 172
MSSM, 49
multifractal, 187, 204

noncommutative Grassmannian, 123
noncommutative torus, 15, 130, 141,
    152
noncommutativity, 123, 146, 151, 211

one-loop, 62, 65, 217
orbifold, 219
orientability, 8, 119
origami curves, 53

p-adic, 202
Packed Swiss Cheese cosmology, 187
Painlevé VI, 147, 163, 215
parametrix, 156, 160
partition function, 225
path integral, 210, 213, 215, 217, 250
Pati-Salaam, 57
pencil of quadrics, 173
periods, 171, 177, 183
Perron-Frobenius, 205, 207
Perron-Frobenius operator, 138
Pfaffian, 31
photon fluid, 93
PL manifolds, 235, 238, 242
Plücker embedding, 116, 123
Planck inflation, 210
Planck mass, 94, 145
Planck scale, 123
Poincaré duality, 179
Poincaré homology sphere, 97, 98,
    100, 195, 218

Poisson summation, 26, 98, 104, 144
poles and residues, 194
Pontecorvo-Maki-Nakagawa-Sakata, 36
power spectrum, 94
primordial black holes, 73, 77
pseudodifferential calculus, 156, 161, 173
pseudodifferential symbol, 156, 159, 174, 177

quadric of null directions, 117
quadrupole suppression, 91
quantized area, 240
quarter-turn space, 102
quaternionic space, 98, 99
quiver, 245

radiation dominated, 93
random geometry, 250
random matrix, 250
rationality, 163, 171
real structure, 4
reconstruction, 8
Reidemeister moves, 234
renormalization group, 62
renormalization group flow, 41, 61, 67, 69
residual gravity acceleration, 91
residual set, 189
Riemann surfaces, 53
right-handed neutrinos, 64

scalar and tensor fluctuations, 94
scalar field, 43, 81, 93, 101, 144
scale factor, 93, 143
Schubert cells, 124
screens and windows, 193
see-saw scales, 66, 78
Seeley-deWitt coefficients, 156, 163
Segre quadric, 182
Selberg trace formula, 55
semialgebraic set, 173, 177
semigroupoid, 220
sensitive dependence, 69
shift operator, 127, 134, 137

Sierpinski dodecahedra, 196
Sierpinski gasket, 19, 197
sixth-turn space, 102
slow-roll inflation, 73, 79, 92, 94, 101, 109, 113, 143, 201
slow-roll parameters, 94, 101, 109, 201
slow-roll potential, 95, 101, 109, 113, 143, 144, 201
small category, 220
spectral action, 19, 27, 28, 39, 55, 92, 156, 157, 165, 199, 217, 243, 247, 249, 252
spectral action with boundary, 211
spectral triple, 4, 6, 12, 18, 152, 213, 243
spectral triple with boundary, 232
spherical space forms, 90, 96, 108
spin foam, 233, 237, 243
spin network, 233, 236, 243
Standard Model, 41
super Riemann surfaces, 54
superalgebras, 51
supersymmetry, 20, 45, 51

Tate motives, 173
temperature, 93
theta characteristics, 149, 163
theta deformation, 16, 123, 139, 141, 151
theta functions, 148
third-turn space, 102
topspin foam, 238
topspin network, 238
toric variety, 123
trees of projective spaces, 120
triangle group, 54
triangulated category, 179, 180
Turaev-Viro invariants, 234
twistor theory, 115
twistor transform, 116, 121

unification, 63, 66, 67
uniformization, 54
universal knots, 218

vertical composition, 227, 240

virtual dimension, 228

Weyl curvature, 22, 25, 147, 164, 214
Wheeler-deWitt equation, 210, 213,
  215
Wick rotation, 23, 93, 150
Wigner semicircle law, 251, 252
Wirtinger presentation, 235
Wodzicki residue, 161, 173

Yang-Mills, 41, 46, 60, 243, 249
Yoneda algebra, 230
Yukawa coupling, 86
Yukawa parameters, 64

zeta function, 21, 191, 193, 199

* 9 7 8 9 8 1 3 2 0 2 8 4 9 *